Guided Wave Optics

Guided Wave Optics

Selected Topics

Editor

Anurag Sharma

MV Learning

3, Henrietta Street
London WC2E 8LU
UK

4737/23, Ansari Road,
Daryaganj, New Delhi 110 002
India

A Viva Books imprint

ISBN: 978-81-309-3216-3

Printed and bound in India.

SUB-0004

We, the authors, dedicate this book to Professor Ajoy K. Ghatak, a teacher par excellence, for initiating us to the exciting field of Guided Wave Optics and for his continued guidance and inspiration over the last three decades.

Authors

I.C. Goyal	*Physics Department, IIT Delhi*
Banshi Dhar Gupta	*Physics Department, IIT Delhi*
Ajit Kumar	*Physics Department, IIT Delhi*
Arun Kumar	*Physics Department, IIT Delhi*
Bishnu Pal	*Physics Department, IIT Delhi*
Anurag Sharma	*Physics Department, IIT Delhi*
Enakshi K. Sharma	*Department of Electronic Science, Delhi University, South Campus*
Sangeeta Srivastava	*Physics Department, Rajdhani College, Delhi University.*
M.R. Shenoy	*Physics Department, IIT Delhi*
K. Thyagarajan	*Physics Department, IIT Delhi*
R.K. Varshney	*Physics Department, IIT Delhi*

CONTENTS

Preface

Under the guidance of Professor Ajoy Ghatak, the activities of the Fiber Optics Group at IIT Delhi have grown over the past 30 years. The nucleus of the group was formed in the early 1970s around a few graduate students by the efforts of Professor M.S. Sodha and Professor Ghatak. The initial efforts were focused on modelling and analysis of propagation effects in optical waveguides. These activities slowly expanded into experimental investigations about a decade later and diversified to include emerging applications like fiber sensors and components, nonlinear guided wave optics, integrated optics and gradient-index optics. Professor Ghatak has created a stimulating and motivating environment in the research group where learning is fun, and has nurtured its growth.

This volume contains articles by authors, who have been mentored by Professor Ghatak. It covers areas of research, which have been actively pursued by our group members in the recent past and in which Professor Ghatak had taken an active role himself. It was conceived as a collective offering to Prof. Ghatak on his 65th birth anniversary as a token of our respect and affection for him. The volume essentially covers three broad areas - physics of guidance in optical waveguides, guided wave optical components and tools of analysis of such waveguides and components, which broadly reflect the current expertise of the group. The first six chapters are devoted to various aspects of propagation in fiber and integrated optical waveguides and include topics like evolution of single-mode fiber designs including photonic crystal fibers, polarization effects in optical fibers, physics underlying fiber amplifiers like EDFA and Raman amplifiers, formation and propagation of spatial and temporal solitons in optical fibers, nonlinear optical effects like Cerenkov radiation and their utility in optical waveguides. The next four chapters cover various guided wave components and sensors. In particular, three versatile technology platforms namely, fused fiber couplers, side-polished fiber half couplers, and in-fiber gratings, which could yield a wide variety of all-fiber components have been discussed in details. These are followed by an exclusive chapter on fiber sensors. The last five chapters focus on analytical and numerical tools to analyse fiber and integrated optical waveguides. These include the approximate techniques such as the perturbation, the MAF and the variational method, and numerical methods for beam propagation.

We hope that this volume would serve a useful introduction to some of the contemporary areas of research and applications on guided wave optics.

We thank Miss Arti Agrawal for her help in formatting and copy editing of the manuscript.

Authors

1

Tailoring Optical Fiber Design for Communication

Bishnu P. Pal

1. Introduction

Spurred by the development of high-quality optical fiber technology, which is considered to be a major driver behind the relatively recent information technology revolution and the phenomenal progress on global telecommunications, the optical communication industry has witnessed an enormous growth in the last two decades. Fiber optics is now taken for granted in view of its wide-ranging application as the most suitable singular transmission medium for voice, video, and data signals. Indeed, optical fibers have now penetrated virtually all segments of telecommunication networks - be it trans-oceanic, transcontinental, inter-city, metro, access, campus, business, or on-premise. The Internet revolution of late 1990s and deregulation of the telecommunication sector from the Government controls all over the world, which took place in the recent past, have substantially contributed to this unprecedented growth. Initial revolution in this field of technology centered on achieving optical *transparency* in terms of exploiting the low-loss and low-dispersion transmission wavelength windows of high-silica single-mode optical fibers. Though the low loss fiber with a loss under 20 dB/km that was reported for the first time was a single-mode fiber at the He-Ne laser wavelength,[1] the earliest fiber optic lightwave systems exploited the first low loss wavelength window centered on 820 nm with graded index multimode fibers as forming the transmission media. However, primarily due to the unpredictable nature of the bandwidth of jointed multimode fiber links, since early-1980s the system focus shifted to single-mode fibers (SMF) by exploiting the zero material dispersion characteristic of silica fibers, which falls at a wavelength of 1280 nm[2] in close proximity to its second low loss wavelength window centered at 1310 nm.[3] Development of *broadband* optical *amplifiers* in the form of Erbium Doped Fiber Amplifiers (EDFA) in 1987,[4] whose operating wavelengths fortuitously coincided with the lowest loss transmission window of silica fibers centered at 1550 nm,[5] heralded the birth of the era of *dense wavelength division multiplexing* (DWDM) technology in the mid-1990s.[6] EDFA ushered in the next revolution in fiber optics. A schematic block diagram of a DWDM optical fiber communication link is shown in Fig.1. The DWDM technology, defined as simultaneous transmission of at least 4 wavelengths in the 1550 nm low-loss wavelength window through one single-mode fiber, has indeed led to an enormous increase in available bandwidth for high-speed telecommunication and data transfer. Recent development of low water peak fibers (LWPF) known as

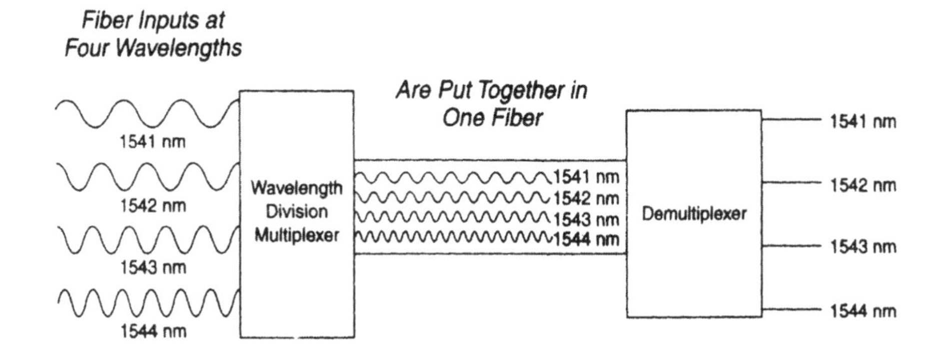

Figure 1 Schematic of a DWDM optical communication system (adapted from [7] ©Prentice Hall 2002)

enhanced SMF (E-SMF) or referred to as G.652.C type of fibers by the International Telecommunication Union (ITU) are characterized with an additional low-loss window in the E-band (1360 ~ 1460 nm), which is about 30% more than the two low-loss windows centered about 1310 and 1550 nm in legacy SMFs namely, G.652 fibers (e.g. SMF-28™) characterized with near-zero dispersion at 1310 nm. With the availability of E-SMFs e.g. AllWave™ and SMF-28e™ fibers devoid of the characteristic OH^- loss peak (centered at 1380 nm) the low loss wavelength window in high-silica fibers now extends from 1280 nm (235 THz) to 1650 nm (182 THz) thereby offering, in principle, an enormously broad 53 THz of optical transmission bandwidth to be potentially tapped through DWDM technique! Classically WDM system implied simultaneous transmission of time division multiplexed (TDM) signals via 1310 and 1550 nm wavelengths. An important attribute of the DWDM technology is that it enables the flexibility of gradual increase in channels as per demand thereby enabling a huge increase in the leveraged capacity of a single fiber. It provides great flexibility in provisioning new services and affords capacity upgrades of existing networks without the need for expensive installation of new cables.

Today the Internet data may take different formats e.g. video clips, audio, large computer files to travel through large distances on a transmission media before being dropped locally. Thus the demand for high-capacity transmission fibers has grown as fast as the Internet.[8] Besides long-haul market, need for high-speed connectivity in metro and access networks have also fuelled demand for application specific fibers. The DWDM technology has already revolutionized backbone networks. In this chapter our focus would be to discuss trends on design tailoring in single-mode optical fibers from the point of view of optical telecommunication. Besides the long-haul market due to growing interest to design fibers with dedicated applications in a metro network to transmit DWDM signals as also deploy coarse WDM (CWDM) techniques in places like university campus local area networks (LANs), some discussions on these emerging applications are also discussed in this chapter. Finally a short discussion on most recent research on dispersion tailoring in Bragg fiber type of photonic bandgap microstructured fiber designs for dispersion compensation and potential applications in metro networks are also presented.

2. Loss spectrum of an optical fiber

Loss and *dispersion spectra* are the two most important propagation characteristics of a single-mode optical fiber. An illustrative example of loss spectrum of single-mode fibers is shown in Fig.2. Except for portion of the loss spectrum near about 1380 nm, which corresponds to characteristic OH^- absorption peak, the rest of the spectrum more or less varies with wavelength as $A\lambda^{-4}$ meaning thereby that signal in

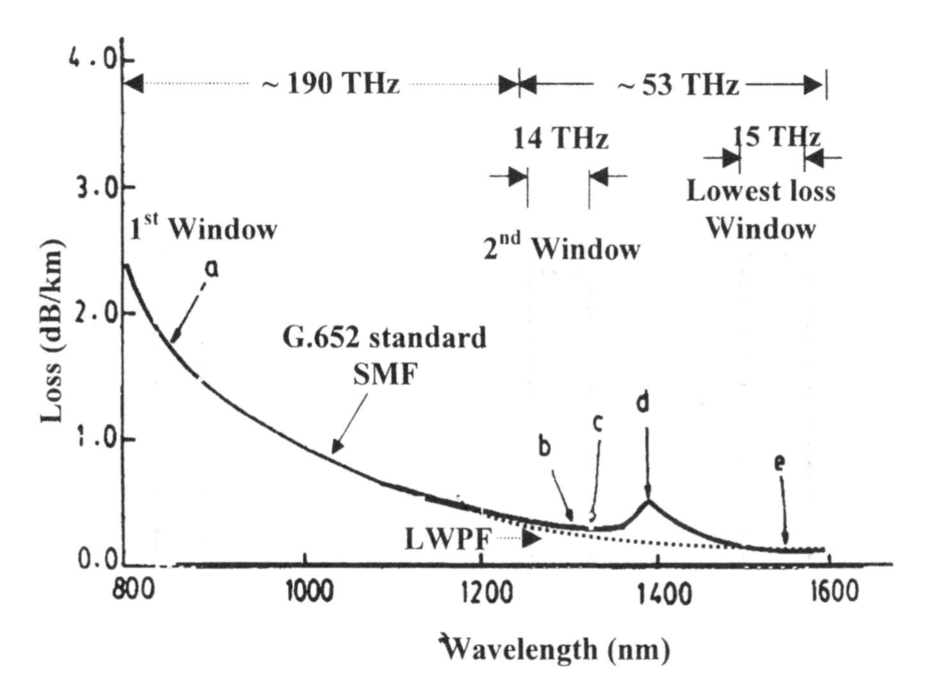

Figure 2 Loss spectrum (full curve) of a commercial state-of-the art G.652 fiber (adapted from a Corning product catalogue ©Corning Inc.); **a**: 1.81 dB/km @ 850 nm, **b**: 0.35 dB/km @1300 nm; **c**: 0.34 dB/km @1310 nm, **d**: 0.55 dB/km @ 1380 nm, **e**: 0.19 dB/km @1550 nm. The dotted curve is a superimposed curve representing typical loss spectrum corresponding to a low water peak fiber (LWPF)/enhanced SMF. The figure also shows a measure of available theoretical transmission bandwidth at different low loss spectral regimes.

a state-of-the-art single-mode fiber is essentially caused by Rayleigh scattering. Rayleigh scattering loss coefficient A in a fiber may be approximately modeled through the relation[9]

$$A = A_1 + A_2\Delta^n \qquad (1)$$

where A_1 = 0.7 dB-μm^4/km in fused silica and $A_2 \approx 0.4 - 1$ dB-μm^4/km while Δ is relative core-cladding index difference expressed in % and exponent $n = 0.7 \sim 1$, whose value depends on the type of the dopant used as refractive index modifier in a high-silica fiber. With GeO_2 as the dopant and $\Delta \sim 0.37\%$, estimated Rayleigh scattering loss in a high-silica fiber is about 0.18 – 0.2 dB/km at 1550 nm. Near about 1400 nm another curve (shown as dotted without the hump) is superimposed on Fig.2 and this corresponds to loss spectrum of a LWPF or E-SMF, examples of which are AllWave™ and SMF-28e™ fibers.

In real-world systems, however, there are other sources of loss, which are required to be budgeted for, on case-to-case basis, and which could add up to more than the inherent loss. Examples of these are cabling-induced losses due to microbending, bend-induced losses along a fiber route, losses due to splices and connectors including those involving insertion of various components in a fiber link. Several of these also depend to a certain extent on the refractive index profile of the fiber in question. Detailed studies have indicated that these extraneous sources of loss could be addressed by optimizing mode field radius (W_P, known as Petermann spot size) and effective cut-off wavelength (λ_{ce}).[10,11] The parameter W_P effectively determines transverse offset-induced loss at a fiber splice as well as sensitivity to microbend-induced losses and λ_{ce} essentially determines sensitivity to bend-induced loss in transmission.[12,14] Both of these are well known important characteristic parameters of a single-mode fiber.[10] For operating at 1310 nm, optimum values of these parameters turned out to be $4.5 < W_P$ (μm) < 5.5 and $1100 < \lambda_{ce}$ (nm) < 1280 .[11] An indirect way to test that λ_{ce} indeed falls within this range is ensured if the measured excess loss of 100 turns of the fiber loosely wound around a cylindrical mandrel of diameter 7.5 cm is below 0.1 dB at 1310 nm and below 1.0 dB at 1550 nm.[12]

In early 1980s it was noticed that many of the installed (multimode) fibers exhibited an increase in transmission loss at $\lambda > 1200$ nm with aging of as little as 3 years. Detailed studies indicated that this phenomenon was attributable to formation of certain chemical bonds between silicon and various dopants with hydroxyl ions in the form of Si-OH, Ge-OH, and P-OH due to trapping of hydrogen at the defect centers formed in the Si-O network caused by incorporation of various dopants. Out of the two most often used dopants namely, GeO_2 and P_2O_5, the latter was found to be more troublesome because defect density formed with it were much more in number due to a difference in valency between Si and P.[11] The phenomenon of hydrogen-induced loss increase with aging of a fiber could be effectively eliminated (at least over the life of a system assumed to be about 25 years) by avoiding phosphorus as a refractive index modifier of silica and through complete polymerization of the fiber coating materials, for which UV-curable epoxy acrylates were found to be the best choice and in fact these are now universally chosen as the optimum fiber coating material before a fiber is sent to a cabling plant.

3. Dispersion spectrum

Chromatic dispersion is the other most important transmission characteristics (along with loss) of a single-mode fiber and its magnitude is a measure of its information transmission capacity. In a fiber optical telecommunication network, signals are normally transmitted in the form of temporal optical pulses (typically in a high speed fiber optical link individual pulse width is about 30 ps),[15] which are replica of electrical TDM signals derived from analog signals e.g. from a telephone. These process steps are schematically depicted in Fig.3. Due to the phenomenon of *chromatic dispersion*, these signal pulses in general broaden in time with propagation through the fiber. Chromatic dispersion arises because of the dispersive nature of an optical fiber due to which the group velocity of a propagating signal pulse becomes a function of frequency (usually referred to as *group velocity*

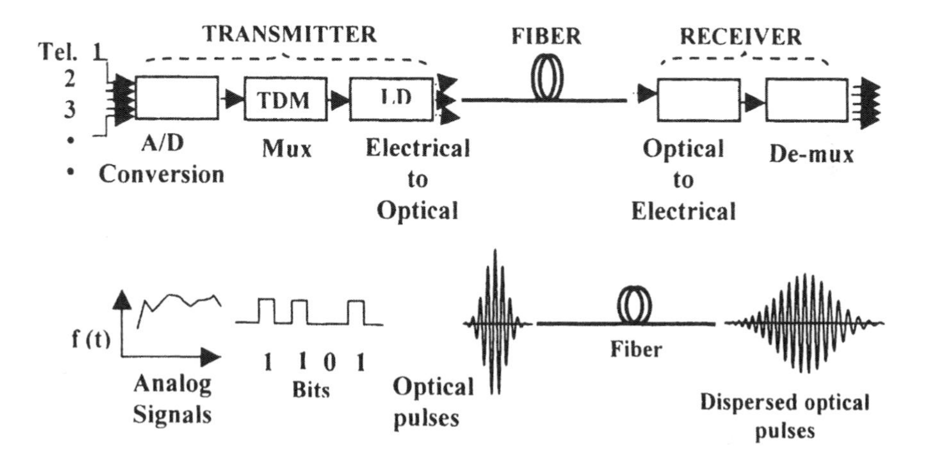

Figure 3 Schematic of an optical telecommunication system; the lower part of the figure depicts sequence of process steps involved and the phenomenon of chirping

dispersion [GVD] in the literature)[14] and this phenomenon of GVD induces *frequency chirp*[13] to a propagating pulse, meaning thereby that the *leading* and the *trailing edges* of the propagating pulse differ in frequency. The resultant frequency chirp [i.e. $\omega(t)$] leads to inter-symbol interference (ISI), in the presence of which the receiver fails to resolve the digital signals as individual pulses when the pulses are transmitted too close to each other. Thus these pulses though started, as individually distinguishable pulses at the transmitter may become indistinguishable at the receiver depending on the amount of chromatic dispersion-induced broadening introduced by the fiber (see Fig.4). In fact the phenomenon of GVD limits the number of pulses that can be sent through the fiber per unit time. For self-consistency of our discussions on pulse dispersion, in the following we outline basic principles that underlay pulse dispersion in a single-mode fiber:[13,14,16,17]

The *phase velocity* of a plane wave propagating in a medium of refractive index n having propagation constant k, which determines the velocity of propagation of its phase front, is given by

$$v_p = \frac{\omega}{k} = \frac{c}{n(\omega)} \tag{2}$$

On the other hand in case of a propagating optical pulse though a *dispersive medium*

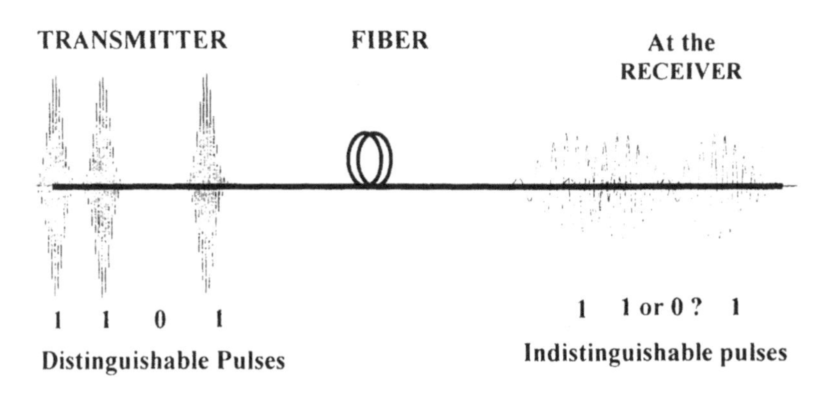

Figure 4 Illustration of loss of signals due to pulse dispersion and consequent inter-symbol interference [16]

like a single-mode fiber via its LP_{01} guided mode, the characteristic velocity at which the pulse energy propagates through the medium is decided by its *group velocity* v_g given by

$$\frac{1}{v_g} = \frac{1}{c}\left[n_{eff}(\lambda_0) - \lambda_0 \frac{dn_{eff}}{d\lambda_0} \right] \tag{3}$$

where $n_{eff}(\omega)$ represents effective index ($= \beta/k_0$; β being propagation constant of the guided mode) of the LP_{01} mode. We now consider a *Fourier transform limited* Gaussian-shaped pulse, which is launched into a single-mode fiber such that at $z = 0$

$$f(x, y, z = 0, t) = E_0(x, y) e^{-t^2/2\tau_0^2} \exp(i\omega_c t) \tag{4}$$

The quantity E_0 (x,y) represents LP_{01} modal field of the fiber, ω_c represents the frequency of the optical carrier, which is being modulated by the pulse and τ_0 corresponds to the characteristic width of the Gaussian pulse, and $\omega_c t$ represents phase (ϕ) of the pulse. By obtaining Fourier transform of this pulse in the frequency domain and incorporating Taylor's series expansion (retaining only terms up to second order) of the phase constant β around $\omega = \omega_c$ (in view of the fact that Fourier transform of such a Fourier transform limited pulse is sharply peaked around a narrow frequency domain $\Delta\omega$ around $\omega = \omega_c$)[13] we would get

$$f(z = L, t) = E_0(x, y) \exp[i(\omega_c t - \beta|_{\omega=\omega_c})]\Psi(z = L, t) \tag{5}$$

where

$$\Psi(z=L,t)=\frac{1}{2\pi}\int_{-\infty}^{+\infty}\left\{F(z=0,\Omega)\exp\left[-i\left(\frac{\Omega}{v_g}+\frac{1}{2}\beta''\Omega^2\right)L\right]\right\}e^{i\Omega t}d\Omega \tag{6}$$

represents the envelope of the dispersed pulse at a fiber length = L, and $F(\Omega)$ is Fourier transform of the pulse in the frequency domain $\Omega = \omega - \omega_c$; the ratio L/v_g simply represents the delay i.e. the time taken by the pulse to propagate through a fiber of length L. In Eq.(6), the quantity $\beta'' = (d^2\beta/d\omega^2)_{\omega=\omega_c}$ is known as the GVD parameter. We may rewrite the Fourier transform [cf. Eq.(6)] in a new time frame, which includes effect of delay due to propagation through $z = L$ i.e. we replace t by $(t - L/v_g)$ and obtain

$$F.T.\ of\ \left\{\Psi\left(z=L,t-\frac{L}{v_g}\right)\right\} = F(z=0,\Omega)\exp\left(-i\frac{1}{2}\beta''\Omega^2 L\right) \tag{7}$$

Thus the RHS of Eq.(7) may be taken as the transfer function of a single-mode fiber of length L. By carrying out the integration in Eq.(6) we would obtain from Eq.(5)

$$f(L,t)=\frac{E_0(x,y)}{\left(\tau(L)/\tau_0\right)^{1/2}}\exp\left[-\frac{\left(t-\frac{L}{v_g}\right)^2}{2\tau^2(L)}\right]\exp[i(\Phi(L,t)-\beta_c L)] \tag{8}$$

where

$$\Phi(L,t)=\omega_c t+\kappa\left(t-\frac{L}{v_g}\right)^2-\frac{1}{2}\tan^{-1}(\bar{\alpha}) \tag{9}$$

$$\kappa=\frac{1}{2\tau_0^2}\frac{\bar{\alpha}}{\left(1+\bar{\alpha}^2\right)} \tag{10}$$

$$\bar{\alpha}=\frac{\beta'' L}{\tau_0^2} \tag{11}$$

and

$$\tau^2(L)=\tau_0^2\left(1+\bar{\alpha}^2\right). \tag{12}$$

Thus the Gaussian temporal pulse remains Gaussian in shape with propagation but its characteristic width increases to $\tau(L)$ [given by Eq.(12)] as it propagates through a fiber length of L. After propagating through a distance of $L_D = \left(\tau_0^2/\beta''\right)$ the pulse assumes a width of $\sqrt{2}\tau_0$. This quantity L_D is a characteristic dispersion-related parameter of a single-mode fiber and is referred to as the *dispersion length*, which is often used to describe dispersive effects of a medium. Smaller the dispersion, longer is this length. Further, it can be seen from Eq.(12) through differentiation and equating the differential to zero that for a given length of a fiber there is an optimum input pulse width $\tau_0^{opt} = \sqrt{\beta'' L}$, for which the output pulse width is minimum given by $\tau_{min} = \sqrt{2\beta'' L}$. The broadening of a Gaussian temporal pulse with propagation could be seen from lower part of Fig.3. The GVD parameter β'' may be alternatively expressed as

$$\beta'' = \left.\frac{d^2\beta}{d\omega^2}\right|_{\omega=\omega_0} = \frac{\lambda_0^3}{2\pi c^2}\frac{d^2 n_{eff}}{d\lambda_0^2} = \frac{\lambda_0^2}{2\pi c}D \tag{13}$$

where D, known as the dispersion coefficient, is given by

$$D = \frac{1}{L}\frac{d\tau}{d\lambda_0} = -\frac{\lambda_0}{c}\frac{d^2 n_{eff}}{d\lambda_0^2} \tag{14}$$

and it is expressed in unit of ps/km.nm. This parameter D is very extensively used in the literature to describe pulse dispersion in an optical fiber. The mode effective index n_{eff} is often expressed through[14]

$$n_{eff} \approx n_{cl}(1 + b\Delta) \tag{15}$$

where $0 < b < 1$ represents the normalized propagation constant of the LP_{01} mode:

$$b = \frac{n_{eff}^2 - n_{cl}^2}{n_c^2 - n_{cl}^2} \tag{16}$$

and $\Delta \approx ((n_c - n_{cl})/n_{cl})$ stands for relative core-cladding index difference in a single-mode fiber. Thus n_{eff} is comprised of two terms – one is purely material related (n_{cl}) and the second arises due to wave guiding effect (b) and hence total dispersion in a single-mode fiber could be attributed to two types of dispersion namely material dispersion and waveguide dispersion. Indeed it can be shown that the total dispersion coefficient (D_T) is given by the following algebraic sum to a very good accuracy[18]

$$D_T \cong D_M + D_{WG} \tag{17}$$

where D_M and D_{WG} correspond to material and waveguide contributions to D, respectively. The material contribution can be obtained from Eq.(15) by replacing n_{eff} in it with n_{cl} while the waveguide contribution would be yielded by[12]

$$D_{WG} = -\frac{n_2\Delta}{c\lambda_0}\left(V\cdot\frac{d^2(bV)}{dV^2}\right) \tag{18}$$

where V is the well known normalized waveguide parameter defined through

$$V = \frac{2\pi}{\lambda_0}an_c\sqrt{2\Delta} \tag{19}$$

where a represents core radius of the fiber. A plot of a typical dispersion spectrum of a single-mode fiber (i.e. D vs λ) along with its components D_M and D_{WG} are shown in Fig.5. It is apparent from the figure that D_{WG} is all along negative and while sign of D_M changes from negative to positive (going through zero at a wavelength of ~ 1280 nm)[16] as wavelength increases – as a result, the two components cancel each other at a wavelength near about 1300 nm, which is referred to as the *zero dispersion wavelength* (λ_{ZD}) and it is a very important design parameter of single-mode fibers. Realization of this fact led system operators to choose the operating wavelength of first generation single- mode fibers as 1310 nm. These fibers optimized for transmission at 1310 nm are now referred to as G.652 fibers as per ITU standards; millions of kilometers of these fibers are laid underground all over the world. Though it appears that if operated at λ_{ZD} one might

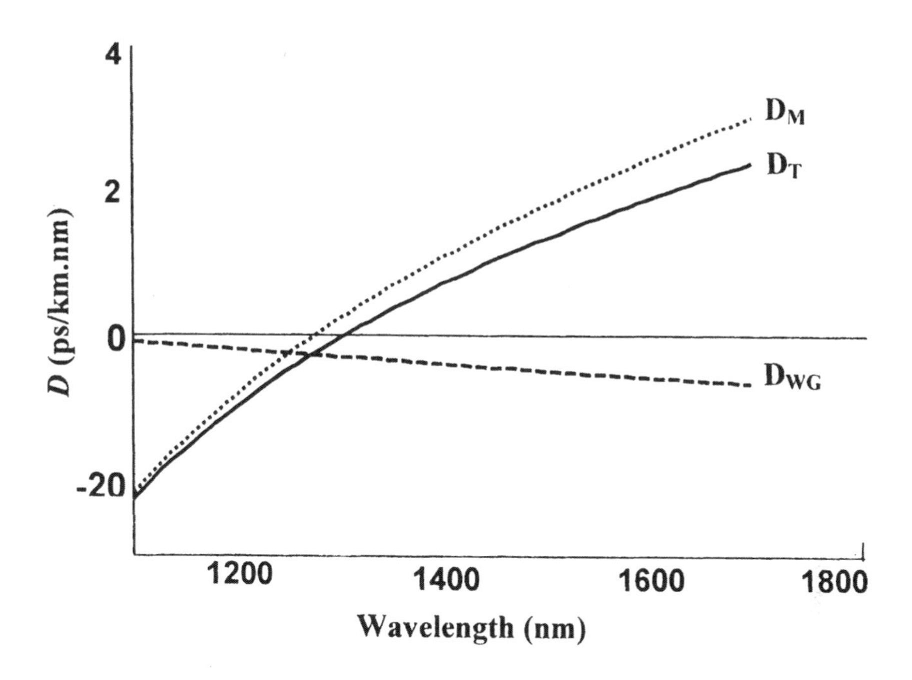

Figure 5 Dispersion (D_T) spectrum of a typical G.652 type single-mode fiber with its material and waveguide components shown separately (Figure courtesy K. Thyagarajan)

get infinite transmission bandwidth, in reality zero dispersion is only an approximation (albeit a very good approximation) because it only signifies that only the second order dispersive effects would be absent. In fact as per ITU recommendations, G.652 fibers qualify for deployment as telecommunication media provided at the 1310 nm wavelength its D_T is < 3.5 ps/km.nm. At a wavelength around λ_{ZD}, higher order dispersion namely third order dispersion characterized by $d^3\beta/d\omega^3$ would determine the net dispersion of a pulse. Thus in the absence of second order dispersion, pulse dispersion is quantitatively determined by the dispersion slope S_0 at $\lambda = \lambda_{ZD}$ through

$$\Delta\tau = \frac{L(\Delta\lambda_0)^2}{2} \cdot S_0 \tag{20}$$

where

$$S_0 = (dD/d\lambda_0)\big|_{\lambda_0 = \lambda_{ZD}},$$

which is measured in units of ps/km.nm^2; S_0 in G. 652 fibers at 1310 nm is ≤ 0.09 ps/km.nm. If third order dispersion is the sole determining factor of pulse dispersion, the output pulse does not remain symmetric and acquires some oscillation at the tails.[19] A knowledge of D and S_0 enables determination of dispersion (D) at any arbitrary wavelength within a transmission window e.g. EDFA band in which D in G.652 fibers varies approximately linearly with λ_0. Utility of this feature often finds applications in component designs and $D(\lambda_0)$ is usually explicitly stated in commercial fiber data sheets as

$$D(\lambda_0) = \frac{S_0}{4}\left[\lambda_0 - \frac{\lambda_{ZD}^4}{\lambda_0^3}\right] \tag{21}$$

The genesis to this relation lies in the following three-term polynomial equation often used as a fit to measured data for delay (τ) vs λ in a single-mode fiber as

$$\tau(\lambda_0) = \frac{A}{\lambda_0^2} + B + C\lambda_0^2 \tag{22}$$

where coefficients A, B, and C are determined through a least square fit to the data for measured $\tau(\lambda_0)$. Equation (22) yields for $D(\lambda_0)$ the following:

$$D(\lambda_0) = \frac{d\tau}{d\lambda_0} = 2C\lambda_0 - \frac{2A}{\lambda_0^3} \tag{23}$$

Setting $D(\lambda_0 = \lambda_{ZD})$ to zero yields $\lambda_{ZD} = (A/C)^{1/4}$. Thus

$$S_0 = 8C \tag{24}$$

Besides pulse broadening, since the energy in the pulse gets reduced within its time slot, the corresponding signal to noise ratio (SNR) will decrease, which could be compensated by increasing the power in the input pulses. This additional power requirement is termed as *dispersion power penalty*.[20] For estimating the dispersion power penalty a more general definition of pulse width in terms of root mean square (*rms*) width becomes more relevant, especially for pulse shapes, which may not necessarily be of a well-defined shape like a Gaussian function.[16] The general definition for *rms* width of a temporal pulse is defined through

$$\tau_{rms} = \sqrt{\langle t^2 \rangle - \langle t \rangle^2} \tag{25}$$

where

$$\langle t^n \rangle = \frac{\int_{-\infty}^{\infty} t^n |f(z,t)|^2 \, dt}{\int_{-\infty}^{\infty} |f(z,t)|^2 \, dt} \tag{26}$$

For a Gaussian-shaped output pulse, its τ_{rms} would be related to its input pulse width τ_0 [cf. Eq.(4)] through $\tau_{rms} = \tau_0/\sqrt{2}$. The bit period of a system operating at a bit rate of B bits per second is $1/B$ s. In order to keep the interference between neighbouring bits at the output below a specified level, one requires τ_{rms} of the dispersed pulse to be kept below a certain fraction ε of the bit period B.[20] Accordingly for maximum allowed *rms* pulse width (assuming it to be Gaussian in shape) of the output pulse, τ_{rms} of the dispersed pulse should satisfy

$$\tau_{rms} = \frac{\tau_0}{\sqrt{2}} < \frac{\varepsilon}{B} \tag{27}$$

which implies

$$B\tau_0 < \sqrt{2}\varepsilon \tag{28}$$

On substitution of the optimum value for output pulse width i.e. $\tau_{min} = \sqrt{2\beta'' L}$ in Eq.(28) and making use of Eq.(13) we get

$$B^2 DL < \frac{2\pi c}{\lambda_0^2} \varepsilon^2 \tag{29}$$

For a 2 dB dispersion power penalty, it is known that $\varepsilon = 0.491$, whereas for 1 dB dispersion power penalty it is 0.306; these values for ε were specified by Bellcore standards (Document no. TR-NWT-000253). For 1 dB dispersion power penalty at the wavelength of 1550 nm, we can write the inequality [Eq.(29)] approximately as

$$D.L \leq \frac{10^5}{B^2} \text{ Gbit}^{-2}\text{ps/nm} \tag{30}$$

where B is assumed to be expressed in Gbits/s, D in ps/km.nm and L in km. Based on Eq. (30), Table I lists the maximum allowed dispersion for different standard bit rates assuming a dispersion power penalty of 1 dB.

Table I

Data rate (B)	Maximum allowed dispersion ($D.L$)
2.5 Gb/s (OC-48)*	~ 16000 ps/nm
10 Gb/s (OC-192)	~ 1000 ps/nm
40 Gb/s (OC-768)	~ 60 ps/nm

*OC stands for Optical Channels

By mid-1980s, proliferation of optical-trunk networks and steady increase in bit transmission rate gave rise to the requirement for a standard transmission format of the digital optical signals so that they may be shared without any need for an interface (this is the so-called "mid-fiber meet") between the American, European and Japanese telephone networks, each of whom until then were following a different digital hierarchy. This lack of standards led to development of synchronous digital hierarchy (SDH), which has now become the global industry standard (in USA this is known as SONET standing for Synchronous Optical Networks) for transmission of digital *optical channels* (OC).[21] A SDH network has a master clock that controls the timing of events all through the network. The base rate (OC-1) in SDH is taken as 51.84 Mbit/s; thus, OC-48 rate corresponds to a transmission rate of 2.488 (≈ 2.5 Gbit/s), which became the standard signal transmission rate at the 1310 nm wavelength window over standard G.652 single-mode fibers. Although in principle very high bit transmission rate was achievable at 1310 nm, maximum link length/repeater spacing was limited by transmission loss (~ 0.34 dB/km at this wavelength) to ~ 40 km in these systems. Thus it became evident that it would be of advantage to shift the operating wavelength to the 1550 nm window, where the loss is lower (cf. Fig.2) in order to overcome transmission loss-induced distance limitation of the 1310 nm window.

4. Dispersion shifted fiber

As could be seen from Fig.2 that the lowest loss in a high-silica fiber appears at 1550 nm, which is almost 40% less than its value at 1310 nm! This motivated system engineers to exploit this lowest loss window by shifting the operating wavelength from 1310 nm to 1550 nm. However simply shifting the operating wavelength to 1550 nm to make use of the already laid G.652 fibers posed a problem because of the large dispersion of 17 ps/km.nm of these fibers at 1550 nm (see Fig.5), which would severely limit dispersion-limited repeater spans of such systems at 1550 nm. By mid-1980s, it was realized that repeater spacing of 1550 nm-based systems could be pushed to a much longer distance if the transmission fiber designs could be so tailored to shift λ_{ZD} to coincide with this wavelength. Such fibers given the name of dispersion-shifted fibers (DSF) indeed got developed, which are generically classified as G.653 fibers by the ITU. A comparative plots of dispersion spectra for G.652 and G.653 type of single-mode fibers are shown in Fig.6. In a DSF typically,

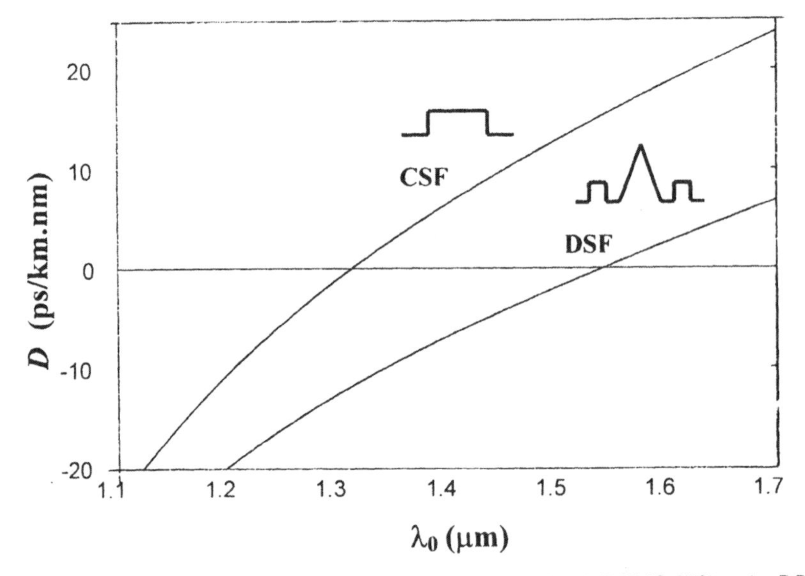

Figure 6 Typical dispersion spectra of a conventional SMF (G.652) and a DSF (G.653); nominal refractive index profiles of these fibers are also shown on the inset as an illustration[22].

D_T is ≤ 2.7 ps/(nm-km) with $S_0 \leq 0.058$ ps/(nm^2-km) at the 1550 nm wavelength while maintaining inherent scattering loss low (~ 0.21 dB/km at 1550 nm). One serious problem with these fibers was their sensitivity to microbend and bend-induced losses at a $\lambda_0 \geq 1550$ nm due to their low cut-off wavelength of about 1100 nm.[11,23] A theoretical design was suggested in 1992 to overcome this issue of bend-loss sensitivity by which the cut-off wavelength was moved up close to about 1550 nm through a spot size optimization scheme.[24] On the laser transmitter front the most common type of laser diodes available for optimum exploitation of the 1550 nm wavelength window operation at 1550 nm were *Fabry-Perot* (FP) and

Distributed Feedback (DFB) lasers, which were based on InGaAsP semiconductors. FP lasers are characterized by spectral width (FWHM = 2 to 5 nm), which is broader than the more expensive DFB lasers (cw spectral width << 1 nm). However, due to chirping and mode hopping, DFB lasers when pulsed exhibit broader dynamic spectral widths (~ 1 to 2 nm) at high modulation rates (≥ 2 Gb/s). In view of these factors, FP lasers in conjunction with dispersion-shifted fibers seemed best *near-term option.*[23] These new generation systems operating @ 2.5 Gb/s became commercial in 1990 having potentials to work at bit rates in excess of 10 Gb/s with careful design of sources and receivers and use of DSF's. These attractive features motivated Japanese telecommunication administration to aggressively deploy DSF in the backbone trunk network in Japan in the early 1990s.[11] This trend persisted for a while before erbium doped fiber amplifiers emerged on the scene and led to a dramatic change in technology trends.

5. Erbium doped fiber amplifiers

In the late 1980s typical state-of-the-art repeater spans were about 40-50 kms @ 560 Mbit/s transmission rate. At a repeater, the so-called 3**R**-regeneration functions (*re-amplification, retiming, and reshaping*) are performed in the electric domain on the incoming to neutralize the effects of both transmission loss as well as dispersion-induced distortions of the signals after these were detected by the photo-detector. However, these complex functions are expensive and required unit replacement in case network capacity is to be upgraded to higher bit transmission rates. Since these units are required to convert photons to electrons and back to photons often at modulation rates approaching the limits of the then available electronic switching technology, a bottleneck was encountered in the late 1980s. *What was needed was an optical amplifier to bypass this electronic bottleneck.* In 1986, the research group at Southampton University in England reported success in incorporating rare earth trivalent erbium ions into host silica glass during fiber fabrication.[25] Erbium is a well known lasing species characterized by strong fluorescence at 1550 nm. Subsequently, the same group demonstrated that excellent noise and gain performance is feasible in a large part of the 1550 nm window with erbium-doped standard silica fibers .[26] The concept of optical amplification in fiber is almost as old as the laser itself. Today erbium-doped fiber amplifiers (EDFA) looks like an outstanding breakthrough, but it is really an old idea. In 1964, Koester and Snitzer had demonstrated a gain of 40 dB at 1.06 μm in a 1-meter long Nd-doped fiber side pumped with flash lamps.[27] The motivation at that time was to find optical sources for communication but impressive development of semiconductor lasers that took place in subsequent years pushed fiber lasers to the background. The operation of an EDFA is very straightforward[13,28] (see also Chapter 3). The electrons in the 4*f* shell of the erbium ions are excited to higher energy states by absorption of energy from a pump. Absorption bands most suitable as pump for obtaining amplification of 1550 nm signals are the 980 nm and 1480 nm wavelengths. When pumped at either of these wavelengths an erbium doped fiber amplifies signals over a band of almost 30 ~ 35 nm at the 1550 nm wavelength region. Typical pump powers required for operating an EDFA as an amplifier could range from 20 to 100 mW depending on the gain required. Absorption of pump energy by the erbium ions leads to *population inversion* of these ions from the ground state ($4I_{15/2}$) to either $4I_{11/2}$ (980 nm) or $4I_{13/2}$ (1480 nm) excited states: $4I_{13/2}$ level effectively acts as a storage of pump power from which the incoming weak signals may stimulate emission and experience amplification.[13] Stimulated events are extremely fast and hence the amplified signal slavishly follows the amplitude modulation of the input signal.[22] EDFAs are accordingly *bit rate transparent*. EDFAs are new tools that system planners now almost routinely use for designing networks.[29] EDFAs can be incorporated in a fiber link with insertion loss ~ 0.1 dB and almost full population inversion could be achieved at the 980 nm pump band. Practical EDFAs with output power of around 100 mW (20 dBm), 30 dB small signal gain, and a noise figure of < 5 dB are now commercially available.

6. Dense wavelength division multiplexing (DWDM)

An attractive feature of EDFAs is that they exhibit a fairly smooth gain versus wavelength curve (especially in case the fiber is doped with Al) almost 30 nm wide

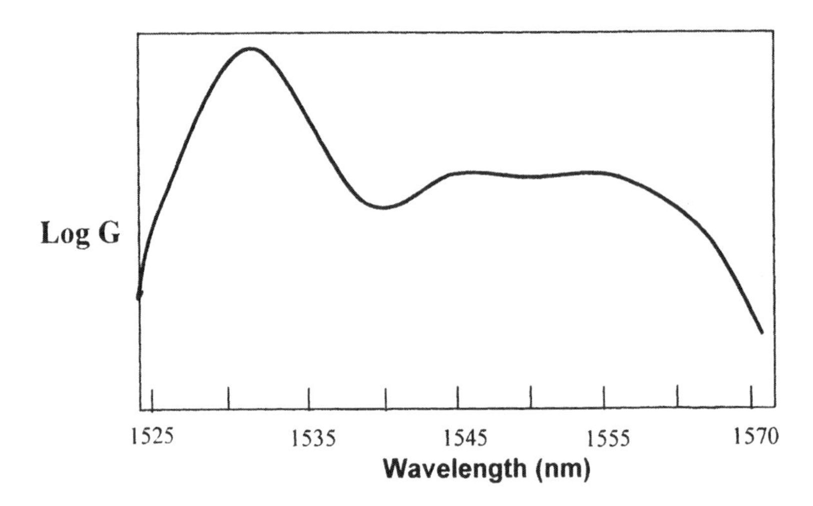

Figure 7 Typical gain spectrum of an EDFA in the C-band.

(C- band: 1535 ~ 1565 nm) (see Fig.7). Thus multi-channel operation via WDM within this gain spectrum became feasible, each wavelength channel being simultaneously amplified by the same EDFA. Relatively long lifetime of the excited state (~ 10 ms) leads to slow gain dynamics and therefore minimal cross talk between WDM channels. In view of this, system designers were blessed with a great degree of freedom to boost the capacity of a system as and when required due to the flexibility to make the network responsive to future demands simply by introducing additional wavelengths through the same fiber, each multiplexed @ e.g. 2.5 Gb/s or 10 Gb/s or even higher. With the development of L-band EDFA characterized by a gain spectrum, which extends from 1570 ~ 1620 nm,[30] the potential for large-scale increase in capacity at will of an already installed link through DWDM became enormous. However, to avoid haphazard growth of multi-wavelength transmitting fiber links, ITU has introduced certain wavelength standards, which are now referred to as ITU *wavelength grids* for DWDM operation. As per ITU standards the reference wavelength is chosen to be 1552.52 nm corresponding to Krypton line, which is equivalent to 193.1 THz (f_0) in the frequency domain; the chosen channel spacing away from f_0 in terms of frequency (Δf) is supposed to follow the relation: $\Delta f = 0.1I$ THz with I = positive/negative integers.[31] Recommended channel spacings are 200 GHz (≡1.6 nm), 100 GHz (≡ 0.8 nm), 50 GHz (≡ 0.4 nm); the quantities within brackets have been calculated by assuming central wavelength as 1550 nm. Today's distributed feedback lasers DFB lasers can be tuned to exact ITU wavelength grids. All the terabit transmission Hero experiments that were reported in recent years took the route of DWDM.[32,33]

The possibility of introducing broadband services in the 1990s led to a boom in the communication industry blurring the distinction between voice, video, and data service providers and the demand for more and more BW grew at a rate that began to strain the capacity of installed communication links.[29] DSF in combination with EDFA appeared to be the ideal solution to meet the demand for high data rate and long-haul applications. However it was soon realized that large optical power throughput that is encountered by a fiber in a DWDM system due to simultaneous amplification of multiple wavelength channels pose problems due to onset of nonlinear propagation effects, which may completely offset the attractiveness of multi-channel transmission with DSF. It turned out that due to negligible temporal dispersion that is characteristic of a DSF at the 1550 nm band, these fibers are highly susceptible to nonlinear optical effects, which may induce severe degradation to the propagating signals.[34]

By late 1990s the demand for bandwidth had been steadily increasing and it became evident that new fibers are required to handle multi-wavelength transmission, each of which was expected to carry a large volume of data especially for under-sea applications and long-haul terrestrial links. A very useful figure of merit for a DWDM link is known as *spectral efficiency*, which is defined as the *ratio* of bit rate to channel spacing. Since bit rate can not be increased arbitrarily due to the constraints with regard to availability of electronic components relevant to bit rate > 40 Gbit/s, decrease in channel spacing appeared to be the best near-term option to tap the huge gain BW of an EDFA for DWDM applications. However it was soon realized that with decrease in channel spacing, the fiber became more strongly sensitive to detrimental nonlinear effects like four-wave mixing (FWM), which could be relaxed by allowing the propagating signals to experience a finite dispersion in the fiber [see Fig.8]. If the number of wavelength channels in a DWDM stream is N, FWM effect, if present, leads to generation of $N^2(N-1)$ number of sidebands! These sidebands would naturally draw power from the propagating signals and hence could result in serious cross talks.[34] Therefore, for DWDM applications, fiber designers came up with new designs for the signal fiber for low-loss dedicated DWDM signal transmission at the 1550 nm band, which were generically named as non-zero dispersion shifted fibers (NZ-DSF). These fibers were designed to meet the requirement of low sensitivity to non-linear effects especially to counter four-wave mixing (FWM), which could be substantially

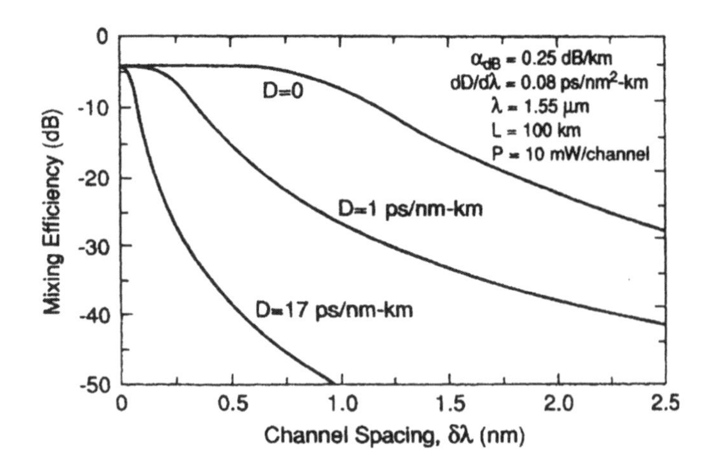

Figure 8 Nonlinearity-induced FWM efficiency with channel spacing for different *D* (Adapted from [34]; © IEEE)

suppressed if each wavelength channel is allowed to experience a finite amount of dispersion during propagation.[30] This requirement is precisely counter to the dispersion characteristics of a DSF – and hence the name of NZ-DSF.[35-37] ITU has christened such fibers as G.655 fibers, which are generally characterized with refractive index profiles, which are relatively more complex with multiple claddings than a simple matched clad step index design of G.652 fibers. The need to deploy such advanced fibers became so imminent that ITU-T Committee had to evolve new standards for G.655 fibers [ITU-T G.655]. As per ITU recommendations, G.655 fibers should exhibit finite dispersion $2 \leq D$ (ps/(nm.km)) ≤ 6 in the 1550 nm band in order to detune the phase matching condition required for detrimental nonlinear propagation effects like four-wave mixing (FWM) and cross-phase modulation (XPM) to take place during multi-channel transmission. ITU arrived at the above range for *D* by assuming a channel spacing of 100 GHz or more. For smaller channel spacing like 50 GHz or 25 GHz, it turned out that the above range for *D* is insufficient to suppress potential nonlinear effects unless i) the power per channel is reduced substantially and ii) number of amplifiers is limited. Unfortunately these steps would amount to a decrease in the repeater spacing(s), which would be counterproductive since this would defeat the very purpose for which G.655 fibers were proposed.[37] In order to overcome this disadvantage of an NZDSF, a more advanced version of NZDSF as a transmission fiber was proposed for super-DWDM, which has been christened by ITU as advanced NZDSF (A-NZDSF)/G.655b fibers. As per ITU recommendation, upper limit on *D* for an A-NZDSF should be 10 ps/(nm.km) at the longer wavelength edge of the C-band i.e. at 1565 nm. For A-NZDSF, its *D* falls in the range $6.8 \leq D$ (ps/(nm.km) ≤ 8.9 (C-band) and $9.1 \leq D$ (ps/nm.km) ≤ 12.0 (L-band).[38] In view of its reduced sensitivity to detrimental nonlinear effects, it can accommodate transmission of as many as 160 channels of 10 Gb/s at 25 GHz channel spacing within the C-band alone and also offers ease in deployment of 40 Gb/s systems. Another attractive feature of A-NZDSF is that it can be used to transmit signals in the S-band as well, for which gain-shifted thulium doped fiber amplifiers (TDFA), and Raman fiber amplifiers (RFA) have emerged as attractive options. There are variations in G. 655 fibers e.g. large effective area fibers (LEAF™),[36] reduced slope (TrueWave RS™),[35] Teralight™,[39] each of which are proprietary products of well-known fiber manufacturing giants. Table II depicts

Table - II

ITU standard fiber type	Dispersion coefficient D (ps/nm.km) @ 1550 nm	Dispersion slope S (ps/nm^2.km)	Mode effective area A_{eff} (μm^2) @1550 nm
G.652: SMF™ -28	~ 17	~ 0.058	~ 80
G.655:			
LEAF™	~ 4.2	~ 0.085	~ 65-80
TrueWave RS™	~ 4.5	~ 0.045	~ 55
Teralight™	~ 8	~ 0.058	-

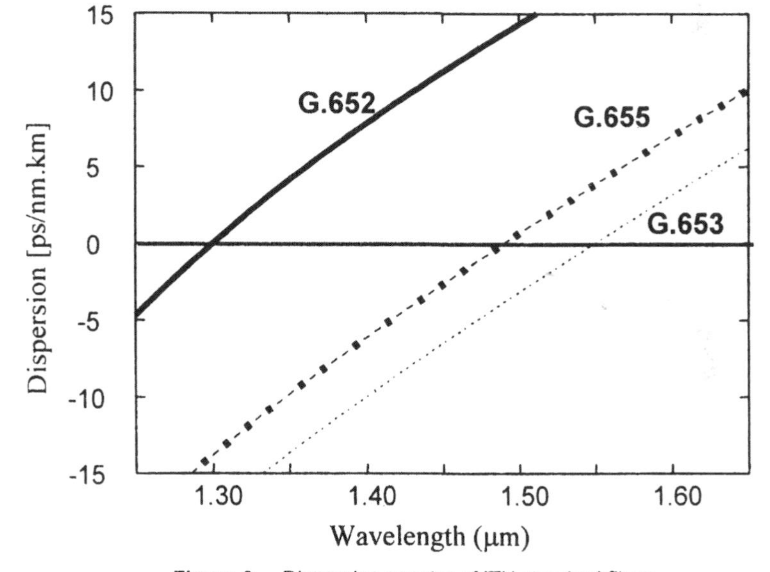

Figure 9 Dispersion spectra of ITU standard fibers

typical characteristics of some of these fibers at 1550 nm along with those of G.652 fibers while Fig.9 shows typical dispersion spectra for G.652, G.653 and G.655 types of transmission fibers. In Table II, the parameter A_{eff} stands for mode effective area, which is defined as[19]

$$A_{eff} = \left[\int_0^\infty \int_0^{2\pi} E_0^2(r)\, r\, \mathrm{d}r\, \mathrm{d}\phi \right]^2 \Bigg/ \int_0^\infty \int_0^{2\pi} E_0^4(r)\, r\, \mathrm{d}r\, \mathrm{d}\phi \qquad (31)$$

where E_0 is modal field of the concerned single-mode fiber.

7. Dispersion compensating fibers

In G.655 fibers for DWDM transmission since a finite though low D is deliberately kept at the design stage itself to counter potentially detrimental nonlinear propagation effects, one would accumulate dispersion between EDFA sites! Assuming a D of 2 ps/km.nm, though a fiber length of about 500 km could be acceptable @ 10 Gbit/s without requiring correction for dispersion, @ 40 Gbit/s corresponding unrepeatered span would hardly extend to 50 km [see Eq.(30)]. The problem is more severe in G.652 fibers for which @ 2.5 Gbit/s though a link length of about 1000 km would be feasible at the 1550 nm window, if the bit rate is increased to 10 Gbit/s, tolerable D in this case over 1000 km would be hardly ~ 1 ps/km.nm! Thus repeater spacing of 1550 nm links based on either of these fiber types would be chromatic dispersion limited for a given data rate.

Around mid-1990s before the emergence of G.655 fibers, it was felt that for network upgrades it would be prudent to exploit the EDFA technology to "mine" the available BW of the already embedded millions of kilometers of G.652 fibers all over the world by switching over transmission through these at the 1550 nm window. However transmission of 1550 nm signals through G.652 fibers would imply accumulation of unacceptably high chromatic dispersion at EDFA sites (ideally spaced at ~ every 80 - 120 kms) due to the fiber's D being +17 ps/km.nm. Hence to reap the benefit of the availability of EDFAs, system upgrade through DWDM technique in the 1550 nm band of the already installed G.652 fiber links would necessarily require some dispersion compensation scheme. Realization of this immediately triggered a great deal of R&D efforts to develop some dispersion compensating modules, which could be integrated to a single-mode fiber optic link so that net dispersion of the link could be maintained/managed within desirable limits. One of the most popular schemes involved inserting a negative dispersion fiber, which are known as *dispersion compensating fibers* (DCF).[40-45]

In order to understand the logic behind dispersion compensation techniques, we notice from Eqs.(9) and (10) that the instantaneous frequency of the output pulse is given by

$$\omega(t) = \frac{d\Phi}{dt} = \omega_c + 2\kappa\left(t - \frac{L}{v_g}\right) \tag{32}$$

The center of the pulse corresponds to $t = L/v_g$. Accordingly the leading and trailing edges of the pulse correspond to $t < L/v_g$ and $t > L/v_g$, respectively. In the normal dispersion regime where $d^2\beta/d\lambda_0^2$ is positive (implying that the parameter κ defined in Eq.(10) is positive), the leading edge of the pulse will be down-shifted i.e *red-shifted* in frequency while the trailing edge will be up-shifted i.e. *blue-shifted* in frequency with respect to the center frequency ω_c. Thus red spectral component i.e. longer wavelength components of the signal pulse would travel faster than its blue spectral component, thereby meaning shorter wavelength components i.e. the group delay would increase with decrease in wavelength. The converse would be true if the signal pulse wavelength corresponds to the anomalous dispersion region where $d^2\beta/d\lambda_0^2$ is negative. Hence as the pulse broadens with propagation due to this variation in its group velocity with wavelength in a dispersive medium like single-mode fiber it also gets chirped meaning thereby that its instantaneous frequency within the duration of the pulse itself changes with time. Chirping of a temporal pulse with propagation in the normal and anomalous dispersion regimes in a fiber are discussed in details in Ref.[13]. If we consider propagation of signal pulses through a G.652 fiber at the 1550 nm wavelength band at which its D is positive (i.e. its group delay increases with increase in wavelength), it would exhibit *anomalous* dispersion. If this broadened temporal pulse were transmitted through a DCF, which exhibits *normal* dispersion (i.e. its dispersion coefficient D is negative) at this wavelength band, then the broadened pulse would get compressed with propagation through the DCF. This could be understood with the help of transfer function of a fiber given by Eq.(7) and studying evolution of the pulse as it propagates through different segments of the fiber link shown in Fig.10. At stage (1), let the F.T. of the

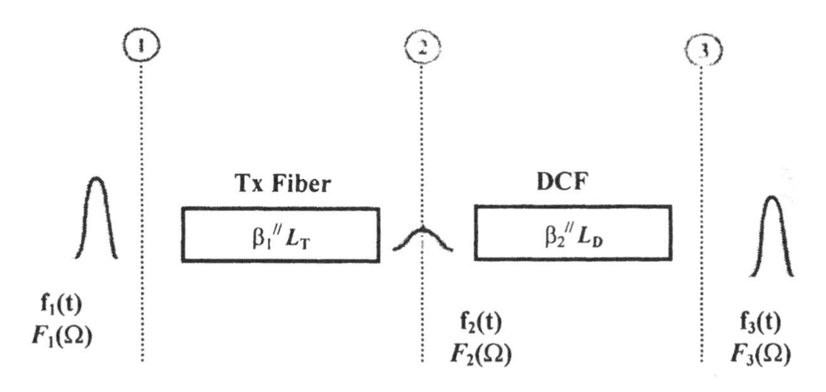

Figure 10 Schematic of the principle behind a DCF (after [15]).

input pulse $f_1(t)$ be $F_1(\Omega)$, which transforms in view of Eq.(7) at subsequent stages to[15]

$$\begin{aligned} &stage(2): F_2(\Omega) = F_1(\Omega)\exp\left(-i\,\beta_T^{\prime\prime} L_T \Omega^2\right) \\ &stage(3): F_3(\Omega) = F_2(\Omega)\exp\left(-i\,\beta_D^{\prime\prime} L_D \Omega^2\right) \\ &= F_1(\Omega)\exp\left[-i\left(\beta_T^{\prime\prime} L_T + \beta_D^{\prime\prime} L_D\right)\Omega^2\right] \end{aligned}$$

where β_T and β_D represent GVD parameters of the transmission fiber and the DCF, respectively, and $L_{T,D}$ refer to corresponding fiber lengths traversed by the signal pulse. It is apparent from the above that if the following condition is satisfied:

$$\beta_T^{\prime\prime} L_T + \beta_D^{\prime\prime} L_D = 0 \qquad (33)$$

the dispersed pulse would recover back its original shape [i.e. evolve back to $f_1(t)$] at stage (3). Thus GVD parameters (and hence coefficients $D_{T,D}$) of the transmission fiber and the DCF should be opposite in sign. Consequently if a G.652 fiber as the transmission fiber is operated at the EDFA band, corresponding DCF must exhibit negative dispersion at this wavelength band. Figure 11 illustrates as an example, the concept of dispersion compensation by a DCF for the dispersion suffered by 10 Gbit/s externally modulated signal pulses over a 600-km span of a G.652 fibers; the full curve represents delay versus length in the transmission fiber and dashed curve corresponds to delay versus length for the DCF[46]. It shows three possible routes for introducing the DCF: i) at the beginning of the span (dashed curve), ii) at the end of the span (full curve), or iii) every 100 km (dotted curve) say at EDFA sites. Further, larger the magnitude of D_D is, the smaller would be the length of the required DCF. This is achievable if D_{WG} of the DCF far exceeds its D_M in absolute magnitude. Large negative D_{WG} is achievable through appropriate choice of the refractive index profile

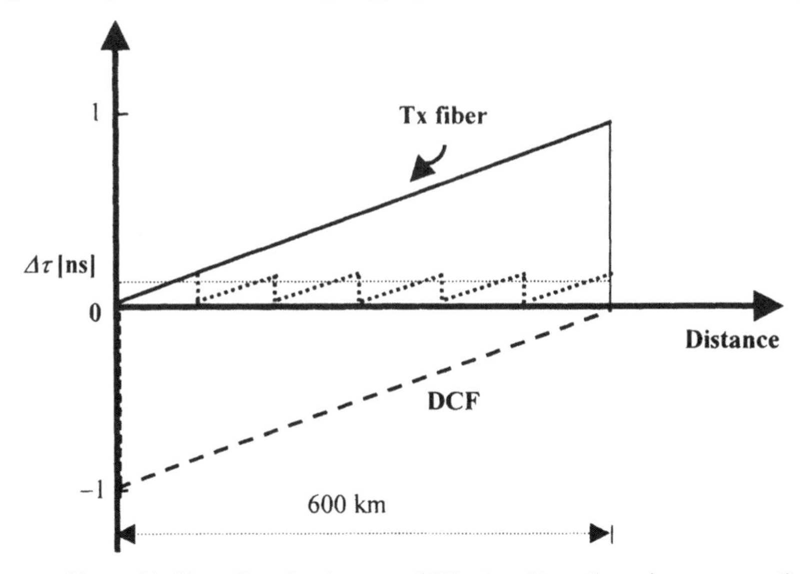

Figure 11 Illustration of various possibilities to achieve dispersion compensation through a DCF (adapted from [46]; ©IEEE).

of the fiber so that at the wavelengths of interest a large fraction of its modal power rapidly spreads into the cladding region for a small variation in the propagating wavelength. In other words, rate of increase in mode size is large for a small increase in λ. Accordingly the modal power distribution is strongly influenced by the cladding, where the mode travels effectively faster due to lower refractive index.[46] In order to maintain transmission loss low and compatibility with telecommunication grade high-silica fibers the material choice was somewhat limited. Thus, the design options were restricted to refractive index distributions alone in order to generate net negative dispersion large. The *first generation* DCFs indeed relied on narrow core and high core-cladding refractive index contrast (Δ) fibers to fulfill this task; in these typically the fiber refractive index profiles were similar to those of the matched clad single-mode fibers with the difference that Δ of the DCF were much larger (≥ 2 %). These DCFs were targeted to compensate dispersion in G.652 fibers at a single wavelength and were characterized with a $D \sim -50$ to -100 ps/(nm.km) and a positive dispersion slope. Due to high Δ and poor mode confinement as explained above, a DCF necessarily involved large insertion loss (typically attenuation in a

DCF could vary in the range 0.5 – 0.7 dB/km)[11] and high sensitivity to bend-induced loss. In order to simultaneously achieve a high negative dispersion coefficient and low attenuation coefficient, α_D, DCF designers have ascribed a *Figure of Merit*, *FOM* to a DCF, which is defined through

$$FOM = -\frac{D_D}{\alpha_D} \tag{34}$$

and expressed in ps/nm.dB. Total attenuation and dispersion in dispersion compensated links would be given by

$$\alpha = \alpha_T L_T + \alpha_D L_D \tag{35}$$

$$D = D_T L_T + D_D L_D \tag{36}$$

It could be shown from Eqs.(34) – (36), that for $D = 0$

$$\alpha = \left(\alpha_T + \frac{D_T}{FOM}\right) L_T \tag{37}$$

which shows that any increase in total attenuation in dispersion compensated links would be solely through *FOM* of the DCF; thus, larger the *FOM* is, smaller would be incremental attenuation in the link due to insertion of the DCF. Dispersion slope (S_D) for the first generation DCF were of the same sign as the transmission fiber and hence perfect compensation was realizable only at a single wavelength in this class of DCFs.

Since DWDM links involve multi channel transmission, it is imperative that ideally one would require a broadband DCF so that dispersion could be compensated for all the wavelength channels simultaneously. Broadband dispersion compensation ability of a DCF is quantifiable through a parameter known as relative dispersion slope (*RDS*) expressed in unit of nm^{-1}, which is defined through

$$RDS = \frac{S_D}{D_D} \tag{38}$$

A related parameter referred to as κ (in unit of nm) also finds reference in the literature and it is simply the inverse of *RDS*.[47] Value of *RDS* for LEAF™, Truewave-RS™, and Teralight™ are 0.0202, 0.01, and 0.0073, respectively. Thus if a DCF is so designed that its *RDS* (or κ) matches that of the transmission fiber then that DCF would ensure perfect compensation for all the wavelengths. Such DCFs are known as dispersion slope compensating fibers (DSCF). A schematic for the dispersion spectrum of such a DSCF along with single-wavelength DCF is shown in Fig.12. Any differences in the value of *RDS* (or κ) between the transmission fiber and the DCF would result in under- or over- compensation of dispersion leading to increased BERs at those channels. In practice a fiber designer targets to match values of the *RDS* for the Tx fiber and the DCF at the median wavelength of a particular amplification band i.e. C- or L- band. In principle this is sufficient to achieve dispersion slope compensation across that particular wavelength band because dispersion spectrum of the transmission fiber is approximately linear within a specific amplification band. However, other propagation issues e.g. bend loss sensitivity, countenance of nonlinear optical effects through large mode effective area etc often demands a compromise between 100% dispersion slope compensation

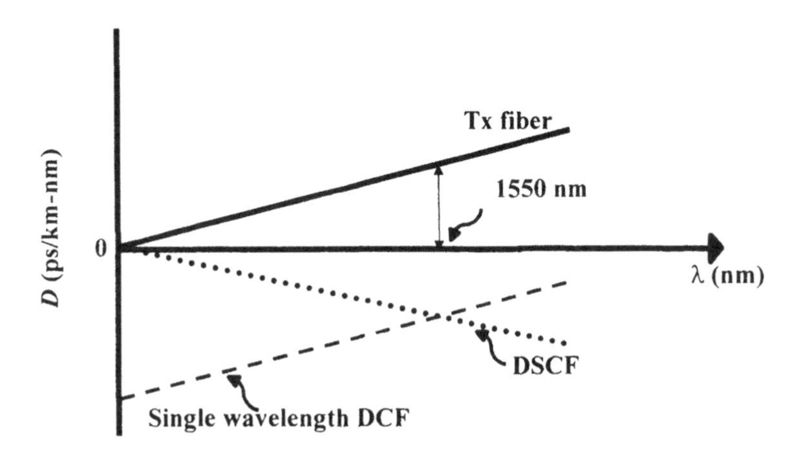

Figure 12 Dispersion slope compensating fiber for broadband dispersion compensation.

and largest achievable mode effective area. In such situations, dispersion slope S_D may not precisely match for those wavelength channels, which fall at the edges of a particular amplification band. However for bit rates up to 10 Gbit/s it does not pose any serious problem because fractional mismatch as much as 20% between S_T and S_D is tolerable though it becomes a much tighter proposition at 40 Gbit/s, in which case only upto 1% mismatch is allowed.[11]

Typical value of *RDS* for G.652 type of SMF at 1550 nm is about 0.00335 nm^{-1} (or κ as 298 nm) while in LEAF and Truewave fibers its values are about 0.026 $nm^{-1}$and 0.01 nm^{-1}, respectively. The insertion loss of commercial DCFs, which approximately compensate for an 80 km transmission span is about 7 dB. Ideally a DCF design should be so optimized that its insertion loss is low, low sensitivity to bend loss, large negative D with an appropriate negative slope for broadband compensation, large mode effective area (A_{eff}) for reduced sensitivity to nonlinear effects and low polarization mode dispersion. Few proprietary designs of DCF index profiles involve multiple claddings in order to achieve better control on the mode-expansion induced changes in the guided mode's effective index. One recent design that yielded the record for largest negative D (– 1800 ps/km.nm at 1558 nm) was based on a dual-core refractive index profile (see Fig.13).[48,49] It can be seen that the two cores are interconnected through a matched cladding in contrast to most of the other proprietary designs, in which there is usually a depressed index clad or a moat region in the cladding. In the fabricated perform (Fig.13b), the central core was more of a triangular shape (shown as dotted line in Fig.13 a) than a step distribution. The two cores essentially function like a directional coupler.[48] Since these two concentric fibers are significantly non-identical, at some desired wavelength (called

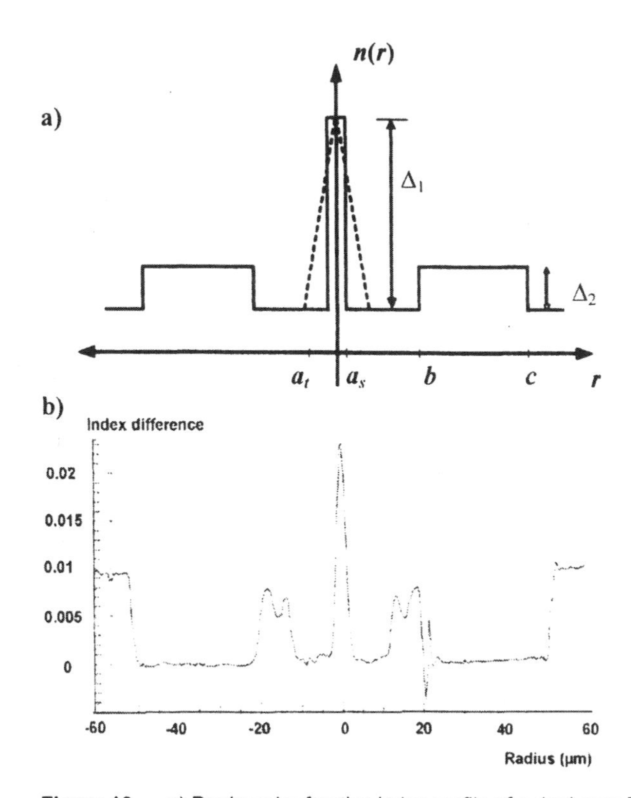

Figure 13 a) Designed refractive index profile of a dual-core DCF; b) Preform r.i. profile of the fabricated dual-core DCF[49]

phase matching wavelength λ_P) their effective indices could be made equal, in which case the effective indices as well as modal field distributions of the normal modes of this dual core fiber would exhibit rapid variations with λ around λ_P.[48] Typical distributions of modal power for three different wavelengths around the phase matching wavelength (λ_p) for one dual-core fiber in which a wide inner core of smaller Δ is coaxially surrounded by a narrow higher index ring are shown in Fig.14. These rapid variations in modal distributions with wavelength is essentially exploited to tailor dispersion characteristics of the DCF e.g. achieve very large negative dispersion or broadband dispersion compensation characteristics.[50-52] Figure 15 shows the measured chromatic dispersion spectrum on the DCF drawn from the above performs. Mode effective area of the commercial DCFs typically range between 15 to 25 μm^2, which make these susceptible to nonlinear effects unless care is taken to reduce launched power (≤ 1 dBm per channel). In contrast, the ones based on dual core DSCFs could be designed to attain A_{eff}, which are comparable to

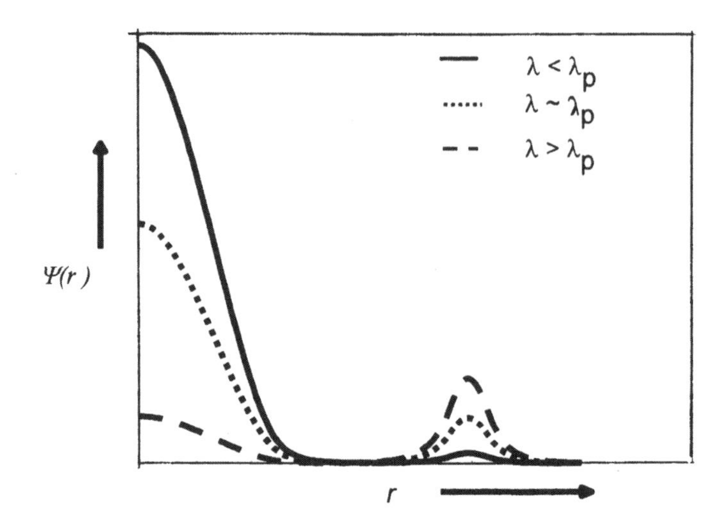

Figure 14 Modal power distributions for different λ relative to λ_P

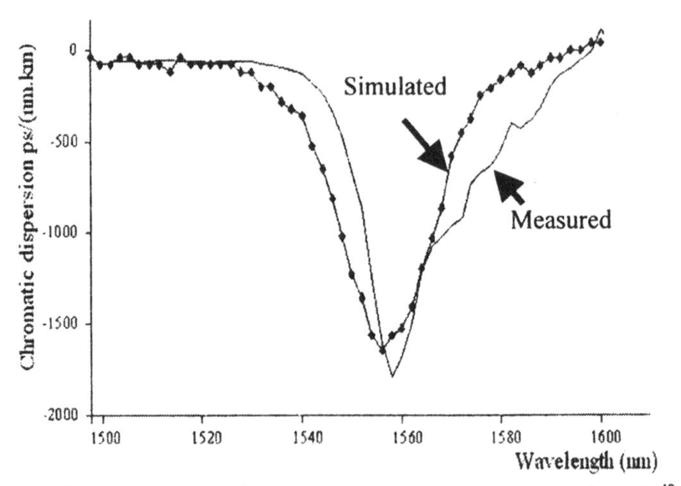

Figure 15 Comparison between simulated and measured results[49].

that of the G.652 fiber (≈ 80 μm^2). The net residual dispersion spectra of a 100 km long G.652 fiber link along with so designed DSCFs (approximately in 10:1 ratio) at each of the amplifier band are shown in Fig. 16. It could be seen that residual average dispersion is well within ± 1 ps/km.nm within all the three bands. Other estimated performance parameters of the DSCFs are shown in Table III. Similar results were obtained for a set of designed DSCFs for standard G.655 fibers for various amplifier bands.[51]

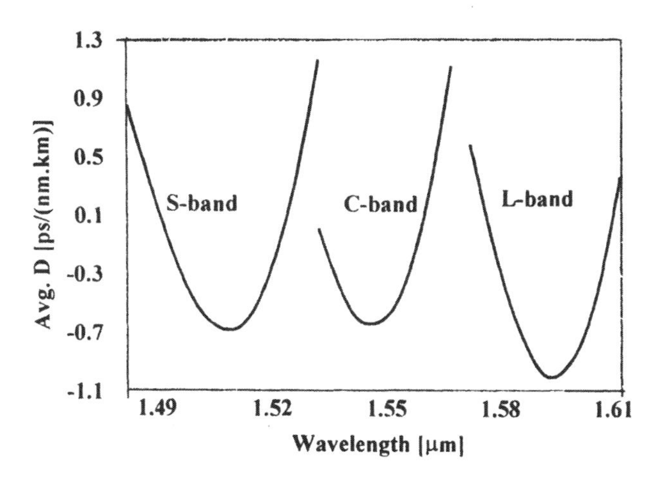

Figure 16 Residual dispersion spectra of the jointed DCF with G.652 transmission fiber in the different amplifier bands

Table III

Important performance parameters of designed DCFs for G.652 fibers[51]

Amplifier band nm	D ps/km.nm	RDS nm^{-1}	A_{eff} μm^2	MFD μm	Estimated FOM ps/dB.nm	Bend loss[#] dB
S (1500)	−182	0.0056	51	7.15	771	0.0221
C (1550)	−191	0.0027	70	7.97	941	0.095
L (1590)	−162	0.0034	70	8.12	837	0.0014

[#]for a single-turn bend of diameter 32 mm

8. Reverse/Inverse dispersion fibers

The above-mentioned DCFs are normally introduced in a link in the form of a spool as a stand-alone module. An alternative scheme involves use of the compensating fiber within the cable itself after having been jointed with the transmission fiber as part of the overall fiber link. Such DCFs are referred as *reverse/inverse* dispersion fibers (RDF/IDF).[53,54] RDFs are used with the transmission fiber almost in the ratio of 1:1 meaning thereby that their characteristic dispersion parameters D and S are almost same as that of the transmission fiber except for the signs. For example, the RDF used in [53], D and S were − 16 ps/km.nm and −0.050 ps/km.nm^2, respectively.

One major attribute of these fibers was that loss in such fibers was low (in contrast to high negative *D* DCFs) ~ 0.25 dB/km. Terabit transmission experiments with such RDFs in conjunction with G.652 fibers over a BW of 50 nm through a repeater span of over 125 km have been reported[55] as well as experiments, in which RDFs were used with NZ-DSFs (in a ratio of 1:2).[56]

9. Fibers for metro networks

In recent years metro optical networks have attracted a great deal of attention due to potentials for high growth. A metro network provides generalized telecommunication services transporting any kind of signal from point to point in a metro. Metro networks with distances less than 200 km, which bridges the gap between local/access networks and long distance telecommunication networks were originally designed essentially to link local switching offices, from which telephone lines branch out to individual customers in the access portion of the network. In larger cities, it may consist of two layers- *metro access or edge loops* (20 to 50 km), which connect groups of switching offices, and these loops, in turn, link to a *metro core* or *backbone* (50 to 200 km) network (having several nodes) that serves the whole metropolitan area.[7] However with the growth in Internet usage and increasing demand in data services a need arose to migrate from conventional network infrastructure to an *intelligent* and *data-centric* metro network. Accordingly the network trend in the metro sector has been to move towards *transparent* rings, in which wavelength channels are routed past or dropped off at the nodes.[57] New services are now fed optically at the network edge; for example, *ESCON* (enterprise system connection, which is a 200 Mb/s protocol meant for providing connectivity of a main frame to other mainframe storage devices), *Gigabit Ethernet* (which is a simple protocol that uses inexpensive electronics and interfaces for sharing a local area network), *FDDI* (a LAN standard covering transmission @ 100 Mb/s, which can support up to 500 nodes on a dual ring network with an inter-node spacing up to 2 km), and digital video. Gigabit Ethernet is fast evolving as a universal protocol for optical packet switching. For example, most of the data on the Internet start as Ethernet packets generated by system servers. In order to offer service providers the flexibility to increase network capacity in response to customer demand, a metro network is required to address unique features like *low first cost*, *high degree of scalability* to efficiently accommodate unpredicted traffic growth, flexibility to add/drop individual signals at any central office in the network, interoperability to carry a variety of signals and provide connectivity to variety of equipment like cell phones, SONET/SDH, legacy equipment, ATM, and IP. Legacy metro networks relied heavily on G.652 fibers exploiting near-zero dispersion at 1310 nm. In these network designs, technological adaptability is extremely important to outweigh obsolescence because outside plant fixed costs like trenching, ducts, etc are almost twice of long-haul networks.[57] DWDM in a metro network is attractive for improved speed in provisioning due to possibility of allocating dynamic bandwidth to customers on demand.[58] With DWDM transmission becoming an attractive option to absorb ever-increasing demand in bandwidth, Corning came out with a MetroCor™ fiber in particular to maximize the uncompensated reach of directly modulated lasers. Directly modulated lasers in contrast to externally modulated lasers is an economic advantage in a metro environment. However directly modulated lasers are usually accompanied with a positive chirp, which could introduces severe pulse dispersion in the EDFA band if the transmission fiber is characterized with a positive *D*. The positive chirp-induced pulse broadening can be countered if the transmission fiber is so designed that it exhibits normal dispersion (i.e. negative *D*) at the EDFA band. This is precisely the design philosophy followed by certain fiber manufacturers.[58] As a typical example, a schematic of the dispersion spectrum of one type of metro fiber relative to that of a standard single-mode fiber (SSMF) is shown in Fig.17. Prior to introduction of metro-specific fibers, for economic reasons

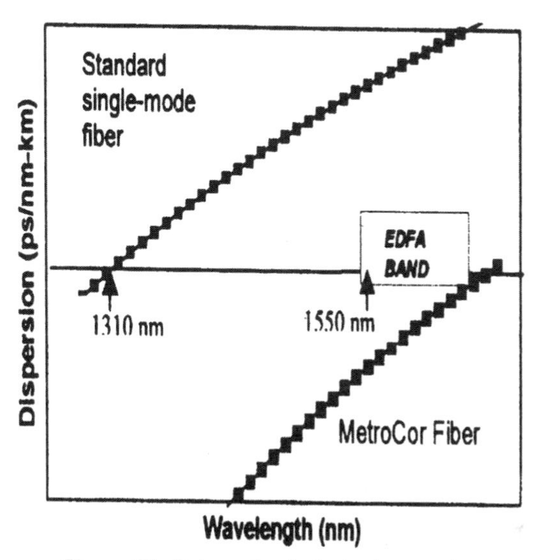

Figure 17 Schematic of dispersion spectrum of a MetroCorfiber relative to SSMF (adapted from [58]; ©Corning Inc 2000)

most of the metro systems till late 1990s involved use of SSMF with multimode Fabry-Perot lasers at 1310, which have little dispersion tolerance; capacity upgrades were based on lighting of new fibers. Deployment of SSMF in contrast to a metro-specific fiber would invariably require DCF thereby requiring additional cost, which could be avoided through a metro-specific negative dispersion fiber. In contrast to this philosophy an alternate school recommends a positive dispersion fiber for metro, details of which are discussed in.[57]

10. Coarse wavelength division multiplexing

With the availability of OH^--free single-mode fibers (see Fig.1), which are sometime referred to as low water peak fiber (LWPF) and christened as G.652C fibers by ITU (e.g. AllWave™ and SMF-28e™), a new market has emerged for configuring short distance communication (up to ~ 20 km, e.g. for campus backbone) networks with these fibers. Due to wider (by as much as 30%) low loss wavelength window in the

E-band offered by these fibers in comparison to G.652 legacy fibers, one could use these to transmit widely spaced wavelengths as carriers with low-cost un-cooled FP-lasers as the light sources. Recently introduced ITU standards allow channel spacing as wide as 20 nm under coarse wavelength division multiplexing (CWDM) scheme. Under CWDM scheme laser cooling becomes unnecessary because environmental temperature-induced wavelength wander would not be an issue, thereby bringing down system complexity and overall network cost because wavelength tolerance demands on other components like filters would also be much less stringent as compared to DWDM systems.

11. Microstructured fibers

Consequent to the mind boggling progress on high speed optical telecommunications that was witnessed in recent times, it appeared for a while that it would be only a matter of time when the huge 53 THz of theoretical bandwidth offered by low-loss transmission windows (extending from 1280 nm (235 THz) to 1650 nm (182 THz)) in OH^--free high-silica optical fibers would be tapped for telecommunication through DWDM techniques! Inspite of this possibility on the horizon, there has been a considerable amount of resurgence of interest amongst researchers to develop specialty fibers, e.g. fibers in which transmission loss of the material would not be a limiting factor, and in which nonlinearity or dispersion properties could be conveniently tailored to achieve transmission characteristics that are otherwise almost impossible to realize in conventional high-silica fibers. Research targeted at such fiber designs gave rise to a new class of fibers known as *microstructured* optical fibers.[59] One category of such microstructured fibers is known as *photonic bandgap* (PBG) fibers (see Chapter 6). In a PBG fiber, light of *certain frequency bands* cannot propagate along directions perpendicular to the fiber axis but, instead, are free to propagate along its length confined to the fiber core. This feature of forbiddance of photons (of certain frequencies) to propagate transverse to the axis in certain type of microstructured fibers led to the christening of such specialty fibers as PBG-guided optical fibers, in analogy with electronic bandgaps encountered by electrons in semiconductors. In contrast to the electronic bandgap, which is a consequence of periodically arranged atoms/molecules in a semiconductor crystal lattice, photonic bandgap arises due to a *periodic distribution of refractive index* in certain dielectric structures, generically referred to as *photonic crystals* (reminiscent of semiconductor crystals in solid-state physics).[60] Photonic bandgap (for photons) in photonic crystals is analogous to electronic bandgap (for electrons) in semiconductor crystals. If the frequency of incident light happens to fall within the photonic bandgap, which is characteristic of a photonic crystal, then light propagation is forbidden in the photonic crystal. Depending on the number of dimensions in which periodicity in refractive index exists, photonic crystals are classified as either one-dimensional or two-dimensional or three-dimensional photonic crystals. A Bragg fiber (as shown in Fig.18) is an example of 1-D photonic crystal and it consists of a core surrounded by a series of periodic layers of alternate high and low refractive index materials (each of which has a refractive index higher than that of the core). The thickness of the

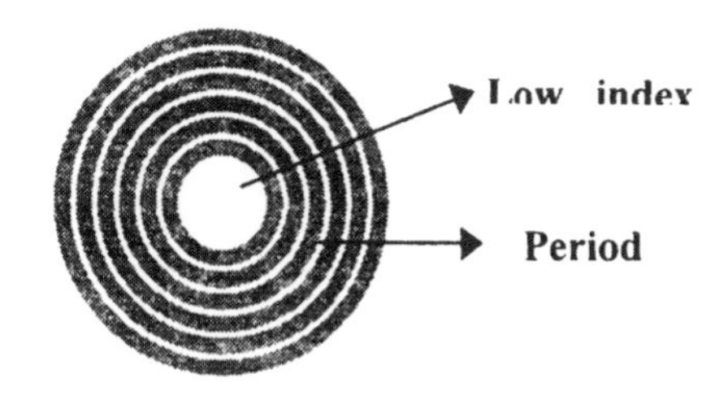

Figure 18a) Cross-sectional view of a Bragg fiber.

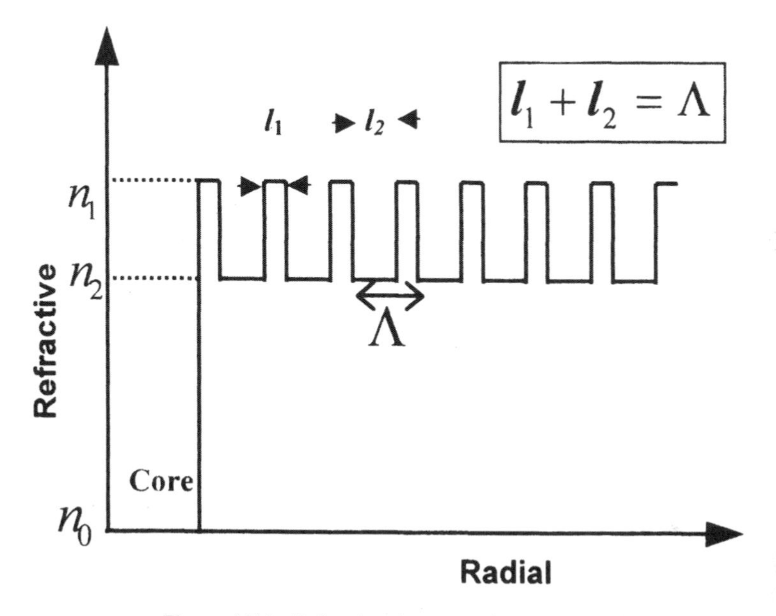

Figure 18b) Refractive index profile of Bragg fiber.

cladding layers is chosen suitably so as to induce Bragg reflections. The periodic layers constituting the cladding form a photonic bandgap (PBG). The PBG, analogous to the electronic bandgap for electrons, does not allow light of certain frequencies to propagate along the direction of periodicity. Thus, in a Bragg fiber, light within a certain frequency range is confined within the core through the mechanism of *photonic bandgap guidance*, in contrast to total internal reflection in conventional fibers. The multilayer cladding of the Bragg fiber acts like a highly reflecting mirror due to *in-phase* multiple reflections that add up from each boundary between the high and the low refractive index layers. The precise nature of the bandgap structure depends on the quantitative values of the refractive index contrast and periodicity of the cladding layers.

Light confinement through photonic bandgap guidance allows the possibility of having an air core. Also, large refractive index contrast in cladding layers allows a large fraction of the guided light to be confined within the air core; thereby enabling low-loss optical transmission and reduced sensitivity to optical nonlinearity. Bragg fibers involve several independent physical parameters e.g. core size, refractive indices and thickness of the cladding layers, and hence offer a wide choice of parametric avenues to tailor their propagation characteristics. As is true with any fiber, loss and dispersion are the two most important propagation characteristics of a Bragg fiber. Owing to their unique guidance mechanism, Bragg fibers can exhibit dispersion characteristics that are otherwise nearly impossible to achieve in conventional silica fibers. Reported designs of Bragg fibers have indicated the potential feasibility of achieving zero dispersion at wavelength ~ 1 μm, multiple zero-dispersion and high negative dispersion ~ – 20,000 ps/km.nm.[61-64] Recently two designs of Bragg fibers: one to achieve a highly efficient (estimated FOM ~ 190000 ps/nm.dB at the C-band) dispersion compensator for long-haul networks (see Fig.19) and the second one to achieve non-zero dispersion-shifted fiber suitable for application as a metro fiber (Fig.20) have been reported, details of which could be found in [65-67].

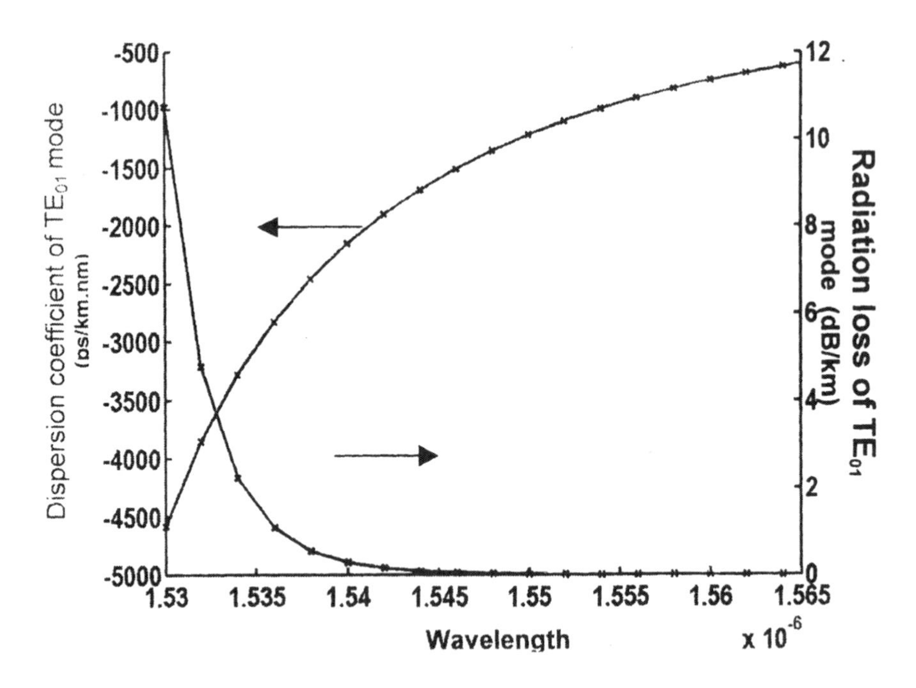

Figure 19 Loss and dispersion spectra of a Bragg fiber-based DCF designed for a ultra-high FOM.[65]

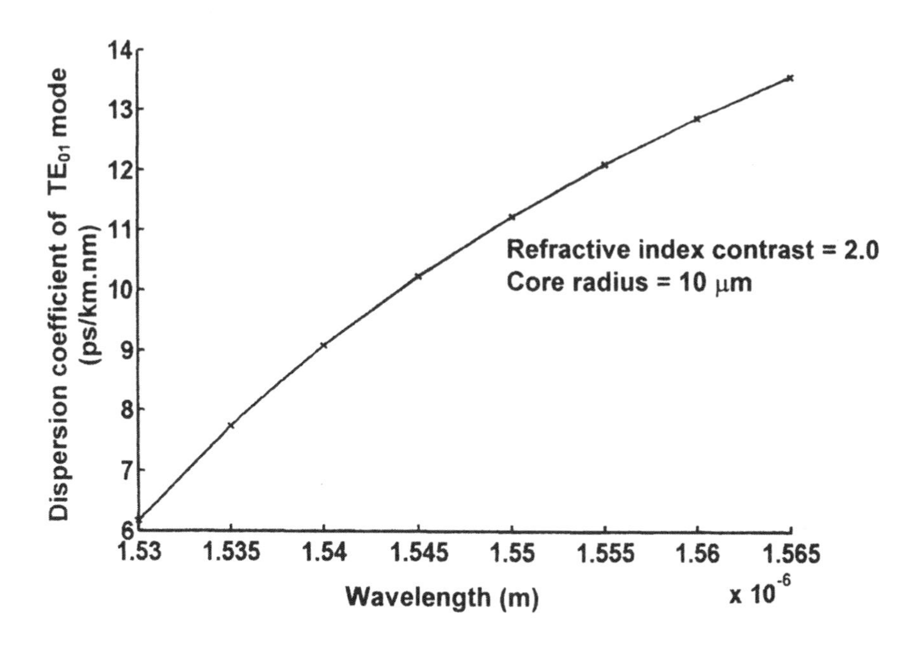

Figure 20 Designed dispersion spectrum of a Bragg fiber-based metro fiber with an average dispersion ~ 9.9 ps/km.nm and average loss ~ 0.04 dB/km[66]

12. Conclusion

In this chapter we have attempted to review the trends in the design of single-mode fibers that have evolved over the years for optical telecommunication. Starting with basic physical principles underlying optical transparency namely, loss and dispersion in a single-mode optical fiber, evolution of optical network based on legacy G.652 fibers and more advanced G.655 fibers have been presented. Short discussions have also been presented on more recent fibers meant for applications in metro networks as well as applications involving low water peak fibers. Finally a short discussion on most recent research on Bragg fiber type of photonic bandgap microstructured fiber designs for dispersion compensation and potential applications in metro networks are also presented.

Acknowledgement: I would like to acknowledge many stimulating and educative discussions that I have had with colleagues of our Fiber Optics group and my Graduate students at various times during the last 30 years, which helped me to enrich my understanding of this exciting field of guided wave optics over all these years.

References

1. F.P. Kapron, D.B. Keck, and R.D. Maurer, "Radiation losses in glass optical waveguides," *App. Phys. Lett.* 17, 423 - 425 (1970).
2. D.N. Payne and W.A. Gambling, "Zero material dispersion in optical fibers," *Electron. Lett.* 11, 176-178 (1975).
3. J.I. Yamada, S. Machida, and T. Kimura, "2 Gb/s optical transmission experiments at 1.3 μm with 44 km single-mode fiber," *Electron. Lett.* 17, 479 - 480 (1981).
4. R.J. Mears, L. Reekie, I.M. Jauncy, and D.N. Payne, "Low-noise erbium-doped fiber amplifier operating at 1.54 μm," *Electron. Lett.* 23, 1026 - 1027 (1987).
5. T. Miya, Y. Terunume, T. Hosaka, and T. Miyashita, "An ultimate low-loss single-mode fiber at 1.55 μm," *Electron. Lett.* 15, pp. 106-108 (1979).
6. S.K. Kartalopoulos, *Introduction to DWDM Technology*, SPIE Press, Bellingham, Washington and IEEE Press, Piscataway, NJ (2000).
7. J. Hecht, *Understanding Fiber Optics* (4th ed.), Prentice Hall, New Jersey (2002).
8. D.J. DiGiovanni, S. das, L.L. Blyler, W. White, and R.K. Boneck, Chapter on "Design of optical fibers for communications systems," in *Optical Fiber Telecommunications IVA*, Academic Press, NY (2002).
9. F.P. Kapron, Chapter on "Transmission properties of optical fibers" in *Optoelectronic Technology and Lightwave Communication Systems*, Ed. Chinlon Lin, Van Nostrand, New York (1989).
10. K. Kitayama, Y. Kato, M. Ohashi, Y. Ishida, and N. Uchida, "Design considerations for the structural optimization of a single-mode fiber," IEEE J. Lightwave Tech. LT-1, 363-369 (1983).
11. N. Uchida, "Development and future prospects of optical fiber technologies," IEICE Trans. Electron. E85 C, 868-880 (2002).
12. B.P. Pal, Chapter on "Transmission characteristics of telecommunication optical fibers" in *Fundamentals of Fiber Optics in Telecommunication and Sensor Systems*, B.P. Pal (Ed.), John Wiley, New York (1992); 2nd Reprint, New Age pub., New Delhi (2002).
13. A. Ghatak and K. Thyagarajan, *Introduction to Fiber Optics*, Cambridge University Press, Cambridge (1998).
14. B.P. Pal, Chapter on "Optical Transmission" in *Perspective in Optoelectronics*, S.S. Jha (Ed.), World Scientific, Singapore (1995).
15. A. Yariv, *Optical Electronics in Modern Communication*, Oxford University Press, New York (1997).
16. K. Thyagarajan and B.P. Pal, Chapter 1 on "Modeling dispersion in optical fibers: applications to dispersion tailoring and dispersion compensation," in *Fiber-based Dispersion Compensation*, Siddharth Ramachandran (Ed.), Springer-Verlag, Berlin (in press, 2005).
17. B.P. Pal, Chapter on "Optical fibers for lightwave communication: evolutionary trends in designs," in *Guided Wave Optical Components and Devices*, B.P. Pal (Ed.), Elesevier, Burlington (in press, 2005).

18. D. Marcuse, "Interdependence of waveguide and material dispersion," *App. Opt.* 18, 2930-2932 (1979).
19. G.P. Agrawal, *Lightwave Technology: Components and Devices*, John Wiley and Sons, New Jersey (2004).
20. R. Ramaswamy and K. N. Sivarajan, *Optical Networks: A practical Perspective*, Harcourt Asia, Singapore (1998).
21. R.J. Boehm, Y.C. Ching, C.G. Griffith, and F.A. Saal, "Standardized fiber optic transmission system – a synchronous optical network view," *IEEE J. Selected Areas in Commn.* SAC-4, 1424-1431 (1986).
22. A. Ghatak and B.P. Pal, Chapter on "Progress in Fiber Optics for Telecommunication" in the book *International Trends in Applied Optics*, Ed. A. Guenther, SPIE, Washington, 359-388 (2002).
23. B.P. Pal, Chapter on "Optical fibers for lightwave communication: Design issues," in *Fiber optics and Applications*, A.K. Ghatak, B. Culshaw, V. Nagarajan and B.D. Khurana (Eds.), VIVA publishers, New Delhi, (1995).
24. R. Tewari, B.P. Pal, and U.K. Das, "Dispersion-shifted dual-shape-core fibers: optimization based on spot size definitions," *IEEE J. Lightwave Tech.* 10, 1 – 5 (1992).
25. R.J. Mears, L. Reekie, S.B. Poole, and D.N. Payne, "Low-threshold tunable cw and Q-switched fiber laser operating at 1.55 μm," *Electron. Lett.* 22, 159-160 (1986).
26. R.J. Mears, L. Reekie, I.M. Jauncy, and D.N. Payne, "Low-noise erbium-doped fiber amplifier operating at 1.54 μm," *Electron. Lett.* 23, 1026 - 1027 (1987).
27. C.J. Koester and E. Snitzer, "Amplification in a fiber laser," *App. Opt.* 3, 1182-1184 (1964).
28. E. Desurvire, *Erbium doped fiber amplifier: Principle and applications*, Wiley-Interscience, New York (1994).
29. K.M. Abe, Optical fiber designs evolve, Lightwave (February 1998).
30. A.K. Srivastava, Y. Sun, J.W. Sulhoff, C. Wolf, M. Zirngibl, R. Monnard, A.R. Chraplyvy, A.A. Abramov, R.P. Espindola, T.A. Strasser, J.R. Pedrazzani, A.M. Vengsarkar, J.L. Zyskind, J. Zhou, D.A. Ferrand, P.F. Wysocki, J.B. Judkins, S.W. Granlund, and Y.P. Li, "1 Tb/s Transmission of 100 WDM 10 Gb/s Channels over 400 km of Truewave™ Fiber," Technical Digest *Optical Fiber Communication Conference*'1998, Opt. Soc. Am., Washington, Post-deadline paper PD10 (1998).
31. ITU-T-Recommendation G.692: *Optical Interfaces for multichannel systems with optical amplifiers.*
32. K. Fukuchi, T. Kasamatsu, M. Morie, R. Ohhira, T. Ito, K. Sekiya, D. Ogasahara, and T. Ono, "10.92-Tb/s (273 x 40-Gb/s) triple-band/ultra-dense WDM optical-repeatered transmission experiment," Technical Digest *Optical Fiber Communication Conference*'2001, Opt. Soc. Am., Washington (2001) Post-deadline paper PD24.
33. S. Bigo, Y. Frignac, G. Charlet, S. Borne, P. Tran, C. Simonneau, D. Bayert, A. Jourdan, J.-P. Hamaide, W. Idler, R. Dischler, G. Veith, H. Gross, and W. Poehlmann, "10.2Tbit/s (256x42.7Gbit/s PDM/WDM) transmission over 100km

TeraLightô fiber with 1.28bit/s/Hz spectral efficiency," – *ibid*, (2001) Post-deadline paper PD25.

34. T. Li, 'The impact of optical amplifiers on long-distance lightwave telecommunication," *Proc. IEEE*, 81, 1568-1579 (1995).
35. D.W. Peckham, A.F. Judy, and R.B. Kummer, "Reduced dispersion slope, non-zero dispersion fiber," in *Proceedings of 24th European Conference on Optical Communication ECOC'98* (Madrid, 1998), 139-140 (1998).
36. Y. Liu, W.B. Mattingly, D.K. Smith, C.E. Lacy, J.A. Cline, and E.M. De Liso, "Design and fabrication of locally dispersion-flattened large effective area fibers," in *Proceedings of 24th European Conference on Optical Communication ECOC'98* (Madrid, 1998), 37-38 (1998).
37. J. Ryan, Special report "ITU G.655 adopts higher disperсion for DWDM," *Lightwave*, 18, no. 10 (2001).
38. Y. Danziger and D. Askegard, "High-order-mode fiber - an innovative approach to chromatic dispersion management that enables optical networking in long-haul high-speed transmission systems," *Opt. Networks Mag.* 2, 40-50 (2001).
39. Y. Frignac and S. Bigo, "Numerical optimization of residual dispersion in dispersion managed systems at 40 Gb/s," in *Proceedings of Optical Fiber Communications Conference OFC 2000* (Baltimore, 2000), 48-50 (2000).
40. Izadpanah H, Lin C, Gimlett H, Johnson H, Way W, and Kaiser P, "Dispersion compensation for upgrading interoffice networks built with 1310 nm optimized SMFs using an equalizer fiber, EDFAs, and 1310/1550 nm WDM," Tech Digest Optical Fiber Communication Conference, Post-deadline paper PD15, 371-373 (1992).
41. Y. Akasaka, R. Suguzaki, and T. Kamiya, "Dispersion Compensating Technique of 1300 nm Zero-dispersion SM Fiber to get Flat Dispersion at 1550 nm Range," in *Digest of European Conference on Optical Communication*, paper We.B.2.4 (1995).
42. A.J. Antos and D.K. Smith, "Design and characterization of dispersion compensating fiber based on LP_{01} mode," *IEEE J. Lightwave Tech.* LT-12, 1739 – 1745 (1994).
43. D.W. Hawtoff, G.E. Berkey, and A.J. Antos, "High Figure of Merit Dispersion Compensating Fiber," in *Technical Digest Optical Fiber Communication Conference* (San Jose), Post-deadline paper PD6 (1996).
44. A.M. Vengsarkar, A.E. Miller, and W.A. Reed, "Highly efficient single-mode fiber for broadband dispersion compensation," in *Proceedings of Optical Fiber Communications Conference OFC'93* (San Jose), 56-59 (1993).
45. L. Grüner-Nielsen, S.N. Knudsen, B. Edvold, T. Veng, B. Edvold, T. Magnussen, C.C. Larsen, and H. Daamsgard, "Dispersion Compensating Fibers," *Opt. Fib. Tech.* 6, 164-180 (2000).
46. B. Jopson and A. Gnauck, "Dispersion compensation for optical fiber systems," *IEEE Comm. Mag.* June, 96-102 (1995).
47. V. Srikant, "Broadband dispersion and dispersion slope compensation in high bit rate and ultra long haul systems," in *Technical Digest Optical Fiber Communication Conference* (Anaheim, CA.), paper TuH1 (2001).

48. K. Thyagarajan, R.K. Varshney, P. Palai, A.K. Ghatak, and I.C. Goyal, "A novel design of a dispersion compensating fiber," *IEEE Photon. Tech. Letts.* 8, 1510-1512 (1994).
49. J.L. Auguste, R. Jindal, J.M. Blondy, M. Clapeau, J. Marcou, B. Dussardier, G. Monnom, D.B. Ostrowsky, B.P. Pal and K. Thyagarajan, "–1800 ps/km-nm chromatic dispersion at 1.55 μm in a dual concentric core fiber," *Electron. Lett.* 36, 1689-1691 (2000).
50. P. Palai, R.K. Varshney, and K. Thyagarajan, "A dispersion flattening dispersion compensating fiber design for broadband dispersion compensation," *Fib. Int. Opt.* 20, 21-27 (2001).
51. B.P. Pal, and K. Pande, "Optimization of a dual-core dispersion slope compensating fiber for DWDM transmission in the 1480-1610 nm band through G.652 single-mode fibers," *Opt. Comm.* 201, 335-344 (2002).
52. K. Pande and B.P. Pal, "Design optimization of a dual-core dispersion compensating fiber with high figure of merit and a large effective area for dense wavelength division multiplexed transmission through standard G.655 fibers," *Appl. Opt.* 42, 3785-3791 (2003).
53. K. Mukasa, Y. Akasaka, Y. Suzuki, and T. Kamiya, "Novel network fiber to manage dispersion at 1.55 μm with combination of 1.3 μm zero dispersion single-mode fiber," Proc. 23rd European Conf on Opt. Commn. (ECOC), Edinburgh, Session MO3C, 127-130 (1997).
54. T. Grüner-Nielsen, S.N. Knudsen, B. Edvold, P. Kristensen, T. Veng, and D. Magnussen, Dispersion compensating fibers and perspectives for future developments," Proc. 26th Europ. Conf. Opt. Comm. (ECOC'200), 91-94 (2000).
55. Y. Miyamoto, K. Yonenaga, S. Kuwahara, M. Tomizawa, A. Hirano, H. Toba, K. Murata, Y. Tada, Y. Umeda, and H. Miyazawa, "1.2 Tbit/s (30x42.7 Gb/s ETDM optical channel) WDM transmission over 376 km with 125-km spacing using forward error correction and carrier suppressed RZ format," Tech. Dig. Opt. Fib. Comm. Conference (OFC'2000), Baltimore, MD, Post-deadline paper, PD26 (2000).
56. L-A. de Montmorrilon, F. Beamont, M. Gorlier, P. Nouchi, L. Fleury, P. Sillard, V. Salles, T. Sauzeau, C. Labatut, J.P. Meress, B. Dany, and O. Leclere, "Optimized Teralight™/reverse Teralight© dispersion-managed link for 40 Gbit/s dense WDM ultra-long haul transmission systems," Proc. 27th Europ. Conf. Opt. Comm. (ECOC'2001), Amsterdam, 464-465 (2001).
57. J. Ryan, "Fiber considerations for metropolitan networks", Alcatel Telecom. Review, 1st Quarter, 52-56 (2002).
58. D. Culverhouse, A. Kruse, C-C. Wang, K. Ennser, and R. Vodhanel, "Corning® MetroCor™ fiber and its applications in metropolitan networks," Corning Inc. White Paper WP5078 (2000).
59. T. Monro, Chapter on "Microstructured Fibers," in *Guided Wave Optical Components and Devices*, B.P. Pal (Ed.), Elesevier, Burlington (in press, 2005).
60. J. D. Joannopoulos, R. D. Meade, and J. N. Winn, *Photonic Crystals: Molding the Flow of Light*, Princeton University Press (1995).

61. F. Brechet, P. Roy, J. Marcou, and D. Pagnoux, "Single-mode propagation into depressed-core-index photonic bandgap fiber designed for zero-dispersion propagation at short wavelengths," *Electron. Lett.*, 36, 514-516 (2000).
62. T. D. Engeness, M. Ibanescu, S. G. Johnson, O. Weisberg, M. Skorobogatiy, S. Jacobs and Y. Fink, "Dispersion tailoring and compensation by modal interactions in omniguide fibers", *Opt. Exp.*, 11, 1175 (2003).
63. G. Ouyang, Y. Xu, and A. Yariv, "Theoretical study on dispersion compensation in air core Bragg fibers," *Opt. Exp.*, **10**, 899 (2002).
64. P. Yeh, A. Yariv, and E. Marom, "Theory of Bragg fiber," *J. Opt. Soc. Amer.*, 68, 1196 (1978).
65. S. Dasgupta, B.P. Pal, and M.R. Shenoy, "Design of a Low Loss Bragg Fiber with High Negative Dispersion for the TE01 Mode", *2004 Frontiers in Optics*, Pres. no. **FWH49,** Rochester, New York, USA (2004)
66. S. Dasgupta, B.P. Pal, and M.R. Shenoy, "Dispersion tailoring in Bragg fibers," International Conference on Photonics (PHOTONICS 2004), Cochin, December 8-11 (2004).
67. S. Dasgupta, B.P. Pal, and M.R. Shenoy, chapter on "Photonic bandgap Bragg fibers," in *Guided Wave Optical Components and Devices*, B.P. Pal (Ed.), Elsevier, Burlington (in press, 2005).

2

Polarization Effects in Single Mode Optical Fibers

Arun Kumar

1. Introduction

The so-called Single Mode Fibers in fact support two orthogonally polarized (say x and y) modes (see Fig.1). In an ideal case these modes should be degenerate, having same propagation constants and group velocities. However, in practice, because of various perturbations such as a small core ellipticity, bends, twists and asymmetrical stresses, etc., this degeneracy is lifted. As a result the fiber becomes birefringent, supporting two orthogonally polarized modes with slightly different propagation constants and, in general, having different group velocities. This has important consequences in various applications of optical fibers. The birefringence so generated leads to a continuous change in the state of polarization (SOP) of the guided light along the fiber if the light coupled at the input is polarized. Further, as the magnitude of the birefringence mentioned above keeps changing randomly with time due to fluctuations in the ambient conditions such as temperature, the output SOP also keeps fluctuating with time. The change in the output SOP is of little consequence in applications where the detected light is not sensitive to the polarization state. However, in many applications such as fiber optic interferometric sensors, coupling between optical fibers and integrated optic devices, coherent communication systems etc, one requires that the output SOP remains stable. The difference in the group velocities of the two polarized modes also plays a very important role as it leads to a temporal broadening of an optical pulse propagating through it, which limits the bit-rates of optical communication systems involving optical fibers and can also cause errors in the transmission of data. This type of dispersion is called Polarization Mode Dispersion (PMD) and is today a serious issue in the field of ultra-high bit-rate (> 10 Gbps) fiber-optic data transmission[1].

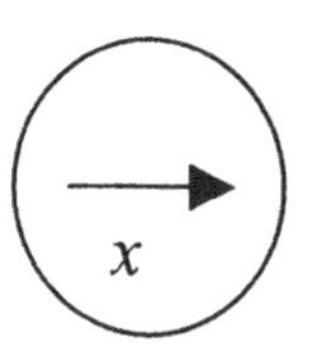

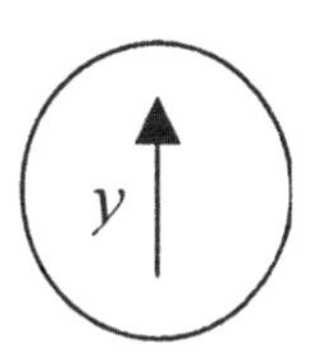

Figure 1 Schematic of orthogonally polarized modes

2. Origins of Birefringence in Optical Fibers

The various reasons due to which an optical fiber may become birefringent are discussed below.

2.1 Core-ellipticity

The geometrical anisotropy due to core-ellipticity introduces a linear birefringence in the fiber, the minor and major axes acting as fast and slow axes respectively. In the absence of analytical solutions of the wave equation for elliptical core fibers, an analytical expression for birefringence in such fibers is not possible. However, for fibers with small core-ellipticity, the birefringence near the first higher mode cutoff is approximately given by[2],

$$B_{ec} = (\beta_x - \beta_y) \approx 0.2 k_0 (\frac{a}{b} - 1)(\Delta n)^2 , \quad (1)$$

where $k_0 = 2\pi/\lambda$ (λ being the free space wavelength), Δn is the core cladding index difference and a and b represents the semi-major and semi-minor axes of the core ellipse. In the above equation β_x and β_y represents the propagation constants of the x and y polarized modes where x and y-axis are taken parallel to the major and minor axis respectively. Equation (1) indicates that the birefringence is higher in fibers with larger core ellipticity and higher core cladding index difference.

2.2 Lateral Stress

An asymmetrical lateral stress on the fiber also produces linear birefringence in the fiber via elasto-optic effect. For example, if the fiber is pressed (say along y-axis) between two parallel plates, the birefringence so generated is given by[3],

$$B_s = - hF \quad , \quad (2)$$

with

$$h = \frac{2 k_0 n_0^3}{\pi r} \frac{(1+\nu)}{E} (p_{11} - p_{12}) \quad (3)$$

Here n_0 is the average refractive index of the fiber, p_{11} and p_{12} are the components of the strain optical tensor, E is the Young's modulus and ν is the Poisson's ratio of the fiber material. Taking $n_0 = 1.46$, $p_{11} = 0.12$, $p_{12} = 0.27$, $E = 7.75\text{x}10^9$ kg/m^2 and $\nu = 0.17$ for fused silica, B_s comes out to be[3],

$$B_s = \frac{2.8\text{x}10^{-10}}{\lambda} \left(\frac{F}{r} \right) \text{ rad/m,} \quad (4)$$

where r represents the outer radius (including the cladding) of the fiber, which for a standard fibers is 62.5 μm and F represents the lateral force per unit length of the fiber in Kg/m.

2.3 Bending

Bending of a fiber introduces different types of stresses in the outer and inner portion of the fiber leading to a linear birefringence, which for silica fibers is given by[4]

$$B_b = \frac{0.85}{\lambda}\left(\frac{r}{R}\right)^2 \text{ rad/m}, \tag{5}$$

where R represents the bend radius. Bending induced birefringence is used to fabricate in-line fiber optic wave plates to control the polarization state of the guided light as discussed later in this chapter.

2.3 Twists

A twist in the fiber produces a circular birefringence given by[5],

$$B_t = g\tau, \tag{6}$$

where τ is the twist rate (in radians/length) and $g = 0.146$ for silica fibers.

2.4 Electric Field

A transverse electric field also produces linear birefringence in the fiber via Kerr electro-optic effect. However, birefringence so produced is extremely small for silica fibers[6].

2.5 Magnetic Field

A longitudinal (*i.e.* axial) magnetic field H produces a circular birefringence in the fiber via Faraday effect and is given by[7],

$$B_H = \beta_{lc} - \beta_{rc} = 2V_d H, \tag{7}$$

where β_{lc} and β_{rc} represents the propagation constants of the left-circular and right-circularly polarized modes and V_d is the Verdet constant of the material and is equal to 4.7×10^{-6} rad/amp for silica fibers.

Birefringence in optical fibers has both advantageous as well as disadvantageous aspects. It can be used to one's advantage if utilized in a proper and controlled way. For example, by imparting a controlled amount of birefringence to an optical fiber, many in-line fiber optic components such as polarizer, all-fiber wave plates (quarter wave plate, half wave plate etc.) can be fabricated. These wave-plates can either be used independently or in a suitable combination to transform the state of polarization (SOP) of incident light from one state to another. Such a device is called a Polarization Controller[8] and is frequently required in various interferometric and polarimetric fiber-optic sensing applications.

In the following sections, these two diametrically opposite aspects of polarization effects in optical fibers, namely (i) the use of a controlled and deliberately introduced birefringence for realizing various in-line fiber optic components and (ii) and the polarization mode dispersion due to uncontrolled and accidentally introduced birefringence in optical fibers are discussed. In order to understand these aspects clearly, some knowledge of the Poincare Sphere representation of polarized light is extremely helpful, which is discussed in the Appendix.

3. Devices using Controlled Birefringence

3.1 Zero-Birefringence Optical Fiber Holder

In many applications such as interferometric and Polarimetric sensors, one requires to hold the fiber without introducing extra birefringence in the fiber. However, in order to fix the fiber securely in its intended position one has to apply lateral forces to the fiber, which produce birefringence in the fiber as discussed in section 2 above. Such a birefringence is highly undesirable as it tends to vary with time e.g. with changing ambient temperature. A possible solution to this problem is to clamp the fiber in a V-groove. In the absence of any friction between the fiber and the groove's walls, a groove angle of 60° presents a symmetry such that the stress distribution is isotropic producing zero birefringence.

Figure 2 shows the cross-sectional view of a bare fiber of radius R pressed in a V-groove of angle 2γ. If a force *F* per unit length is applied on the fiber in the *y* direction, it causes the reactional forces F_1 and F_2 at the points of contact. Kumar and Ulrich[9] have studied the birefringence so produced both theoretically as well as experimentally.

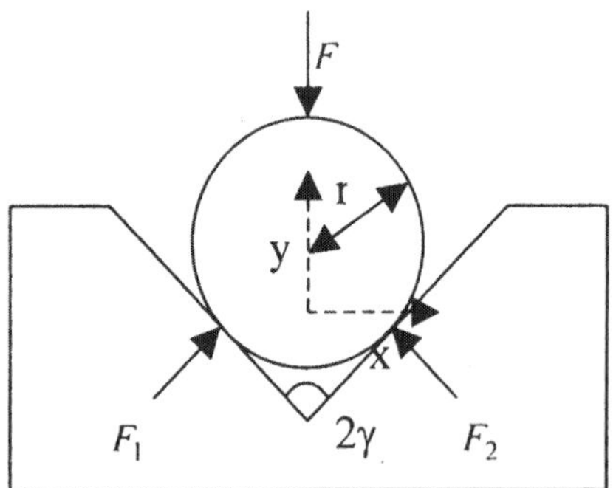

Figure 2 Fiber pressed in a V-groove.

In an ideal case of zero friction in the groove, the birefringence is given by

$$B_s = \beta_x - \beta_y = -\left(h/2\right)\left(1 - \frac{\cos 2\gamma}{\sin \gamma}\right)F, \tag{8}$$

In the case of a silica fiber with an outer diameter of 125 μm, one gets $h = -41^\circ / \mathrm{N}$ at He-Ne wavelength λ = 633 nm. In practice, however, the friction in the groove is not zero and the above equation takes the following form:

$$B_s = -\left(h/2\right)\left(1 - \frac{\cos 2\gamma}{\mu \cos\gamma + \sin \gamma}\right)F, \tag{9}$$

where μ is the coefficient of friction between the fiber and the groove's walls. The above equation reduces to Eq.(8) in the limiting case of zero friction i.e. μ = 0. In the other limiting case of complete friction, forces F_1 and F_2 are exactly in the opposite direction to the applied force F and the corresponding value of the coefficient of the static friction is[10],

$$\mu = 1/\tan\gamma \tag{10}$$

and the induced birefringence is given by:

$$B_s = -\left(h/2\right)\left(1 - \cos 2\gamma \sin\gamma\right)F \tag{11}$$

Figure 3 shows the normalized measured birefringence as a function of groove angle. Theoretical results for the two limiting cases, i.e., zero (dashed curve) and complete friction (solid curve) are also shown. which are obtained using Eqns.(8) and (11) respectively. In the zero-friction case, the birefringence should become zero

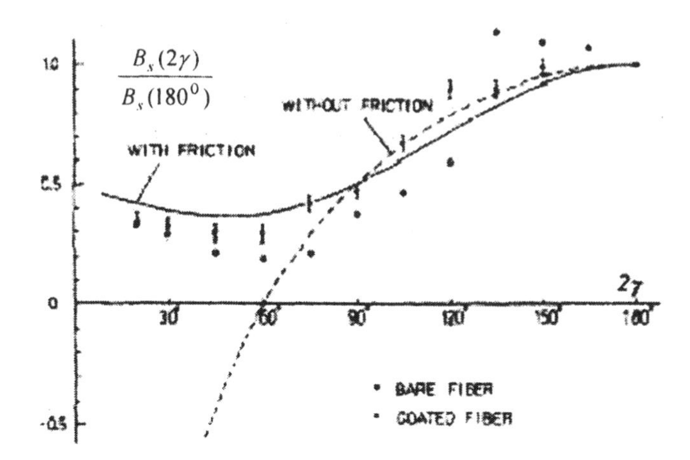

Figure 3 Measured birefringence of a bare fiber (filled circles) and a coated fiber (open circles). Solid and dashed curves correspond to Eqs.11 and 8 respectively. Adapted from from A Kumar, R Ulrich, Optics Letters, **12**, 644-646, 1981. (© Optical Society of America)

and change sign at $2\gamma = 60^0$. However, in the case of complete friction the birefringence would never be zero but has a minimum value at $2\gamma = 48.2^0$. A comparison of the theoretical and experimental results indicates a very large friction in the groove. Maystre and Bertholds[10] have repeated such measurements by preparing different fiber holders with good surface quality. Their studies shows that V-grooves made with hard steel produces zero birefringence at an angle $\gamma \approx 55^0$.

3.2 In-line Fiber optic wave-plates

As mentioned earlier bending in the optical fiber leads to a linear birefringence i.e the two eigen SOPs of the system are linearly polarized, the fast SOP oriented in the plane of the bend (say y-axis) and the slow one perpendicular to it (say x-axis). The amount of this birefringence can be controlled by varying the bend radius and the number of bends or loops in the fiber. It can easily be shown using Eq.(5), that for a typical silica fiber, the phase difference accumulated between the two eigen SOPs for N fiber loops of radius R (see Fig. 4) is given by:

$$\Delta\phi = 0.133x\frac{4\pi^2 Nr}{\lambda R} \tag{12}$$

Thus, a single fiber loop (N=1) would act as a quarter-wave plate (producing $\Delta\phi = \pi/2$) if the bend radius is given by:

$$R = 0.133 \times \frac{8\pi r^2}{\lambda} \tag{13}$$

Similarly two loops (N=2) instead of one with the same bending radius would give rise to a phase difference of π and would act as a half wave plate. Substituting the parameters with typical values for a standard single-mode fiber at He-Ne laser wavelength (632.8 nm), one gets the above bend radius value of ~2.1 cm.

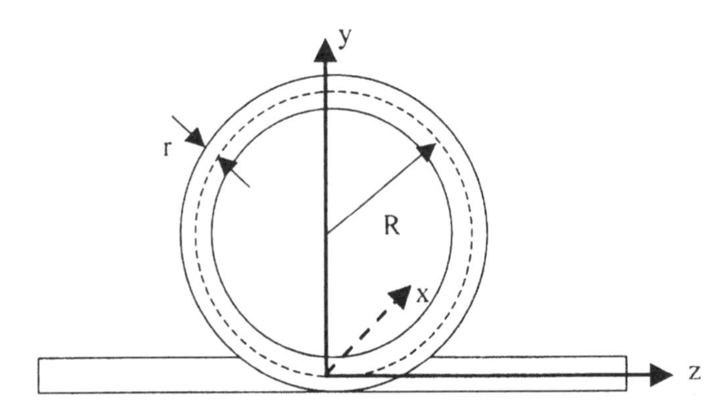

Figure 4 A fiber bent in the form of a loop.

3.3 The all-fiber Polarization Controller

A proper sequence of three wave plates – a quarter wave plate (QWP) followed by a half wave plate (HWP) and then followed by another quarter wave plate with proper orientations can convert the SOP of the input light to any desired output SOP. The resulting device is known as a Polarization Controller[8] (see Fig. 5).

The working of this device can be clearly understood by using the Poincare Sphere. Referring to Fig. 6, let P_i (α_i, γ_i) be the input SOP to the controller which we

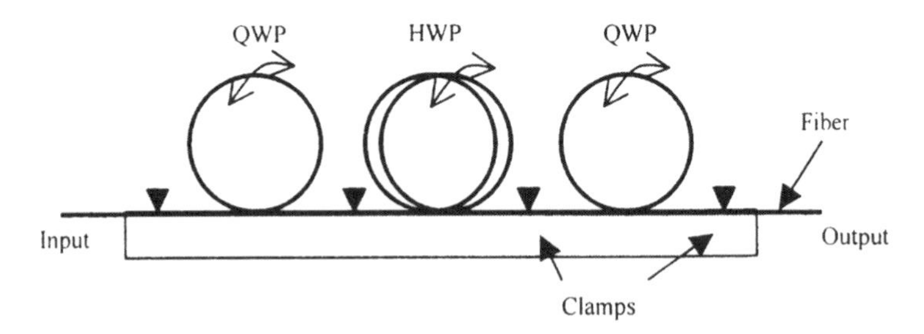

Figure 5 The polarization controller.

wish to convert to another SOP P_f (α_f, γ_f) at the output. The orientation of the first QWP is adjusted in such a way that its fast axis is oriented along the major axis of P_i represented by the point where the meridian passing through P_i meets the equator as shown in Fig.6(a). The half wave-plate is adjusted in such a way that its output SOP (which would also be a linear represented by P_2) can be converted to the required SOP, P_f by the second QWP. It is clear from Fig. 6(c) that to accomplish this, the fast

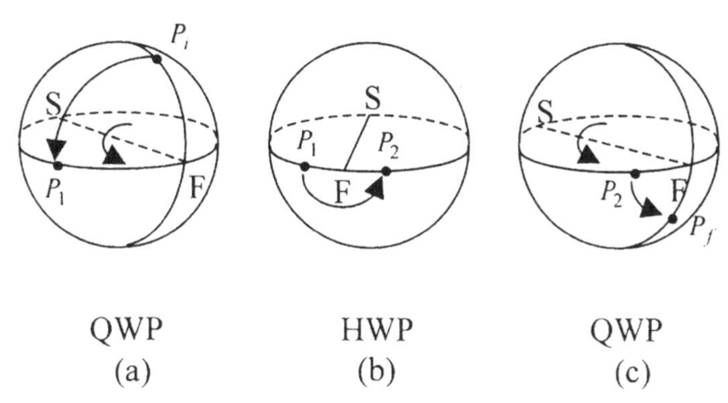

Figure 6 Schematic actions of wave-plates in a Polarization Controller.

axis of the second QWP should be represented by the point where the meridian passing through P_f meets the equator and the point P_2 should be such that arc P_2F should be equal to arc FP_f. Thus the fast axis of the HWP should be so oriented that it is represented by a point midway between P_1 and P_2 as shown in Fig.6(b).

To sum up, the operation of the device is in three steps – convert the given elliptic SOP to a linear SOP, rotate the linear SOP to another appropriate linear SOP, and finally impart to it the required ellipticity to obtain the final desired SOP. The advantage of this device is that it is in an "all-fiber" form. In certain interferometric and polarimetric fiber sensors, one requires control over the SOP of light propagating in the fiber. In those cases, this Polarization Controller finds immense use, as it has to be just spliced at two ends in between a fiber. One does not need to depend on bulk-optic components for these purposes, which are lossy as well as difficult to integrate with fibers.

3.4 Polarization Maintaining Fibers

Polarization maintaining fibers is another application of intentionally induced birefringence in optical fibers. In such fibers the difference between the propagation constants of the two orthogonally polarized modes (i.e. birefringence) is made quite large so that the coupling between the two modes is greatly reduced. Thus if light is coupled into one of the polarization modes of such a fiber then its SOP is maintained as it propagates through the fiber. Such fibers, also known as High-Birefringence (Hi-Bi) fibers, are characterized by their beat length L_b, which is defined as,

$$L_p = \frac{2\pi}{(\beta_x - \beta_y)} = \frac{2\pi}{B}, \tag{14}$$

where $B \; (= \beta_x - \beta_y)$ represents the birefringence in the fiber. A lower value of L_b means a higher value of $(\beta_x - \beta_y)$ and hence high polarization holding capacity of the fiber. Physically L_b represents the distance along the fiber during which the phase difference developed between the two polarized modes is 2π. Thus if the light is coupled in both the modes of the fiber, the SOP of the guided light will repeat itself after a distance L_b.

Figure 7(a) shows the evolution of SOP along a Hi-Bi fiber when a left circularly polarized light is coupled at the input end and the two polarization axes of the fiber are taken along x and y directions. The evolution of the SOP along the fiber can

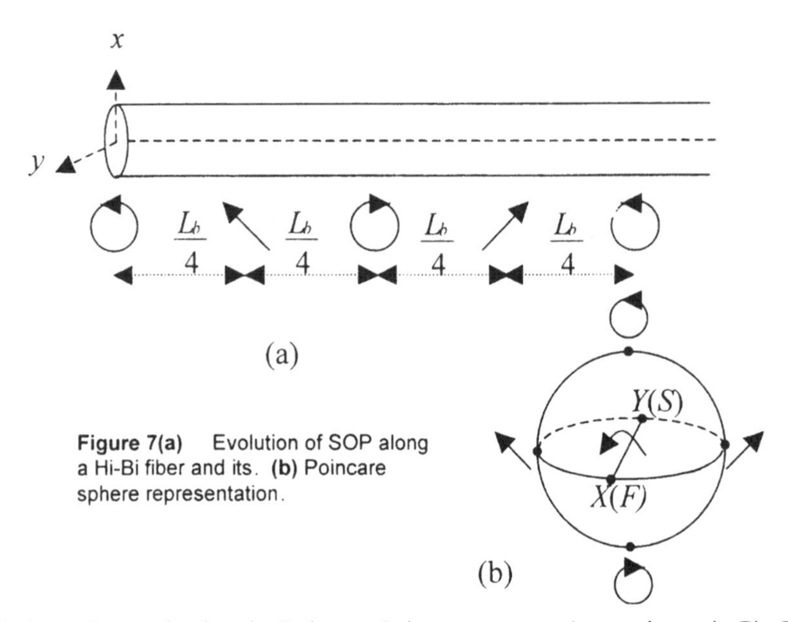

Figure 7(a) Evolution of SOP along a Hi-Bi fiber and its. **(b)** Poincare sphere representation.

easily be understood using the Poincare Sphere representation as shown in Fig 7(b).

On the Poincare Sphere the fiber is represented by the diameter XY and the SOP along the fiber length would be obtained by rotating the sphere anti-clockwise around XY. Thus if the input SOP is left circular it will be converted to linear Polarisation oriented at 45^0, right circular, linear Polarisation oriented at $+45^0$ and again to left circular after a disdance of $L_b/4$, $L_b/2$, $3L_b/4$ and L_b respectively.

Two types of High-Birefringent fibers are most popular, which are discussed below:

3.4.1 Elliptical Core Fibers

These fibers are fabricated with an elliptical core embedded in a circular cladding as shown in Fig.8. The elliptic core creates both geometrical anisotropy as well as asymmetrical stress in the core. As a result the propagation constants of the two fundamental modes polarized along the major (x-axis) and the minor axis (y-axis) become different making the fiber birefringent. The total birefringence is thus a sum of geometrical birefringence B_g (due to non-circular shape) and stress-induced birefringence B_s (due to asymmetrical stress in the core).

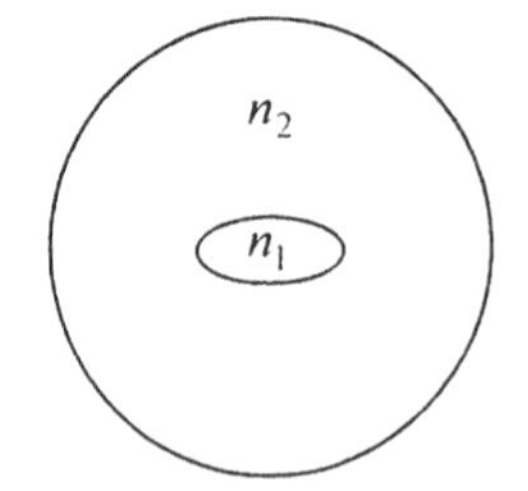

Figure 8 Transverse cross-section of an elliptic core fibre.

Geometrical birefringence B_g

The exact calculation of B_g in such fibers is difficult as the wave equation has to be solved in the elliptic co-ordinates, and the eigenvalue equation so obtained is in the form of an infinite determinant[11]. Dyott et al.[12] has obtained the birefringence of different elliptical core fibers by truncating the infinite determinant to finite order. Numerical methods like finite element[13] and point matching method[14] have also been reported by various authors, however, these methods involve time consuming numerical calculations. A number of approximate methods[15-17] have also been reported, amongst which the one proposed by Kumar et al[17] is relatively simple and is discussed here briefly. According to this method the birefringence of an elliptical-core fibre as shown in Fig. 9 matches very well with that of a rectangular-core waveguide having the same core area, aspect ratio and core-cladding refractive indices. Thus the birefringence of an elliptical core fibre with major and Minor axes as $2a$ and $2b$ respectively, is approximately equal to that of a rectangular core waveguide with major and minor axes given by,

$$2a' = \sqrt{\pi}a \quad \text{and} \quad 2b' = \sqrt{\pi}b \tag{15}$$

and with the same core and cladding indices (n_1 and n_2) respectively. The propagation constants β_x and β_y for the two fundamental modes of this rectangular core waveguide can be obtained using an accurate perturbation approach[18] as discussed in Chapter 12.

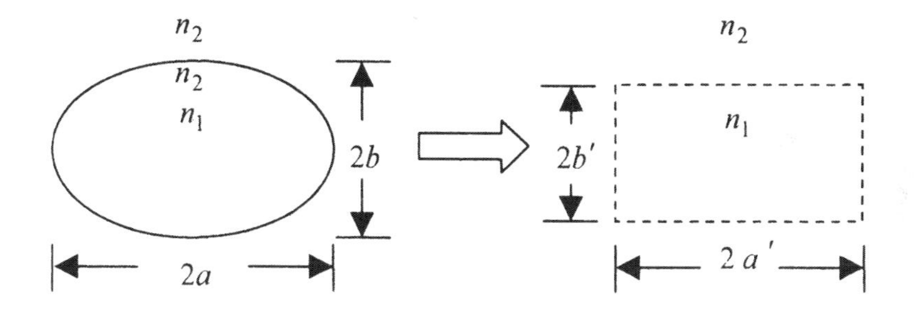

Figure 9 Cross-section of an elliptic-core fiber and its equivalent rectangular-core waveguide.

Stress Birefringence B_s

An asymmetrical stress is frozen in elliptical core fibres during their fabrication due to different thermal expansion coefficients of the core and the cladding materials. This stress introduces a linear birefringence B_s in the fibre through the elasto-optic effect. The amount of this birefringence can be obtained by calculating the difference between the stresses generated at the fibre center along the major and minor axes of the fibre and is given by[19-22]

$$B_s = f(V)\frac{c_0}{1-\nu}\Delta\alpha.\Delta T\frac{a-b}{a+b}, \tag{16}$$

where $c_0 = (n_0^3/2)(p_{11} - p_{12})(1+\nu)$ and $\Delta\alpha$ is the difference between the thermal expansion coefficients of the core and cladding materials, $\Delta T = T_r - T_s$; T_r and T_s being the room temperature and the softening temperature of the fibre material and $f(V)$ represents the fractional modal field power in the core of the fibre. The factor $f(V)$ gives the wavelength dependence of B_s. Approximating the given elliptic core of the fibre by a circular core of radius $d\{=(a+b)/2\}$, $f(V)$ is approximately given by,

$$f(V) = W^2 / V^2, \tag{17}$$

with $W = k_0\, d\sqrt{n_{eff}^2 - n_2^2}$ and $V = k_0\, d\sqrt{n_1^2 - n_2^2}$,where n_{eff} represents the average effective index of the two fundamental modes.

3.4.2 Stress Induced High Birefringence Fibers

A more effective method of introducing the high birefringence in optical fibres is through freezing a transverse stress with two-fold symmetry, in the core of the fibre, which makes the fibre linearly birefringent as discussed earlier. The required stress is obtained by introducing two identical stress applying parts (SAPs) centered on a diameter, one on each side of the core. The SAPs have different thermal expansion coefficient (α_3) than that of the cladding material (α_2) due to which a transverse stress is applied on the fibre core after it is drawn from the perform and cooled down. Two different shapes of SAPs are generally used (i) Bow-tie[23] and (ii) circular[24-26]. The corresponding fibers are known as the ' Bow-tie' and PANDA (Polarization-maintaining AND Absobsion reducing) fibres respectively. The cross-sections of these two types of fiber are shown in Fig.10.

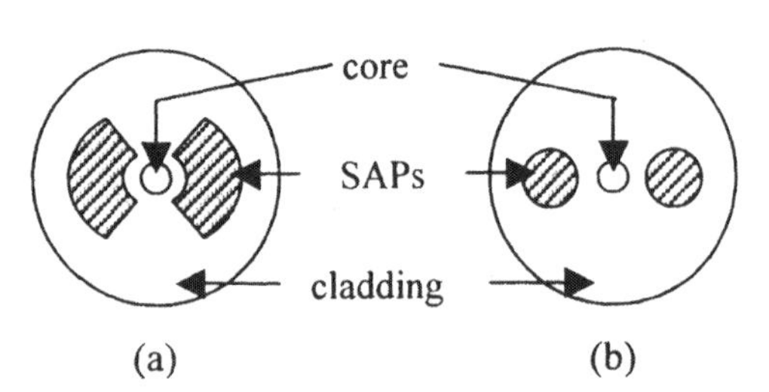

Figure10 Transverse cross-sections of (a) Bow-tie and (b) PANDA Hi-Bi fibers.

The stress distribution due to SAPs in these fibers is quite cumbersome to obtain. Okamoto and co-workers[27] have given a finite-element method to obtain this complicated stress distribution while Varnham et al.[28] and Chu and Sammut[29] have given analytical methods for calculation of thermal stress distribution and birefringence in such fibers. The birefringence in such fibers can be increased by placing the SAPs closer to the core. However, they can not be placed arbitrarily closed since it increases the loss of the fiber as SAPs are doped with materials other than silica. Stress-induced fibers are superior to any other type of birefringent fibres in terms of loss and polarization holding capacity. Fibers with very low-loss and cross-talk have been reported using optimum design parameters. For example, Sasaki[30] have reported the fabrication of a PANDA fibre with 0.22 dB/km loss with birefringence $B = 3.3\text{x}10^{-4}$ and cross-talk of –27 dB in a 5 km length.

3.3.3 In Line Fiber-Polarizers

An optical fiber can be made highly polarization selective by side-polishing it and depositing an appropriate material (birefringent crystal or metal) on the polished surface as discussed in Chapter 7 also. Such a device acts as an in-line fiber polarizer. Fiber polarizers using thin metal coatings are more popular and works through coupling between fiber TM mode to the surface-plasmon mode supported by the cladding-metal interface. Thus a metal-coated side-polished fiber acts as a TE-pass (electric field parallel to the polished surface) polarizer[31].

4. Polarization Mode Dispersion

In the previous section, we discussed devices based on deliberately introduced birefringence in a fiber, thus utilizing the polarization effects to our advantage. In the following, we discuss the detrimental aspect of birefringence in optical fibers namely the Polarization Mode Dispersion (PMD).

The existence of birefringence in a fiber implies that the fiber supports two orthogonal polarized modes that have different effective indices and hence propagate with different group velocities in the fiber. An optical pulse launched into such a fiber would be split up into two orthogonal polarized pulses, which would then propagate with different group velocities and would superimpose with a finite phase difference at the output (see Fig.11). This superimposition of the two phase-shifted pulses leads to the generation of an optical pulse that is now temporally more broadened as compared to the input pulse. Thus the pulse gets "dispersed" due to the effect of fiber birefringence, and the phenomenon is called Polarization Mode Dispersion (PMD). It is a serious limitation in the area of ultra-high bit-rate fiber communication links, as it puts a cap on the bit-rate of the link as well as causes errors in data transmission. The fibers comprising a real long distance link experience stress, bends, temperature changes, twists etc. in a random fashion along the length of the link, and as a result the temporal broadening becomes a randomly varying quantity that makes PMD compensation schemes difficult to implement. Unlike normal dispersion compensation, PMD has to be compensated in "real time", and so the devices have to be made "intelligent" for that purpose.

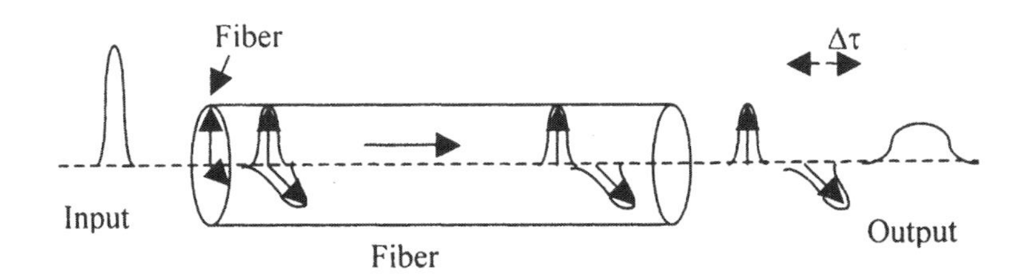

Figure 11 A schematic of pulse broadening due to PMD.

Polarization mode dispersion is measured in terms of the difference in propagation times experienced by the two orthogonally polarized modes of the fiber at a given wavelength, which is called the Differential Group Delay (DGD) of the fiber. It is denoted by $\Delta\tau$ and is measured in picoseconds (psec). Let us assume that the two eigen SOPs are fixed, so that PMD is completely deterministic and not random. If n_{ex} and n_{ey} are the effective indices of the two eigen SOPs, the corresponding propagation constants are then given by:

$$\beta_x = \frac{\omega}{c} n_{ex} \ ; \ \beta_y = \frac{\omega}{c} n_{ey}, \tag{10}$$

where ω is the angular frequency. The corresponding group velocities would be,

$$v_{gx} = \frac{d\beta_x}{d\omega} \ and \ v_{gy} = \frac{d\beta_y}{d\omega} \tag{11}$$

and the DGD in a length L of the fiber would then be given by:

$$\Delta\tau = \frac{L}{v_{gx}} - \frac{L}{v_{gy}} = L\frac{d}{d\omega}[\beta_x - \beta_y] = \frac{d}{d\omega}[L(\beta_x - \beta_y)] = \frac{d\delta}{d\omega} \tag{12}$$

where $\delta = L\ [\beta_x - \beta_y]$ is nothing but the phase difference introduced between the two eigen-polarizations, which, as we know decides the output SOP. Thus by knowing the rate of change of the output SOP with respect to the frequency (or wavelength) of input light, one can determine the PMD, which is the working principle of a simple method to measure PMD, known as the Wavelength Scanning method as discussed below.

4.1 Wavelength Scanning Method for measurement of PMD

In the case of fixed eigen-axes, the fiber will be represented by a fixed diameter on the Poincare Sphere and the output SOP can be obtained from the input SOP by rotating the Poincare Sphere around this diameter. Thus the output SOP of a device with fixed eigen SOPs, such as a length of a High-Birefringent (Hi-Bi) fiber, traces a circle on the Poincare Sphere as the wavelength is changed (Figure 12a). However, in a more general case, the birefringence may be due to a variety of causes like stress, twist, bends etc., whose direction may change randomly along the fiber length and with time. In such a case, the diameter representing the birefringent fiber, rate of

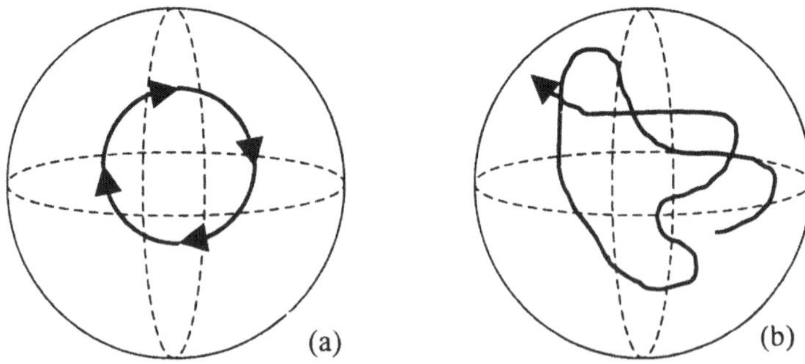

Figure 12 Evolution of the output SOP as the input wavelength is changed in the case of (a) Hi-Bi and (b) practical single mode fiber.

rotation and hence the arc on the Poincare Sphere changes with wavelength, and the output SOP wanders on the sphere in a random fashion as shown in Fig.12b.

The analysis of PMD in practical fibers is thus difficult. Poole and Wagner[32] have proposed a phenomenological model of PMD in long lengths of fibers according to which, in the absence of polarization dependent loss, there exists two orthogonal input states of polarization for which the corresponding output states of polarization are also orthogonal and show no dependence on wavelength to first order. Thus in practical fibers also for a small change in the wavelength, one can assume as the eigen-polarizatin axes are fixed. The DGD will thus be given as $\Delta\tau = \Delta\theta / \Delta\omega$, where $\Delta\theta$ is the arc length along a great circle joining the output SOPs on the sphere corresponding to angular frequencies ω and $\omega + \Delta\omega$. It may be mentioned that $\Delta\theta$ is nothing but the phase difference $\Delta\delta$ introduced between the two eigen-polarization states due to the incremental change $\Delta\omega$ in frequency. As discussed in the Appendix, the effect of a birefringent medium on a given SOP is to rotate the Poincare sphere by an amount equal to the phase difference between the two eigen-polarizations. In the case of a Hi-Bi fibre, since the axes are fixed, the SOP moves on a circle as frequency or wavelength changes; one full circle corresponds to $\Delta\theta = 2\pi$. The change in the output SOP is transformed into output intensity oscillations by passing the output beam through a fixed linear analyzer. A schematic setup for wavelength scanning method for the measurement of PMD is shown in Fig. 13.

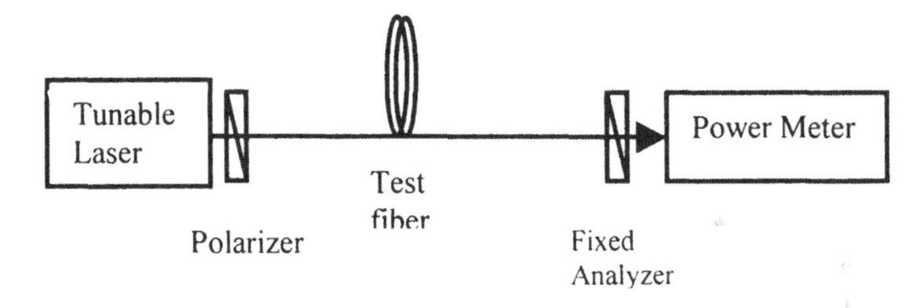

Figure 13 Schematic setup of the wavelength scanning method.

In the case of a Hi-Bi fiber, there is no polarization mode coupling and the output intensity variation is periodic with a fixed amplitude and period. However, in practical single mode fibers there is finite coupling between orthogonal polarization states and output intensity variation looks something like as shown in Fig.(14). By counting the number of oscillations in intensity about a mean threshold level, the value of DGD can be estimated. If λ_1 and λ_2 are the endpoints of the wavelength scan and N represents the number of output polarization cycles that occur across the scan, the DGD is given by[33],

$$\Delta\tau = \frac{Nk\lambda_1\lambda_2}{c(\lambda_1 - \lambda_2)}, \tag{13}$$

where constant k depends on the mode coupling in the fiber. In the case of no mode-coupling such as a Hi-Bi fiber or a short length standard fiber, $k = 1$, while for long lengths of standard fibers, there is a significant mode coupling and k=0.82 is recommended. It may be mentioned that, due to mode coupling, PMD in long lengths of fibers is proportional to the square root of fiber length. Such dependence can be proved using a statistical treatment of the randomly varying nature of PMD, wherein the DGD is modeled as a Maxwellian distribution[34]. For standard communication grade fibers, the PMD is typically 0.01-0.1 ps/km$^{1/2}$.

5. Summary

In summary in this chapter, we discussed the origin and implications of birefringence in single-mode optical fibers. A controlled birefringence in optical fibers can be used to realize various in-line fiber polarization components, On the other hand uncontrolled and random birefringence leads to polarization mode dispersion which limits the information carrying capacity of ultra-high bit-rate fiber optic communication systems. A few examples of polarization components, PMD and a simple method of measuring PMD in optical fibers are discussed.

Acknowledgement: The author would like to thank Mr. Deepak Gupta for his help in preparation of this manuscript.

Appendix: Poincare Sphere Representation of polarized light

Poincare Sphere representation is due to H. Poincare who was a French physicist. This gives a very simplified picture of the SOPs and their change due to birefringence. As we know, any given general elliptic SOP can be specified completely by three quantities:

1. The orientation of the major axis of the ellipse, which may be specified by an angle α that it makes with a given direction in the plane of the wavefront (here x-axis).
2. The ratio of the axes of the ellipse (b/a ; $b < a$) (refer Fig. A1)
3. The sense of rotation (left/right) of the tip of the electric field on the periphery of the representative ellipse. Here the left/right rotations are the anticlockwise/clockwise rotations as seen by an observer looking into the source of the light.

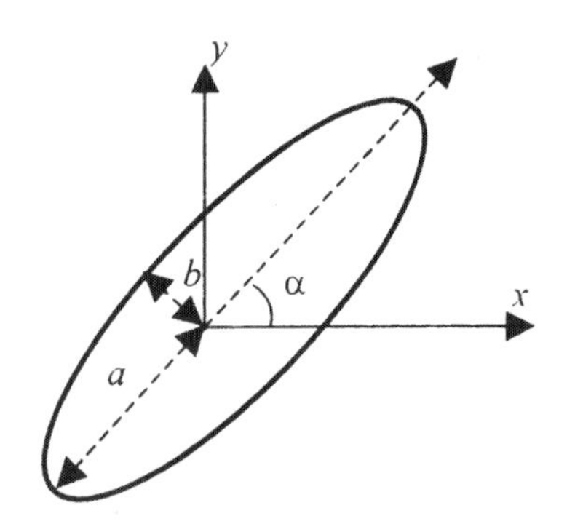

Figure A1 A general elliptic state of polarization.

According to the Poincare sphere representation, the above three quantities are reduced to two parameters namely α as defined above and $\gamma = \pm \tan^{-1}(b/a)$, where + and – signs are used to represent the sense of rotation: + for left and – for right rotations. These two angles, called azimuth and ellipticity respectively, uniquely specify the SOP of light. All possible SOPs are covered by the range of 0 to π of α and $-\pi/4$ to $+\pi/4$ of γ (taken together). Such an SOP can be uniquely represented by a point on the surface of a sphere of unit radius, whose longitude and latitude

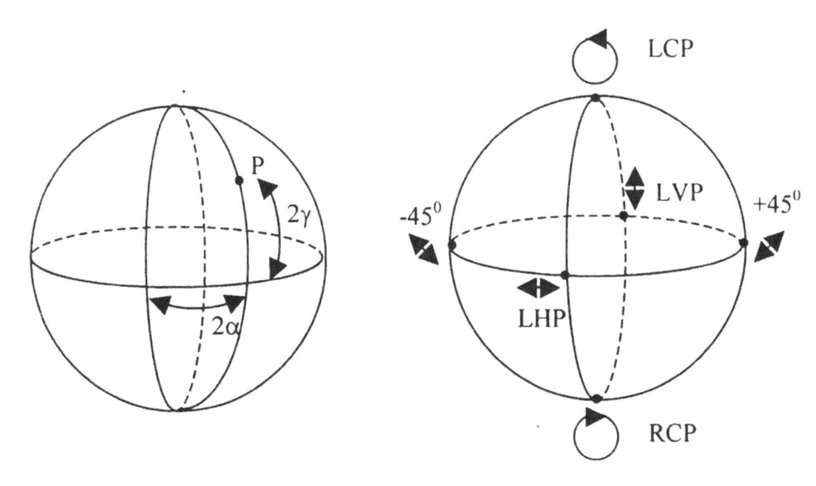

Figure A2 Poincare sphere representation of polarization States.

Figure A3 Some common SOPs on Poincare sphere.

have the values 2α and 2γ respectively (see Figure A2). This representation is called the Poincare Sphere representation and the sphere is called the Poincare Sphere. Figure A3 enlists some of the standard SOPs and their corresponding Poincare Sphere representations.

A1. Representation of a Polarizer/Analyzer

A polarizer is a device which produces a certain SOP say of parameters α_P and γ_P and is represented by a point on the Poincare Sphere having the longitude and latitude as $2\alpha_P$ and $2\gamma_P$. Similarly an analyzer is a device which completely passes a certain SOP say of parameters α_A and γ_A and is represented by the point having the longitude and latitude as $2\alpha_A$ and $2\gamma_A$. If a given SOP represented by a point P is passed through an analyzer represented by a point A on the Poincare sphere, then the fractional intensity transmitted by the analyzer is given by

$$I = \cos^2 \frac{\overline{PA}}{2} \qquad (7)$$

where $\overline{PA}$ is the arc length along the great circle connecting the two points on the Poincare sphere.

A2. Representation of a Birefringent Medium

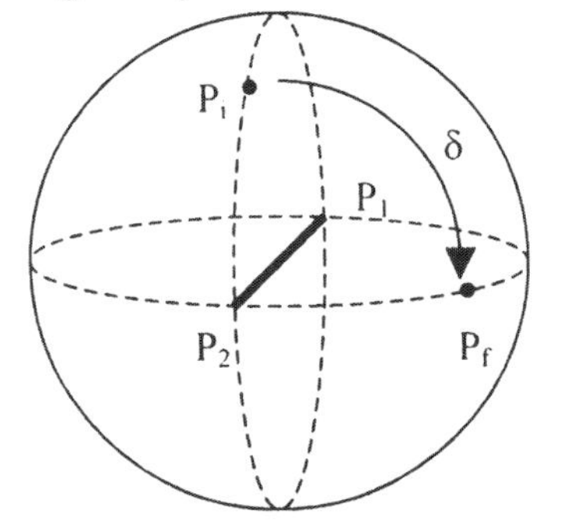

Figure A4 Representation of a birefringent medium on the sphere

A birefringent medium is characterized by two eigen SOPs and is represented by a diameter, the end points $P_1(\alpha_1, \gamma_1)$ and $P_2(\alpha_2, \gamma_2)$ of which represent the two eigen SOPs on the Poincare Sphere (see Fig. A4). If a SOP given by $P_i(\alpha_i, \gamma_i)$ is passed through such a birefringent medium, then the output SOP $P_f(\alpha_f, \gamma_f)$ can be obtained by rotating the Poincare sphere by an angle δ about anti-clockwise looking from P_1 to P_2 ; P_1 leading P_2 by a phase difference of δ.

References

1. C.D. Poole, R.W. Tkach, A.R. Chraplyvy and D.A. Fishman, "Fading in lightwave systems due to polarization-mode dispersion", *IEEE Photon. Technol. Lett.*, **3**, 68-70 (1991).
2. S.C. Rashleigh, " Origins and control of polarization effects in single-mode fibers", *J. Lightwave Technology*, **1**, 312-331 (1983).
3. D.N. Payne, A.J. Barlow and J.J. Ramskov Hansen, "Development of low and high-birefringence optical fibers", *IEEE J. Quantum Elect.*, **18**, 477-488 (1982).
4. R. Ulrich, S.C. Rashleigh and W. Eickhoff, "Bending-induced birefringence in single-mode fibers", *Opt. Lett.*, **5**, 273-275 (1980).
5. R. Ulrich and A. Simon, " Polarization optics of twisted single-mode fibers," *Appl. Opt.*, **18**, 2241-2251 (1979).
6. A. Simon, and R. Ulrich, "Evolution of polarization along a single- mode fiber", *Electron. Lett.*, **25**, 290-292 (1989).
7. A.M. Smith, "Polarization and magneto-optical properties of single- mode fibers", *Appl. Opt.*, **17**, 52-56 (1978).
8. H.C. Lefevre, "Single mode fiber fractional wave devices and polarization controllers", *Electron. Lett.*, **16**, 778-780 (1980).
9. A. Kumar and R. Ulrich, "Birefringence of optical fiber pressed into a V-groove", *Opt. Lett.*, **6**, 644-646 (1981).
10. F. Maystre and A. Bertholds, "Zero-birefringence optical fiber holder", *Opt. Lett.*, **12**, 126-128 (1987).
11. C. Yeh, "Elliptical dielectric waveguides", *J. Appl. Phy.*, **33**, 3235-3245 (1962).
12. R.B. Dyott, J.R Cozens and D.G. Morris, "Preservation of ploarisation in optical fiber waveguides with elliptical cores", *Electron. Lett.*, **15**, 380-382 (1979).
13. C. Yeh, K. Ha, S.B. Dong and W.P. Brown, "Single-mode optical waveguides", *Appl. Opt.*, **18**, 1490-1504 (1979).
14. K. Okamato, T. Hosaka and Y. Sasaki, "Linearly single polarization fibers with zero polarization mode dispersion", *IEEE J. Quantum. Electron.*, **18**, 496-503 (1982).
15. J.D. Love, R. A. Sammut and A.W. Synder, "Birefringence in elliptically deformed optical fibers", *Electron. Lett.*, **15**, 615-616 (1979).
16. R.A. Sammut, "Birefringence in slightly elliptical optical fibers", *Electron. Lett.*, **16**, 728-729 (1980).
17. A. Kumar, R.K. Varshney and K. Thyagarajan, "Birefringence calculations in Elliptical-Core Fibers", *Electron. Lett.*, **20**, 112-113 (1984).
18. A. Kumar, K. Thyagarajan and A.K. Ghatak, "Analysis of rectangular core dielectric waveguide; an accurate perturnbation approach", *Opt. Lett.*, **8**, 63-65 (1983).

19. W. Eickhoff, "Stress induced single-polarisation single-mode fiber", *Opt. Lett.*, 7, 629-631 (1982).
20. P.L. Chu, "Thermal stress induced birefringence in single-mode elliptical optical fiber", *Electron. Lett.*, **18**, 45-47 (1982).
21. J. Sakai and T. Kimura, "Birefringence caused by thermal stress in elliptically deformed core optical fibers", *IEEE J. Quantum. Electron.*, **18**, 1899-1909 (1982).
22. M.P. Varnham, D.N. Payne, A.J. Barlow and R.D. Birch, "Analytic solution for the birefringence produced by thermal stress in polarization-maintaining optical fibers", *J. Lightwave Technology*, **1**, 332-339 (1983).
23. R.D. Birch, D.N. Payne, M.P. Varnham, "Fabrication of polarization-maintaining fibres using gas-phase etching", *Electron. Lett.*, **18**, 1036-38 (1982).
24. T. Hosaka, K. Okamoto, T. Miya, Y. Sasaki and T. Edahiro, "Low loss single polarization fibers with asymmetrical strain birefringence", *Electron. Lett.*, **17**, 530-531 (1981).
25. Y. Sasaki, K. Okamoto, T. Hosaka and N. Shibata, "Polarisation maintaining and absorption reducing fibers", in *Tech. Dig. Topical Meeting Opt. Fibre Commun., Phoenix*, Az (1982).
26. N. Shibata, Y. Sasaki, K. Okamoto and T. Hosaka, "Fabrication of polarization maintaining and absorption reducing fibers", *J. Lightwave Technology*, **1**, 38-43 (1983).
27. K. Okamoto, T. Hosaka and T. Edahiro, "Stress analysis of optical fibers by a finite element method", *IEEE J. Quantum Elect.*, **17**, 2123-2129 (1981).
28. M.P. Varnham, D.N. Payne, A.J. Barlow and R.D. Birch, "Analytic solution for the birefringence produced by thermal stress in polarization-maintaining optical fibers", *J. Lightwave Technology* **1**, 332-339 (1983).
29. P.L. Chu and R.A. Sammut, "Analytical method for calculation of stresses and material birefringence in polarization maintaining optical fiber", *J. Lightwave Technology*, **2**, 650-662 (1986).
30. Y. Sasaki, T. Hosaka, M. Horiguchi and J. Noda, " Design and fabrication of low-loss and low cross-talk polarization-maintaining optical fibers", *J. Lightwave Technology*, **4**, 1097-1102 (1983).
31. A. Kumar, D. Uttamchandani and B. Culshaw, "Relative transmission loss of TE and TM-like modes in method-coated coupler halves", *Electron. Lett.*, **25**, 310-302 (1989).
32. C.D. Poole and R.E. Wagner, "Phenomenological approach to polarization dispersion in long single-mode fibers," *Electron. Lett.*, **22**, 1029-1030, (1986).
33. C.D. Poole, " Mesurements of polarization-mode dispersion in single-mode fibers with random mode coupling", *Opt. Lett.*, **14**, 523-525, (1989).
34. C.D. Poole, " Statistical treatment of polarization dispersion in single mode fibers", *Opt. Lett.*, **13**, 687-689, (1988). Correction in *Opt. Lett.*, **16**, 372-374, (1991).

3

Optical Fiber Amplifiers

K. Thyagarajan

1. Introduction

When information carrying optical signals propagate through an optical fiber, they suffer from *attenuation* leading to reduction in power, *dispersion* leading to pulse broadening and *nonlinear effects* leading to coupling between various frequencies. In order to achieve greater bandwidths and longer propagation distances, these effects have to be overcome or compensated for in the fiber optic link. In traditional links, compensation of loss and dispersion is usually accomplished by using electronic regenerators in which the optical signals are first converted into electrical signals, then processed in the electrical domain and then reconverted into optical signals. In situations wherein the primary issue is loss, optical amplifiers can indeed be used to amplify the signals without conversion into electrical domain. Such optical amplifiers and have truly revolutionized long distance fiber optic communications.

In comparison to electronic regenerators, optical amplifiers do not need any high speed electronic circuitry, are transparent to bit rate and format, and most importantly can amplify multiple optical signals at different wavelengths simultaneously. Thus their development has ushered in the tremendous growth of communication capacity using wavelength division multiplexing (WDM) in which multiple wavelengths carrying independent signals are propagated through the same single mode fiber, thus multiplying the capacity of the link. Unlike electronic regenerators, optical amplifiers do not compensate for dispersion accumulated in the link and they also add noise to the optical signal.

Optical amplifiers can be used at many points in a communication link. A booster amplifier is used to boost the power of the transmitter before launching into the fiber link. The increased transmitter power can be used to go farther in the link. The preamplifier placed just before the receiver is used to increase the receiver sensitivity. In line amplifiers are used at intermediate points in the link to overcome fiber transmission and other losses.

The three main types of optical amplifiers are the erbium doped fiber amplifier (EDFA), the Raman fiber amplifier (RFA) and the semiconductor optical amplifier (SOA). The primary amplifying bands of EDFAs are the C-band (1530 – 1565 nm) and the L-band (1570-1610 nm); however there has been recent reports of extending the operation range of EDFAs to the S-band (1460-1530 nm). On the other hand RFAs can be made to operate in any band. Today most optical fiber communication systems use EDFAs due to their advantages in terms of bandwidth, high power output and noise characteristics. RFAs and SOAs are also becoming important in many applications. In the following we will discuss the characteristics of EDFAs and RFAs; detailed discussions on EDFAs and RFAs can be found in many texts, e. g., Refs. [1-4].

2. Erbium Doped Fiber Amplifier (EDFA)

Optical amplification by EDFA is based on the process of stimulated emission, which is the basic principle behind laser operation. In stimulated emission, an atom occupying a higher energy state can be stimulated to emit radiation by an incident radiation of appropriate frequency. The radiation coming out of the atom is coherent with respect to the incident radiation. This is in contrast to spontaneous emission wherein an atom in the upper state can emit radiation spontaneously and get de-excited. The radiation coming out of spontaneous emission is incoherent with respect to other existing radiation.

Consider two energy levels of an atomic system: the ground level with energy E_1 and an excited level with energy E_2. Under thermal equilibrium, most of the atoms are in the ground level. Thus if light corresponding to an appropriate frequency ($\nu = (E_2-E_1)/h$; h being the Planck's constant) falls on this collection of atoms then it will result in a greater number of absorptions than stimulated emissions and the incident light beam will suffer from attenuation. On the other hand, if the number of atoms in the upper level could be made more than in the lower level, then an incident light beam at the appropriate frequency could induce more stimulated emissions than absorptions, thus leading to optical amplification. This is the basic principle behind optical amplification by EDFA.

Figure 1 shows the three lowest lying energy levels of Erbium ion in silica matrix. Light from a pump laser at 980 nm excites the erbium ions from the ground level E_1 to the level marked E_3. The level E_3 is a short-lived level and the ions jump down to the level marked E_2 after a time lasting less than a microsecond. The

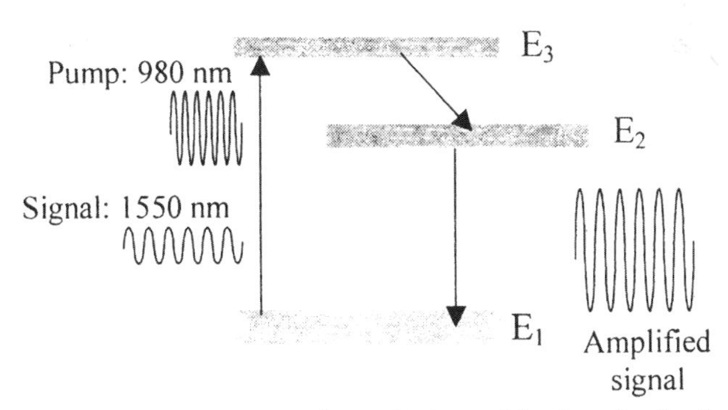

Figure 1 Energy levels corresponding to the lowest lying three levels of erbium ion. Population inversion between E_2 and E_1 brought about by the 980 nm pump leads to optical amplification.

lifetime of level E_2 is much larger and is about 12 milliseconds. Hence ions brought to level E_2 stay there for a long time. Thus by pumping hard enough, the population of ions in the level E_2 can be made larger than the population of level E_1 and thus we can achieve population inversion between levels E_1 and E_2. In such a situation, if a light beam at a frequency $\nu_0 = (E_2 - E_1)/h$ falls on the collection of erbium ions, it will get amplified through the process of stimulated emission. For erbium ions, the frequency ν_0 falls in the 1550 nm band and thus it is an ideal amplifier for signals in the 1550 nm window the lowest loss window of silica based optical fibers. In the case of erbium ions in silica matrix, the energy levels are not sharp but are broadened due to interaction with other ions in the silica matrix. Hence the system is capable of amplifying optical signals over a band of wavelengths.

3. Optical amplification in EDFAs

In order to study amplification we need to consider the evolution of pump and signal as they propagate through the erbium doped fiber. Let us consider an erbium doped fiber and let $I_p(z)$ and $I_s(z)$ represent the variation of intensity of the pump at frequency ν_p (assumed to be at 980 nm) and the signal at frequency ν_s assumed to be in the region of 1550 nm. Let N_1 and N_2 represent the number of erbium ions per unit volume in E_1 and E_2 respectively. As the beams propagate through the fiber, the pump would induce absorption from E_1 to E_3 while the signal would induce absorption and stimulated emissions between levels E_2 and E_1. The population of the various levels would depend on the z value since the intensities of the beams would be z-dependent. We will assume that the lifetime of level E_3 is very small so that we can neglect the population density N_3 of level E_3 and put $N_3 = 0$. Thus the ions are either in level E_1 or in level E_2. We can write for the rate of change of population of level E_2 as[3]

$$\frac{dN_2}{dt}\left(=-\frac{dN_1}{dt}\right)$$
$$=-\frac{N_2}{t_{sp}}+\frac{\sigma_{pa}I_p}{h\upsilon_p}N_1-\left(\sigma_{se}N_2-\sigma_{sa}N_1\right)\frac{I_s}{h\upsilon_s} \tag{1}$$

Here σ_{sa} and σ_{se} are respectively the absorption and emission cross sections at the signal wavelength while σ_{pa} and σ_{pe} are respectively the absorption and emission cross sections at the pump wavelength. t_{sp} is the spontaneous life time of the level E_2. Subscripts p and s correspond to pump and signal respectively. The absorption and emission cross sections depend on the frequency, the specific ion as well as on the pair of levels for a given ion[5].

In Eq. (1) the first term on the right hand side corresponds to spontaneous emission, the second term to pump absorption while the third term corresponds to signal transitions. The pump and signal intensity variation with z are caused due to absorption and stimulated emission and can be described by the following equations:

$$\frac{dI_p}{dz}=-\sigma_{pa}N_1I_p \tag{2}$$

$$\frac{dI_s}{dz}=-\left(\sigma_{sa}N_1-\sigma_{se}N_2\right)I_s \tag{3}$$

In the case of optical fibers since the pump and signal beams propagate in the form of modes, we should describe amplification in terms of powers rather than in terms of intensities. By expressing the intensity patterns in terms of powers we can transform the above equations in terms of powers and solve them to obtain the variation of gain with various parameters such as doped fiber length, pump power, input signal power etc.[3, 6]

A typical erbium doped fiber would have a high numerical aperture of 0.2 and a cut off wavelength around 900 nm so that at the pump wavelength of 980 nm the fiber would be single moded. The erbium concentration is usually about 100 to 500 parts per million (concentration of 5.7×10^{24} m^{-3} to 2.9×10^{25} m^{-3}).

We now present results of simulations of an EDFA with the erbium doped fiber of length 7 m having a core radius of 1.64 μm, an NA of 0.21 and an erbium concentration of 0.68×10^{25} m^{-3}. We assume a single pump at 980 nm and a single signal channel at 1550 nm. The spontaneous lifetime is taken to be 12 ms.

Figure 2 shows the variation of gain versus length of EDFA for different pump powers. As the pump propagates through the fiber, it gets absorbed along the length and reaches a value of z for which the pump power is just enough to equalize the

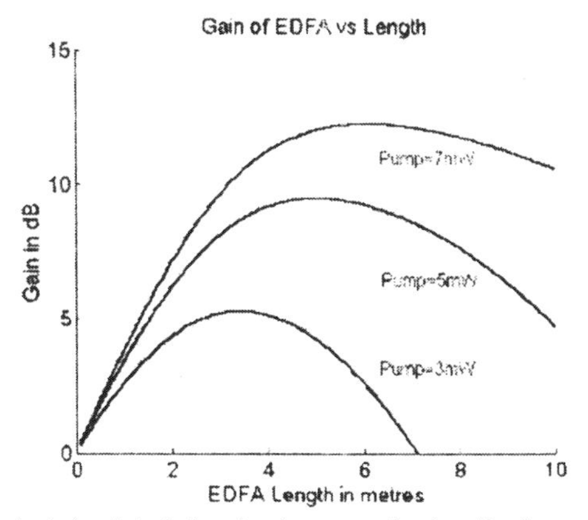

Figure 2 A typical simulated signal gain versus the length of an erbium doped fiber.

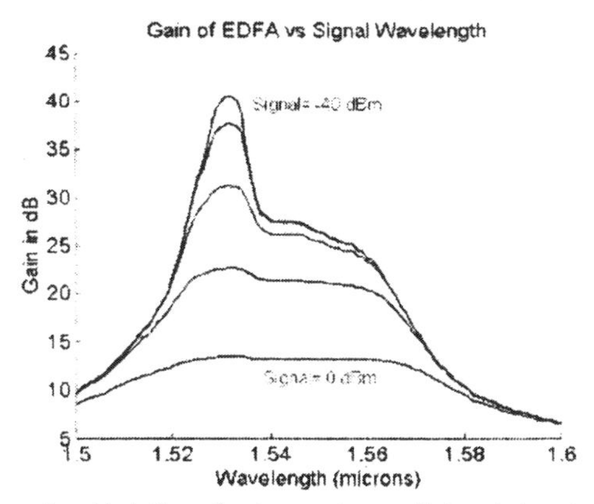

Figure 3 Variation of gain spectrum with input signal power.

populations of the two levels. At this point, referred to as the optimum length for maximum gain, the fiber is neither absorbing nor amplifying and corresponds to the peak in the curve. Beyond this length the signal gets attenuated rather than amplified.

Figure 3 shows the gain spectra at various input signal power levels. It can be seen that as the signal power increases the gain decreases. This is referred to as signal gain saturation. The gain spectrum is also in general not flat and gets modified as the input signal power changes. Non flat gain can lead to problems in their applications in wavelength division multiplexing wherein multiple signals are carrying independent signals.

Figure 4 shows typical measured gain spectra of an EDFA for two different pump powers. As can be seen in the figure, EDFA can provide amplifications of greater than 20 dB over the entire band of 40 nm from 1525 nm to about 1565 nm. This wavelength band is referred to as the C band (conventional band) and is the most common wavelength band of operation.

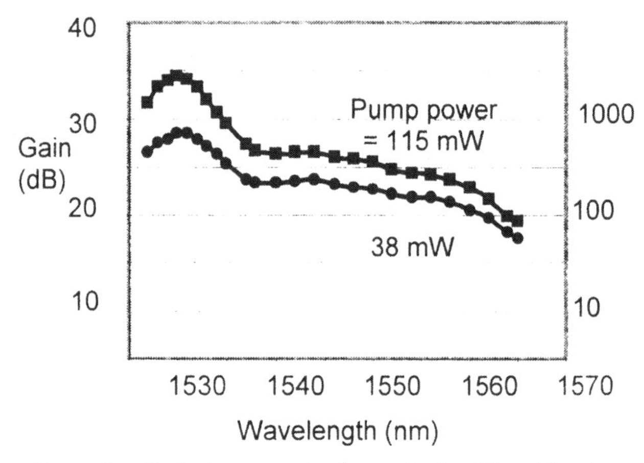

Figure 4 Typical measured gain spectra for different input pump power levels of an EDFA.

With proper amplifier optimization, EDFAs can also amplify signals in the wavelength range of 1570 to 1610 nm; this band of wavelengths is referred to as the L-band (long wavelength band). By controlling the average inversion in the fiber it is possible to achieve amplification in the L band. Recently there has also been activity in realizing EDFAs operating in the S-band (1460 – 1520 nm). This will be discussed in Sec. 6 .

4. Gain flattening of EDFAs

We have seen that although EDFAs can provide gains over an entire band of 40 nm, the gain spectrum is in general not flat, i. e., the gain depends on the signal wavelength. Thus if multiple wavelength signals with same power are input into the amplifier, then their output powers will be different. In a communication system employing a chain of amplifiers, a differential signal gain among the various signal wavelengths (channels) from each amplifier will result in a significant difference in signal power levels and hence in the signal to noise ratio (SNR) among the various channels. In fact, signals for which the gain in the amplifier is greater than the loss suffered in the link, will keep on increasing in power level while those channels for which the amplifier gain is less than the loss suffered will keep on reducing in power. The former channels will finally saturate the amplifiers and will also lead to increased nonlinear effects in the link while the latter will have reduced SNR leading to increased errors in detection. Thus such a differential amplifier gain is not desirable in a communication system and it is very important to have gain flattened amplifiers.

There are basically two main techniques for gain flattening: one uses external wavelength filters to flatten the gain while the other one relies on modifying the amplifying fiber properties to flatten the gain.

4.1 Gain flattening using external filters

The principle behind gain flattening using external filters is to use with the amplifier an external wavelength filter whose transmission characteristic is exactly the inverse of the gain spectrum of the amplifier. Thus channels which have experienced greater gain in the amplifier will suffer greater transmission loss while channels which experience smaller gain will suffer smaller loss. By appropriately tailoring the filter transmission profile it is possible to flatten the gain spectrum of the amplifier. Transmission filters with specific transmission profiles can be designed and fabricated using various techniques. These include thin film interference filters and filters based on long period fiber gratings (LPG). Gain flattening using LPGs will be discussed in Chapter 9.

4.2 Intrinsically flat gain spectrum

We note that the gain of the EDFA is not flat due to the spectral dependence of the absorption and emission cross sections and also due to the variation of the modal overlap between the pump, signal and the erbium doped region of the fiber. Thus it is in principle possible to flatten the gain of the amplifier by appropriately choosing the transverse refractive index profile and the doping profile of the fiber to achieve flatter gain. Figure 5 shows a schematic of a staircase refractive index profile distribution and the corresponding erbium doped region which can provide gain

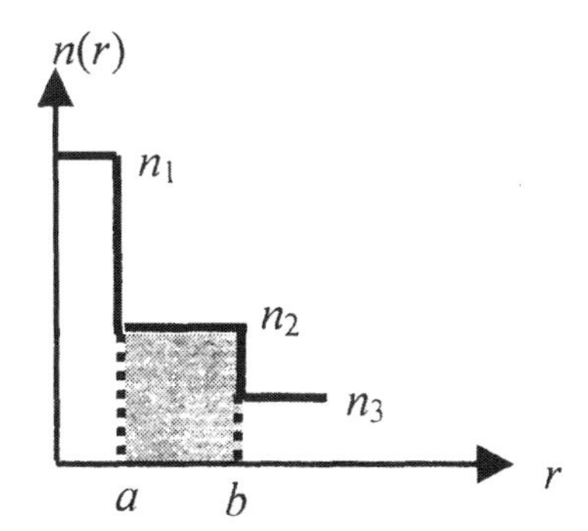

Figure 5 A schematic of a refractive index profile of an erbium doped fiber which exhibits intrinsic gain flattening. The shaded region corresponds to the doped region.

flattening by appropriately optimizing the various parameters. Figure 6 shows the comparison of the gain profile of an EDFA with a conventional fiber and of an optimized EDFA with the staircase design[7,8]. As is evident, much flatter gain profiles can be achieved using proper optimization of the refractive index profile and the doping profile of an EDF.

5. Noise in amplifiers

In an EDFA erbium ions occupying the upper energy level can also make spontaneous transitions to the ground level and emit radiation. This radiation appears over the entire fluorescent band of emission of erbium ions and can travel in all directions. Some of the emitted light can get coupled into the guided mode of the fiber and propagate in the forward as well as in the backward direction. This spontaneously emitted light can then get amplified just like the signal as it propagates through the population inverted fiber. The resulting radiation is called amplified spontaneous emission (ASE). This ASE is the basic mechanism leading to noise in the optical amplifier (see e.g., Ref [1]).

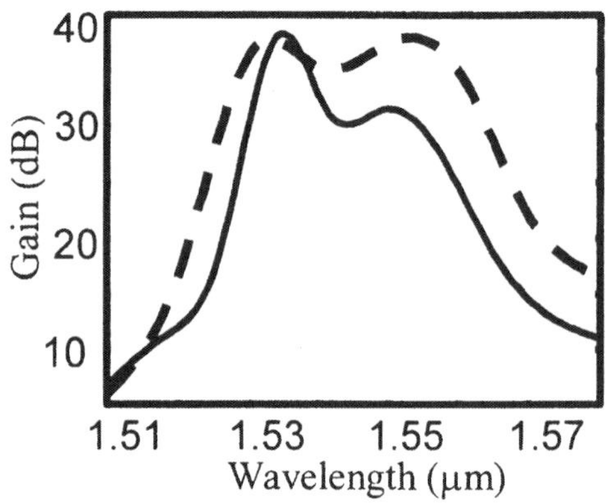

Figure 6 Comparison of the gain spectrum of a EDFA with a conventional erbium doped fiber (———) and the staircase erbium doped fiber (— — —).

ASE appearing in wavelength region not coincident with the signal can be filtered using an optical filter. On the other hand, the ASE that appears in the signal wavelength region cannot be separated and constitutes the minimum added noise from the amplifier.

If P_{in} represents the signal input power (at frequency ν) into the amplifier and G represents the gain (in linear units) of the amplifier then the output signal power is given by $G\,P_{in}$. Along with this amplified signal, there is also ASE power which can be shown to be given by (see e.g., Ref. [1])

$$P_{ASE} = 2n_{sp}(G-1)h\upsilon B_o \tag{4}$$

where B_o is the optical bandwidth over which the ASE power is being measured (which must be at least equal to the optical bandwidth of the signal),

$$n_{sp} = \frac{N_2}{(N_2 - N_1)} \tag{5}$$

Here N_2 and N_1 represent the population densities in the upper and lower amplifier energy levels of erbium in the fiber. Minimum value for n_{sp} corresponds to a completely inverted amplifier for which $N_1 = 0$ and thus $n_{sp} = 1$.

The optical signal to noise ratio (OSNR) is defined as the ratio of the output optical signal power to the ASE power and is given by:

$$OSNR = \frac{P_{out}}{P_{ASE}} = \frac{GP_{in}}{2n_{sp}(G-1)h\nu B_o} \tag{6}$$

where P_{in} is the average power input into the amplifier. For large gains G >> 1 and assuming B_o = 12.5 GHz (=0.1 nm at 1550 nm), for a wavelength of 1550 nm, we obtain

$$OSNR(\text{dB}) \approx P_{in}(\text{dBm}) + 58 - F \tag{7}$$

where

$$F(\text{dB}) = 10\log(2n_{sp}) \tag{8}$$

is the noise figure of the amplifier (for large gains). As an example, for an input power of 1 μW (=-30 dBm) and a noise figure of 5 dB, the estimated OSNR is about 23 dB.

In a long distance fiber optic communication system EDFAs are used periodically to compensate for the loss of each span of fiber. The gains of the amplifiers are chosen to compensate for the loss suffered by the signal. Thus the signal amplitude is maintained along the link at the end of each span. On the other hand, each amplifier in the chain adds noise and thus the noise power keeps increasing after every span resulting in a falling OSNR (see Fig. 7). At some point in the link when the OSNR falls below a certain value, the signal would need to be regenerated. If we assume a noise of 0.6 μW added by each amplifier then after say 5 amplifiers, the signal power would still be the same as at the beginning (assuming) the amplifier gain exactly compensates for the attenuation in the span) but the noise power would be 3 μW. Thus as the signal passes through multiple spans and amplifiers, there is a reduction in the optical signal to noise ratio. Hence there is a

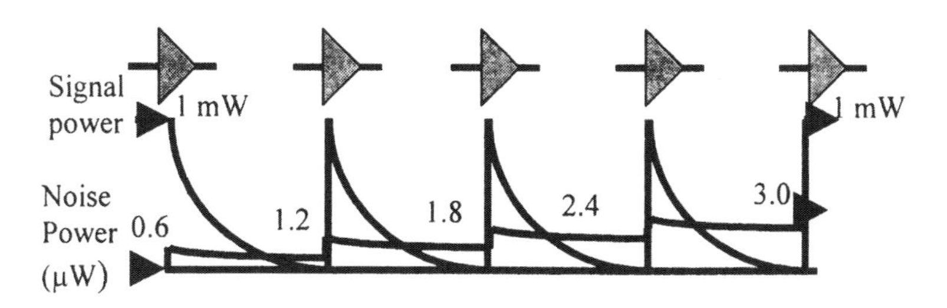

Figure 7 A long fiber optic link with EDFAs for compensation of loss of each span. The signal power returns to the original value after each span while the ASE power keeps on building up leading to reduction in OSNR with increased number of spans.

maximum number of amplifiers that can be placed in a link beyond which the signal needs to be regenerated.

The above discussion was based on the *optical* signal to noise ratio of the amplifier. When the amplified output is received by a detector, then the detector converts the optical signal into an electric current and the noise characteristics of the generated electrical signal are of importance. Apart from the optical signal, the amplified spontaneous emission within the bandwidth of the signal also falls on the photodetector. However, the ASE noise is completely random and contains no information. The photodetector would convert the total optical power received into electrical current; if the electric fields of the signal and noise are E_s and E_n respectively then the generated electric current would be proportional to $(E_s + E_n)^2$. The expansion of this would consist of terms proportional to $(E_s)^2$, $(E_n)^2$ and $E_s E_n$. The first term leads to the signal current while the other two terms correspond to noise. The second term leads to a beating between noise components at various frequencies lying within the signal bandwidth and is referred to as spontaneous-spontaneous beat noise. The last term leads to beating between the signal and spontaneous emission and is referred to as signal-spontaneous beat noise. Under normal circumstances, the signal-spontaneous noise term and the signal-shot noise terms are the dominant noise terms and assuming the input to the amplifier to be shot noise limited we can calculate the output signal to noise ratio (SNR) from the noise terms. We define the noise figure of the amplifier by the following relation:

$$F = \frac{(\mathrm{SNR})_{\mathrm{in}}}{(\mathrm{SNR})_{\mathrm{out}}} \tag{9}$$

By calculating the input and output SNR we can obtain an expression for the noise figure of the amplifier which is given by

$$F = \frac{1 + 2n_{sp}(G-1)}{G} \tag{10}$$

Thus the noise figure depends on the inversion through n_{sp} and on the amplifier gain through G. For large gains $G >> 1$, the noise figure is approximately given by 2 n_{sp}. Since the smallest value of n_{sp} is unity, the smallest noise figure is given by 2 or in decibel units as 3 dB. This implies that there is always a deterioration of the signal to noise ratio due to amplification.

Noise figure is a very important characteristic of an amplifier and determines the overall performance of any amplified link. Noise figures of typical commercially available EDFAs are about 5 dB.

6. EDFAs for the S-band

Recently there has been activity in realizing erbium doped silica based optical amplifiers for the S-band[9]. It has been shown that efficient S-band EDFAs require high inversion levels along the fiber and C-band ASE suppression, which otherwise depletes the population inversion. An efficient design for single stage S-band EDFA based on a co-axial core fiber has recently been proposed[10]. In the proposal distributed ASE filtering is achieved by winding the fiber with an optimally chosen bend radius. Bend loss usually has a strong spectral variation because of the variation of mode field diameter (MFD) with wavelength. Co-axial fiber design provides extra degrees of freedom in terms of tuning the bend loss variation. Hence, by optimising the fiber parameters and the bend radius, high bend loss is ensured at wavelengths above 1525 nm (> 6 dB/m), whereas the wavelengths below 1525 nm suffer minimal loss. For this design, the bend loss at 1530 nm is about 100 times greater than that at 1490 nm, which leads to a high net gain in the S-band. Figure 8 shows the gain spectrum of the S-band amplifier (designed with a bend radius of 3 cm) under unsaturated operating conditions. As can be seen an average gain of 26 dB with a gain variation of ± 2.9 dB over the wavelength region 1495 nm to 1525 nm is achievable.

7. Raman fiber amplifier (RFA)

In the earlier sections we discussed an optical amplifier based on population inversion created between pairs of energy levels of erbium ion. Optical amplifiers based on stimulated Raman effect in optical fibers have also gained importance in recent years. This is primarily due to the fact that unlike EDFAs, RFAs can be made to operate at any wavelength band and they have also a broad bandwidth. Apart from this, the link fiber can itself be used as the amplifier and thus the signal gets amplified as it covers the distance along the communication link itself. Such amplifiers are also referred to as distributed amplifiers.

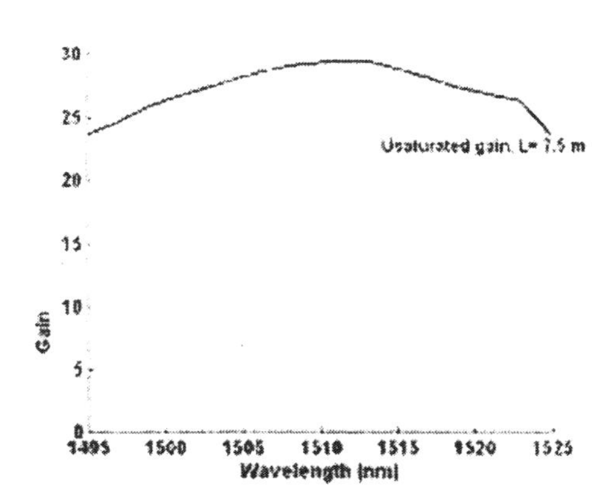

Figure 8 Gain spectrum (in dB) of a novel S-band EDFA

When a light beam interacts with a medium consisting of molecules then apart from Rayleigh scattering, which appears at the same frequency as the incident light, one also observes very weak scattering at a frequency smaller than the incident frequency. This is the result of Raman scattering discovered by Sir C.V. Raman for which he received the Nobel Prize in 1930. The incident photon interacts with the molecules which take up a part of the energy of the incident photon to get excited to a higher vibrational level and thus leading to a scattered photon with a lower energy or frequency (see Fig.9). The difference in frequency between the incident photon and the scattered photon is a characteristic of the molecule. In the case of an optical fiber the vibrational modes (optical phonons) of the glass matrix leads to Raman scattering. When we send a high power laser beam at 1450 nm (say into a long single mode fiber, then at the output apart from the attenuated light at 1450 nm, we would also see light emerging at longer wavelengths. This is due to spontaneous Raman scattering (see Fig. 10). As can be seen the scattered radiation appears over a large band and the peak of the scattered radiation lies at about 100 nm away from the pump wavelength. Indeed Raman scattering in silica leads to a peak Raman shift of between 13 and 14 THz which corresponds to about 100 nm at the wavelength of 1550 nm. The same frequency shift at 1310 nm would correspond to about 70 nm

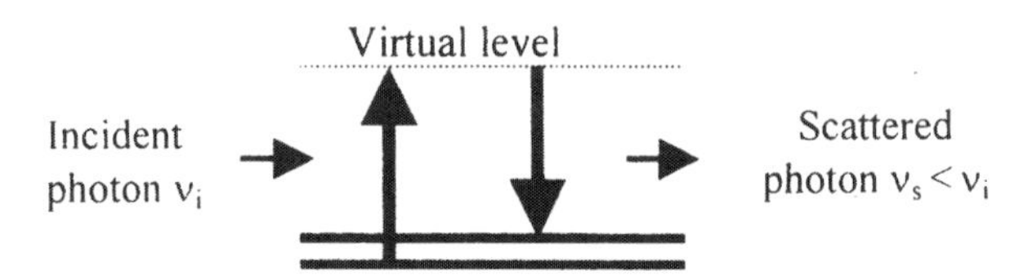

Figure 9 In Raman scattering an incident photon gets scattered in to a lower energy photon leaving the molecule into a higher vibrational energy state.

wavelength shift. When we launch a pump at 1450 nm and simultaneously a signal lying within the band of spontaneous Raman scattering, then it leads to stimulated Raman scattering. In this case, the pump and signal wavelengths are coherently coupled by the Raman scattering process and the scattered photon is coherent with the incident signal photon much like stimulated emission discussed earlier. It is this process that is used to build Raman fiber amplifiers.

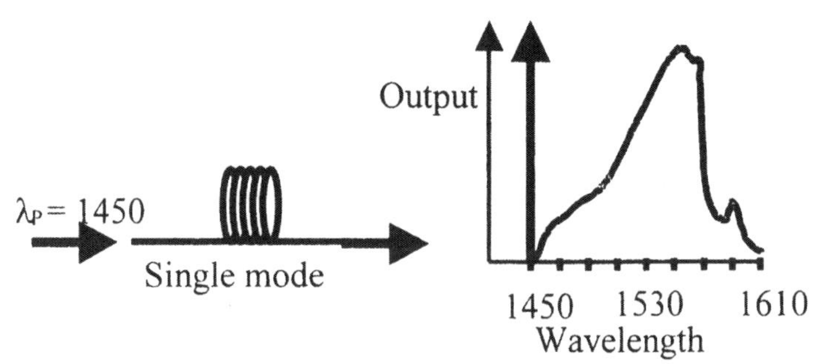

Figure 10 When light at 1450 nm is sent through an optical fiber, then at the output apart from the attenuated 1450 nm signal, we also get light at higher wavelengths via spontaneous Raman scattering.

The equations describing the evolution of signal power (P_s) and pump power (P_p) due to Raman scattering are given by[4, 11]

$$\frac{dP_s}{dz} = \frac{\gamma_R P_p P_s}{K_{eff}} - \alpha_s P_s \tag{11}$$

$$\pm\frac{dP_p}{dz} = -\frac{\nu_p}{\nu_s}\frac{\gamma_R P_p P_s}{K_{eff}} - \alpha_p P_p \tag{12}$$

where α_s and α_p represent the background attenuation of the fiber at the signal and pump wavelengths. Here γ_R is called the Raman gain efficiency and is given by

$$\gamma_R = \frac{g_R}{A_{eff}} \tag{13}$$

where g_R is the Raman gain coefficient and A_{eff} is the effective area defined by

$$A_{eff} = 2\pi \frac{\int \psi_p^2 r dr \int \psi_s^2 r dr}{\int \psi_p^2 \psi_s^2 r dr} \tag{14}$$

with ψ_p and ψ_s representing the modal field distributions at the pump and signal frequencies. In Eqs. (11) and (12) K_{eff} represents the polarization factor; if the signal and pump are co-polarized then it has a value of unity and if the polarizations are completely scrambled then $K_{eff} = 2$. This factor takes account of the strong polarization dependence of Raman amplification. The positive and negative signs in Eq. (12) correspond respectively to co-directional pumping (i.e., pump propagating along with the signal in the $+z$ direction) and contra-directional pumping (i.e., pump propagating along the $-z$ direction).

The second terms on the right hand sides of Eqs. (11) and (12) correspond to attenuation of the fiber and the first terms to Raman scattering. Since the fiber lengths involved in Raman amplification are large (unlike EDFAs), the background attenuation plays a significant role here. The extra factor of ν_p/ν_s in the first term on the right hand side of Eq. (12) comes about due to the fact that for every photon generated in the signal, one photon at the pump frequency must be lost.

Typical values of Raman gain efficiency γ_R for different fibers are:

Standard SMF: ~ 0.7 W^{-1} km^{-1}

Dispersion compensating fiber: 2.5 – 3 W^{-1} km^{-1}

Highly nonlinear fibers: 6.5 W^{-1} km^{-1}

As can be seen dispersion compensating fibers possessing very small effective areas have large values of γ_R.

7.1 Approximate solutions

If the pump depletion due to signal amplification can be neglected (i.e., the first term on the right hand side of Eq. (12)), then the pump equation can be easily solved. Assuming contra directional pumping we obtain the pump power variation with z as

$$P_p(z) = P_p(L)e^{-\alpha_p(L-z)} \tag{15}$$

where $P_p(L)$ is the input pump power at $z = L$. Substituting this solution in Eq. (11) for the signal, we obtain for the signal power at the output of the fiber of length L as

$$P_s(L) = P_s(0)\, e^{-\alpha_s L} \exp[\gamma_R\, P_p(L)\, L_{eff}] \tag{16}$$

where L_{eff} is the effective length of the fiber defined through

$$L_{eff} = \frac{1-e^{-\alpha_p L}}{\alpha_p} \tag{17}$$

We now define the on-off gain as the ratio of signal power at the exit in the absence of the pump and in the presence of the pump which for $\alpha_p L >> 1$ is given by:

$$\tilde{G}(\text{dB}) = G_{on-off}(\text{dB}) \approx 4.34 \frac{\gamma_R}{\alpha_p} P_p(L) \tag{18}$$

The net gain of the Raman amplifier is defined as the ratio of the output signal power to the input signal power (see Fig. 11).

The noise in Raman amplifier is caused due to spontaneous Raman scattering which is independent of the signal and also due to double Rayleigh scattering. Equations describing the evolution of the spontaneous power can also be written

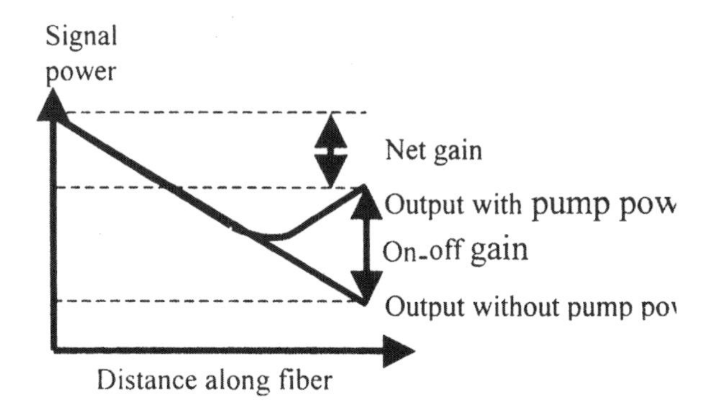

Figure 11 Figure showing on-off gain and net gain of a Raman amplifier.

down and one can then calculate the OSNR of the output signal.

We will now present some results obtained by numerically resolving the Raman amplifier equations in the presence of multiple signals taking into account pump depletion also.

Figure 12 shows the variation of the signal power at 1550 nm as a function of distance along a standard single mode fiber in the presence of a pump at 1450 nm with a power of 400 mW, which is propagating along the fiber in the backward

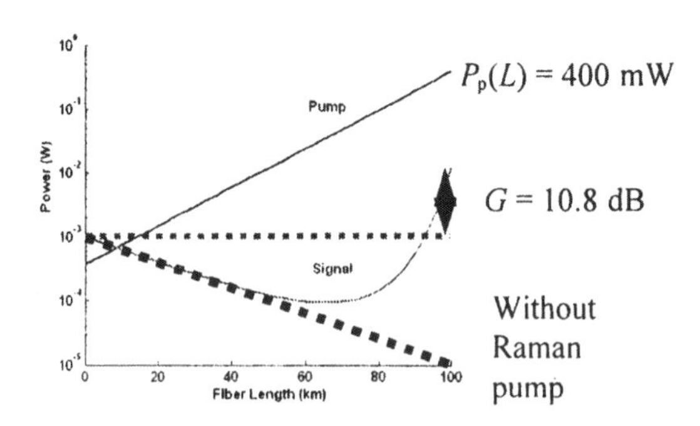

Figure 12 In a contra directionally pumped Raman amplifier, the signal power reduces first and then gets amplified as it encounters the pump.

direction. In the absence of the pump, the signal gets attenuated to 10 μW at the end of the link (corresponding to a loss of 20 dB). On the other hand, in the presence of the pump, the signal initially gets attenuated but as it propagates towards the end of the link it then encounters the high power pump and gets amplified via stimulated Raman scattering and exits with a power of 12 mW leading to a net gain of 10.8 dB instead of a loss of 20 dB.

As an example of how Raman amplifiers can be used to increase span length of a fiber optic link, consider a 16-channel system with an EDFA booster amplifier with an output power of 15 dBm. This implies that each channel has a power of 2 mW. If the receiver sensitivity is –28 dBm, then assuming a fiber loss of 0.25 dB/km, the maximum length of the span would be 124 km corresponding to a loss of 31 dB. If we now launch a 400 mW pump propagating in the backward direction, then this would result in a Raman gain of 15 dB and the total link length that is now possible is 180 km. Thus Raman amplifiers can be used to increase span length.

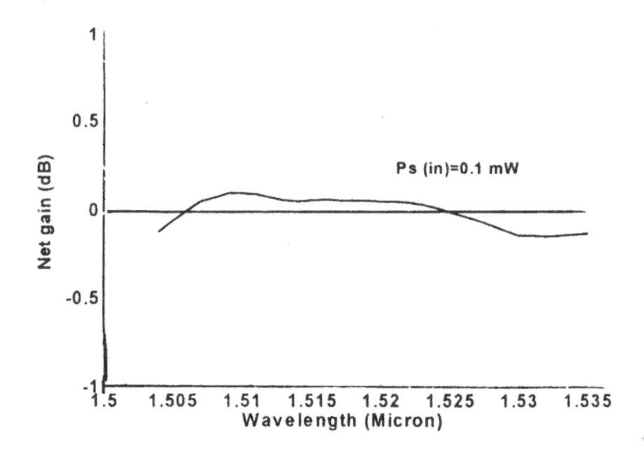

Figure 13 Spectral variation of net gain Raman gain for an ATC fiber [15].

Since the gain spectrum depends on the pump wavelength, it is indeed possible to achieve large flat gain using multiple pumps. Thus using 12 pumps with wavelengths lying between 1410 nm and 1510 nm, a total flat gain bandwidth of 100 nm from 1520 nm to 1620 nm (covering the C-band and the L-band) has been demonstrated[12].

Since the gain coefficient depends on the effective area of the fiber, the Raman gain spectrum could be significantly modified by proper fiber designs having appropriate spectral dependence of effective area. Recently a novel coaxial fiber design and a novel twin core fiber design have been proposed that can give flat Raman gain with just a single pump[13-15]. For example, Fig. 13 shows the gain spectrum of the novel asymmetric twin core fiber design with a single Raman pump at 1450 nm. Both these fiber designs possess high negative dispersion values and thus they can act as zero loss dispersion compensating fibers over a band of 90 nm. Although the simulations have been carried out for the S-band wavelengths, since Raman gain spectrum depends on the pump wavelength, similar results could also be achieved for other bands of amplification.

Compared to EDFAs, Raman amplifiers require much higher pump powers and also the gain is polarization dependent. The polarization dependence can be overcome either by using depolarized pumps or by launching pumps corresponding to two orthogonal polarization directions. Since the amplification is taking place over the entire length of the fiber, double Rayleigh scattering gets amplified and adds to the noise of the amplifier. For nice reviews of fiber Raman amplifiers, readers are also referred to Refs. [4,16,17].

8. Conclusions

Optical fiber amplifiers have truly revolutionized optical fiber communications and have made possible practical implementation of dense wavelength division multiplexing for increased bandwidth. EDFAs can be made to operate in S-, C- and L-bands and although EDFAs are the most popular amplifiers, Raman fiber amplifiers and semiconductor optical amplifiers are also becoming very important. With the availability of optical amplifiers capable of spanning the entire wavelength region of 1250 nm to 1650 nm, the entire low loss window of optical fibers will become available for the telecommunication engineers.

Acknowledgements: The author thanks Ms. Charu Kakkar and Mr. Deepak Gupta for their help in simulations of many of the results presented here.

References

1. P.C. Becker, N.A. Olsson, and J.R. Simpson, *Erbium doped fiber amplifiers*, Academic Press, San Diego (1999).
2. E. Desurvire *Erbium doped fiber amplifiers*, Academic Press, New York (1994).
3. A.K. Ghatak and K. Thyagarajan, *Introduction to fiber optics*, Cambridge University Press, UK (1998).
4. M.N. Islam (Ed.), *Raman amplifiers for telecommunications*, Vols 1 and 2, Springer, New York, (2004).
5. W.L. Barnes, R.I. Laming, E.J. Tarbox, P. Morkel, "Absorption and emission cross section of Er^{3+} doped silica fibers", *IEEE J Quant. Electron.* **27**, 1004-1010 (1991).
6. Y. Sun, J.L. Zyskind and A.Srivastava, "Average inversion level, modeling, and physics of erbium doped fiber amplifiers", *IEEE J. Sel. Topics Quant. Electron.* **3**, 991-1007 (1997).
7. K. Thyagarajan.and J.Kaur, "A novel design of an intrinsically gain flattened erbium doped fiber", *Optics Communications*, **183**, 407-413 (2000).
8. K. Thyagarajan and J.Kaur, "Intrinsically gain-flattened staircase profile erbium doped fiber amplifier", *Optics Communications*, **222**, 227-233 (2003).
9. M.A. Arbore, Yidong Zhou, Hans Thiele, Jake Bromage and Lynn Nelson, "S-band Erbium doped fiber amplifiers for WDM transmission between 1488 and 1508 nm, in *Proceedings of Optical Fiber Communications*, Georgia, USA, paper WK 2, (2003).

10. K. Thyagarajan and Charu Kakkar, "S-band single stage EDFA with 25 dB gain using distributed ASE suppression", *IEEE Photonics Tech. Letts.*, **16,** pp 2448 – 2450 (2004).
11. G.P. Agarwal, *Nonlinear fiber optics*, Academic Press, NY, (1995).
12. Y. Emori and S. Namiki, "100 nm bandwidth flat gain Raman amplifiers pumped and gain-equalized by 12-wavelength-channel WDM high power laser diodes," Proc. Optical Fiber Communication Conference, Post-deadline Paper PD19, (1999).
13. K. Thyagarajan and Charu Kakkar, "Fiber design for broadband, gain-flattened Raman fiber amplifier ", *IEEE Photonics Tech. Letts.,* **15**, 1701-1703 (2003).
14. K. Thyagarajan and Charu Kakkar, "Novel fiber design for flat gain Raman amplification using single pump and dispersion compensation in S-band", *J. Lightwave Tech.* **22**, 2279-2286 (2004).
15. Charu Kakkar and K. Thyagarajan, "Broadband lossless DCF utilizing single pump Raman amplification", CLEO/IQEC 2004, San Francisco, May 2004
16. M. Islam, "Raman amplifiers for telecommunications", *IEEE J. Sel. Topics in Quant. Electron.*, **8**, 548-559 (2002).
17. J. Bromage, "Raman amplification for fiber communication systems", *J. Lightwave Tech.*, **22**, 79-93 (2004).

4

Temporal and Spatial Optical Solitons

Ajit Kumar

1. Introduction

The importance of optical solitons in modern nonlinear guided wave optics is well known to the optics community. After the inginious idea of Hasegawa and Tappert[1], who proposed to use temporal optical solitons in a fiber-optic communication link in order to tackle the long-standing problem of fiber dispersion, the field of optical solitons has progressed rapidly. The experimental work of Mollenauer, Stolen and Gordon[2] on soliton generation and propagation in an optical fiber using picosecond laser pulses immensely contributed to this new area and further accelerated its growth. Beautiful physics with added advantage of technological applications motivated physicists and technologists to look for similar phenomena in other optical systems and it led to the discovery of new types of solitonic structures with exciting properties, useful for various technological applications[3–8].

The main aim here is to introduce the reader to optical solitons and their possible applications. We start with the basic concept of a temporal soliton in an optical fiber with Kerr nonlinearity. Then, we move on to a brief description of a soliton-based all-optical communication system where we focus on the practical problems encountered in a long-haul system and discuss about the ways to overcome them. Finally, we take up spatial optical solitons and discuss their properties and possible applications. In this connection, we deal with the physical concepts in detail using the minimum necessary mathematics.

In general, a soliton is a localized propagating wave solution of a nonlinear dispersive or diffractive partial differential equation which preserves its shape and velocity, in a lossless medium, under a collision with a similar localized propagating wave. The collision property distinguishes a soliton from a so-called solitary wave which may not fulfil this requirement of elastic collision. Depending on the localization property and the spatial dimension in which they exist, solitons are classified into three major categories : (i) temporal solitons, (ii) spatial solitons and (iii) spatio-temporal solitons.

2. Temporal solitons

They can exist in a nonlinear and dispersive medium and are excited when the dispersive boradening of an optical pulse is exactly balanced by nonlinear response of the fiber due to which the pulse tries to collapse on itself. This balance is a dynamical one and depends on the nature of dispersion and the nonlinear properties of the medium. In an optical fiber, they are formed due to mutual cancellation of phase modulations induced by dispersion and nonlinearity in the anomalous dispersion region[1,3-9]. In the normal dispersion region, these phase modulations assist each other and lead to greater pulse broadening as compared to the case of dispersive propagation alone.

The partial differential equation, that governs pulse evolution in a nonlinear and isotropic fiber with circular cross-section, is given by the so-called nonlinear Schroedinger equation (NLSE) :

$$i\,\frac{\partial u}{\partial \xi} + \frac{1}{2}\frac{\partial^2 u}{\partial \tau^2} + n_{nl}(I)\; u = 0 \tag{1}$$

where n_{nl} is the intensity dependent nonlinear refractive index change and I is the intensity of light. u, τ and ξ are the dimensionless pulse envelope amplitude, the dimensionless time (which is measured in a frame moving with the group velocity of the pulse) and the dimensionless distance of propagation along the fiber, respectively. The envelope amplitude, u, is normalized to the square root of the pulse peak power P_0, while the time, τ, is normalized to the initial pulse width T_0. The distance along the fiber, ξ, is normalized to the characteristic dispersion length $L_D = T_0^2/|\beta_2|$, where β_2 is the second-order dispersion coefficient. For the simplest case of Kerr nonlinearity, $n_{nl}(I) = n_2 I$, where n_2 is the nonlinear Kerr coefficient and, consequently, Eq.(1) takes the form[3-8]

$$i\,\frac{\partial u}{\partial \xi} + \frac{1}{2}\frac{\partial^2 u}{\partial \tau^2} + |u|^2\; u = 0. \tag{2}$$

This equation, for an arbitrary localized input condition

$$u(\xi, 0) = \psi(\tau), \quad lim_{|\tau|\longrightarrow\infty}\psi(\tau) = lim_{|\tau|\longrightarrow\infty}\frac{\partial \psi(\tau)}{\partial \tau} = 0, \tag{3}$$

can be integrated by the inverse scattering transform method (ISTM)[10,11]. The general asymptotic solution consists of solitons and localized radiations.

2.1 Bright Soliton

Consider the following initial condition

$$\psi(\tau) = N\; sech(\tau) \tag{4}$$

where N^2 is the ratio of the characteristic dispersion length L_D to the characteristic nonlinear length L_N, and is given by

$$N^2 = \frac{2\pi n_2 P_0 T_0^2}{\lambda|\beta_2|A_{eff}}. \tag{5}$$

Here, in equation (5), A_{eff} is the effective core area of the fiber and λ is the carrier wavelength. N is called the order of the soliton solution. In this case the stationary bright fundamental soliton solution ($N = 1$) of the NLSE (2) is given by[3]

$$u(\xi, \tau) = sech(\tau)\ e^{i\xi/2} \tag{6}$$

The above soliton solution reperesents a lump of light and is called a bright soliton. Soliton solutions of Eq.(2) corresponding to $N > 1$ are called higher-order solitons. When a fundamental soliton is launched into a dispersive and lossless Kerr fiber the shape of the input pulse does not change during propagation (in principle for an infinite length). On the other hand, the propagation of a higher-order soliton is characterized by a continuous change in shape in such a way that the original shape is recovered at $\xi = \frac{m\pi}{2}$, where m is an integer. This happens because of the variations in the relative strength of dispersive and nonlinear effects which depend on the distance of propagation. Since $\xi = \frac{z}{L_D}$, the soliton period z_0, defined as the characteristic distance over which the higher-order solitons recover their initial shape, is given by[4]

$$z_0 = \frac{\pi}{2} L_D = \frac{\pi T_0^2}{2|\beta_2|} \tag{7}$$

The soliton period z_0 and the soliton order N play an important role in quantifying temporal solitons in an optical fiber[5].

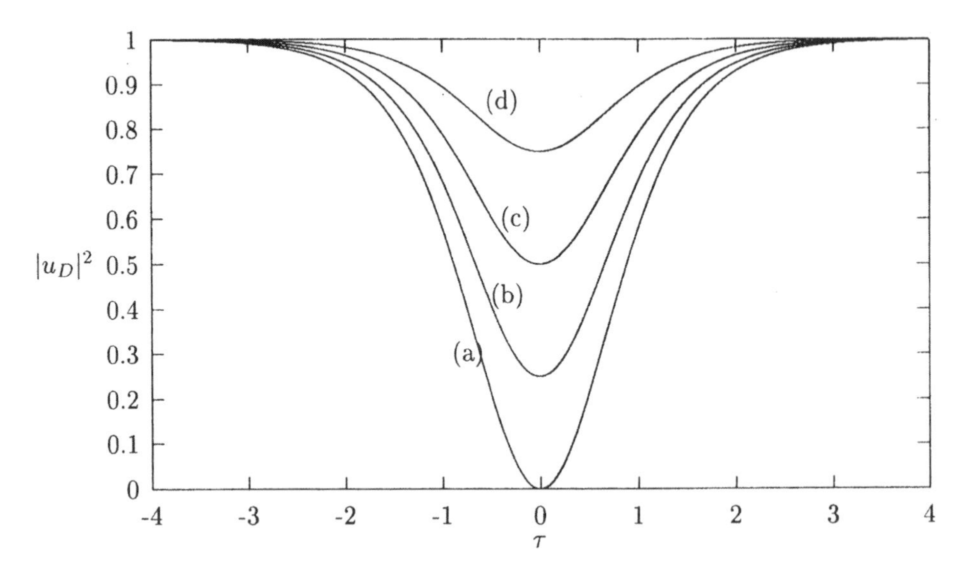

Figure 1 Dark soliton intensity for various values of the dip parameter σ. The curves, labelled (a), (b), (c), and (d) correspond to σ=0.0, 0.25, 0.50 and 0.75, respectively.

2.2 Dark solitons

Besides the bright solitons, discussed above, the NLSE also admits the so-called dark soliton solutions, in the normal dispersion regime. A dark soliton

represents a dip in the light intensity profile against a uniform background, and it is the dip that remains unchanged during propagation along the fiber. Since the dip is characterized by the absence of light in it, the soliton is called a dark soliton.
The dark soliton solution is given by[4]:

$$u_D(\xi,\tau) = \sqrt{(1-\sigma^2\ sech^2\tau)}\ e^{i(\kappa\xi+\phi(\tau))} \tag{8}$$

where the time dependent phase ϕ is

$$\phi(\tau) = sin^{-1}\left(\frac{\sigma\ tanh\tau}{\sqrt{(1-\sigma^2\ sech^2\tau)}}\right). \tag{9}$$

Here κ is a constant and σ is called the dip parameter or blackness and varies in the range [0,1]. For $\sigma = 1$ the intensity at the dip center, $\tau = 0$, is zero, where as for other values of $\sigma \neq 1$ it is given by $(1-\sigma^2)$. The solution with $\sigma = 1$ is called the fundamental dark soliton.

It has been shown by numerical simulations[12] that the dip can propagate as a dark soliton even against a non-uniform background, provided the light intensity is uniform in the vicinity of the dip. The intensity profile of a dark soliton are depicted in Fig.1 for several values of the dip parameter σ. Dark temporal solitons have been experimentally observed[13,14] and are also expected to play important role in future communication systems.

2.3 Bright soliton generation

In a communication system it is proposed to use a fundamental bright soliton in each bit slot, representing 1 in a bit stream. The first and the foremost requirement, in this connection, is the generation of a soliton. The average power needed for creating a soliton can be calculated as follows. The width parameter T_0 of the input pulse is related to the FWHM, T_s, of the soliton by[10]

$$T_s = 2T_0\ ln(1+\sqrt{2}) = 1.763\ T_0. \tag{10}$$

Therefore, the peak power of the soliton is given by

$$P_0 = \frac{(1.763)^2|\beta_2|}{\gamma T_s^2}; \qquad \gamma = \frac{2\pi n_2}{\lambda A_{eff}} \tag{11}$$

and the average pulse energy is

$$E_s = \int_{-\infty}^{+\infty} |A(0,t)|^2\ dt = 3.526\ P_0\ T_s. \tag{12}$$

Since both 1 and 0 bits are supposed to be equally likely, the average power of a RZ signal is given by

$$\bar{P}_s = \frac{1}{2}\ E_s B = \frac{P_0}{2d_0}, \tag{13}$$

where $2d_0$ is the separation between the neighbouring solitons in the soliton units and B is the bitrate of the system related to the duration T_B of the bit slot by

$$B = \frac{1}{T_B} = \frac{1}{2d_0 T_0}. \tag{14}$$

For instance, for a 10 Gb/sec system with a soliton of FWHM of 17.6 ps ($T_0 \sim 10$ ps), the required peak power[4] of the input pulse is as low as 5 mW, if $\beta_2 = -1$ ps^2/km and $\gamma = 2\ W^{-1}/km$. This can easily be done with the available lasers.

2.4 Properties of the fundamental bright soliton

One of the most important properties of the fundamental soliton is its stability against small perturbations. In spite of the fact that a fundamental soliton requires a specific input shape, sech(τ), and a definite input power, given by

$$P_0 = \frac{\lambda|\beta_2|A_{eff}}{2\pi n_2 T_0^2} = \frac{|\beta_2|}{\gamma T_0^2}; \qquad \gamma = \frac{2\pi n_2}{\lambda A_{eff}} \tag{15}$$

for its creation, it can still be generated even if the initial conditions deviate a little from the above mentioned values for a given carrier wavelength and fixed parameters of the fiber.

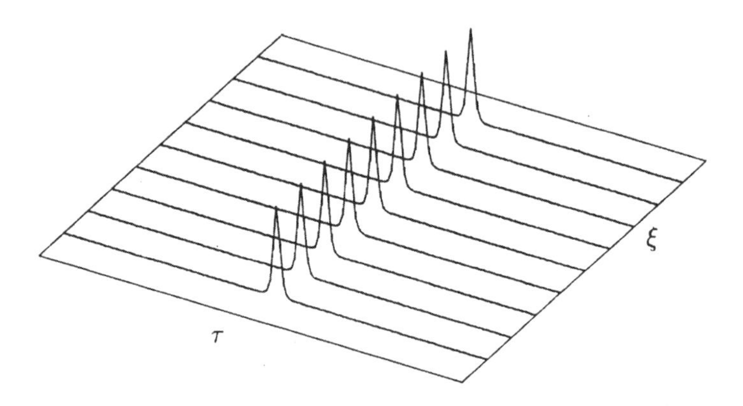

Figure 2 Spatio-temporal evolution of the fundamental bright soliton with N=1.

In general, the ISTM rigorously predicts that the $N-th$ order soliton can be formed when the parameter N lies in the following range[7]

$$N - \frac{1}{2} \leq N \leq N + \frac{1}{2}. \tag{16}$$

So, the fundamental soliton can be created by an input for which $N \in [0.5, 1.5]$. The concrete form of the input shape is not important, so long as it is localized in τ. In other words any localized input with N in the above specified range will evolve into a fundamental soliton during propagation. This can be understood by invoking the concept of the temporal modes[9] of a nonlinear waveguide. Higher intensities in the central part of a localized pulse increase the refractive index there and create a temporal waveguide. The waveguide, thus created, supports temporal modes just as the core-cladding index difference leads to spatial modes in a fiber. Therefore, even if the input pulse does not exactly match the temporal mode of this waveguide, the major part of the pulse will still be coupled to the temporal mode. The left-over energy spreads as a dispersive wave. When such a fundamental soliton is launched into a lossless fiber, it propagates without any distortion. This has been depicted in Fig. 2.

Another important property of the fundamental soliton follows from its form given by Eq.(6). It clearly shows that the soliton pulse acquires a phase shift of $\frac{\xi}{2}$ during propagation, however, its amplitude remains unchanged. Since this phase modulation is uniform over the entire wave form, such solitons are ideal candidates for routing switches, like, a nonlinear directional coupler, because they switch as a unit and help in avoiding the phenomena of output pulse distortion and pulse break-up which invariably occur with any kind of non-solitonic pulses.

The above mentioned properties (along with some other important characteristics related to stability with respect to randomly varying polarization and self-similarity owing to which the initial pulse shape can be restored by proper amplification in a lossy fiber), make the fundamental temporal solitons very useful and attractive for future all-optical communication systems[3,36], switches and logic gates[4–6].

3. Fiber-optic Communication

Since the most important application of temporal solitons is in a fiber-optic communication link, let us discuss, in brief, the basic features of such a communication system.

3.1 Soliton amplification

The first in the list of design issues is loss management. Due to the intrinsic fiber loss, a soliton loses energy and broadens due to the weakening of nonlinear effects. Perturbative calculations yield an exponential growth of the soliton width given by[3–8]

$$T_s(\xi) = T_{s0}\ e^{2\Gamma\xi} = \ T_{s0}\ e^{\alpha z}, \tag{17}$$

where $\Gamma = \alpha L_D/2$ is the normalized loss coefficient and α is the usual power loss coefficient. In reality, however, the increase in the pulse width is not exponential but much slower, as has been shown by several authors using direct numerical calculations of the corresponding damped NLSE. In any case, for a long-haul system, this pulse broadening due to fiber loss must be taken care of by periodic amplification[15]. For this purpose either lumped or distributed amplification scheme is used[5–8].

3.1.1 Lumped amplification

In the simplest possible scheme, an optical amplifier is placed periodically along the fiber link. The amplifier gain G_0 is chosen such that

$$G_0 e^{-\alpha L_A} = 1, \tag{18}$$

in order to exactly compensate for the fiber losses between any two amplifiers.

In this scheme the soliton is boosted up to its initial level locally over a short distance without allowing for a gradual recovery of the fundamental soliton. After amplification, the soliton adjusts its width dynamically shedding a part of its energy as dispersive wave. Therefore, large and rapid variations in the soliton peak power in this process,can destroy the soliton pulse due to rapid variations in its width accompanied by the emission of continuous radiation. A way to avoid this is to keep the amplifier spacing L_A much smaller than the dispersion length L_D. In this case the soliton does not get distorted despite energy losses and can be amplified hundreds of times while keeping its amplitude and width intact. This regime of soliton propagation is known as guiding center soliton regime[21] or path-averaged soliton regime.

3.1.2 Distributed amplification

If the rate of information transfer exceeds 10 Gb/sec, it becomes very difficult to satisfy the above mentioned restriction on the amplifier spacing. This condition can be relaxed by using the distributed amplification scheme. The stimulated Raman scattering[15] is used for this purpose, which provides Raman gain all along the transmission line when the fiber is pumped optically by injecting the pump light of appropriate wavelength. The main advantage of this scheme is that its use leads to nearly lossless fiber by compensating for loss locally at every point inside the transmission fiber. In practice, however, the pump power does not remain constant due to fiber loss and pump depletion and hence the gain $G_0(z)$ can not be made constant along the fiber. As a result, the pump light must be injected to boost the pump power up for amplification. In spite of that the fiber loss can be fully compensated for over the distance L_A, if

$$\int_0^{L_A} G(z)\, dz = \alpha L_A. \tag{19}$$

The experimental feasibility of this scheme and its usefulness for a long-haul system has been demonstrated by several workers[16,17].

3.2 Soliton interaction

So, creating a soliton and compensating for energy dissipation during its propagation down a lossy fiber is not a problem. The next issue which we have to take up is related to the separation between neighbouring solitons in a soliton stream. The reason is that solitons interact[18,19] when in close proximity : If they are in phase (phase difference equals zero), they attract each other, where as, when out of phase (phase difference equals π) they repel each other. This process of attraction and repulsion occurs periodically as the soliton stream moves along the fiber link.

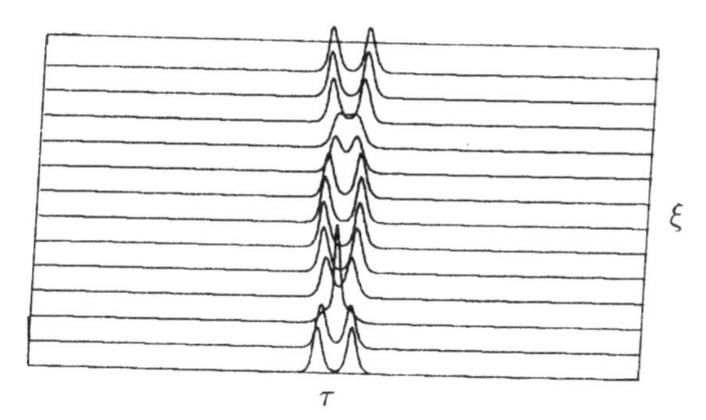

Figure 3 Spatio-temporal evolution of the interaction process of two fundamental bright solitons given by equation 6 with initial separation $2d_0 = 5$.

In Fig. 3 we have shown this process of collision of two fundamental solitons with initial phase difference zero. It is clearly seen that as the pulses propagate down the fiber, they start attracting each other and come closer and closer to form a transient bound state and finally separate again. This process repeates itself periodically. This repeated collision of solitons in a fiber link leads to observable effects which put considerable restrictions on system performance. Recent studies have shown that if the initial spacing between the adjacent solitons is of the order of 5 times the soliton width, soliton interaction negligibly affects system performance.

3.3 Amplifier noise and the Gordon-Haus jitter

In a long-haul system the soliton has to be amplified a large number of times. At every stage of amplification noise is added to the system in the form of amplified spontaneous emission (ASE). It affects the system performance in several ways including degradation of the signal to noise ratio and gain saturation. In addition to these adverse effects, which are present even for the non-solitonic systems, a soliton system suffers from yet another problem known as Gordon-Haus jitter[4–7]. The physics of this effect can be understood as follows.

It was shown by Gordon and Haus[20] that addition of ASE to the signal leads to fluctuations in the amplitude and the phase of the soliton. The amplitude fluctuation results in width fluctuation which is small in magnitude and does not pose any serious problems. However, the phase fluctuation leads to random variations in the soliton carrier frequency due to which the velocity of the soliton gets randomised. As a result the arrival time of the soliton at the end of the amplifier chain varies, from bit to bit, in a random manner. This arrival time jitter induced by the ASE noise is called the Gordon-Haus jitter.

3.4 Reduction of Gordon-Haus jitter

Quite a few methods have been proposed to reduce the Gordon-Haus jitter[4–7] in a communication link with periodic amplification. The most common in all the experimental works is the method in which a passive filter, with a bandwidth roughly ten times larger than the soliton bandwidth, is inserted after every amplifier. This was theoretically proposed by several authors[21-23], who showed that such filters push the solitons, that may have strayed from the carrier frequency, back to the filter center frequency which is set to match the soliton carrier frequency. The physical phenomenon involved can be understood as follows[6]. A spectrum of a pulse which is not centered with the filter center frquency experiences more losses in the part of the spectrum which is farther away from the filter center frequency than in the part which is closer to it. As a consequence the pulse feels a restoring force in the frequency domain. The method has been successfully used to demonstrate long distance stable soliton propagation with periodic amplification in laboratory environment[24,25].

4. Spatial solitons

Spatial solitons are self-guided optical beams with finite cross-section that remain confined in transverse directions orthogonal to the direction of propagation.

We know that when a narrow optical beam propagates in a linear medium, it undergoes diffractive broadening and that the narrower the beam, faster it diverges. In its essense, diffraction creates a wavefront similar to that due to a concave lens and spreads the beam in a wider region. In a nonlinear medium, however, a high intensity beam modifies the refractive index of the medium in such a way that, if the medium has positive nonlinear coefficient n_2, the index increases in the center of the beam, while it remains unchanged in the tails. The region thus created acts like a convex lens which tends to focus the beam toward the center[26]. This phenomenon is known as self-focusing. Therefore, by using a high intensity beam in a proper nonlinear medium, we can counter diffractive broadening by self-focusing. If the parameters of the medium and the beam are properly selected, one can achieve a complete cancellation of these two opposite and competing effects leading to a stationary beam propagation without any change in the shape of the beam. This regime of beam propagation is called the soliton regime and the resulting beam is called a bright spatial soliton[5,26]. In, general, a spatial soliton is formed when the self-focusing of the beam balances its natural diffractive broadening. The above mentioned mechanism of soliton formation is also called self-trapping because the beam gets itself trapped by modifying the refractive index of the medium.

4.1 Types of spatial solitons

Depending on the medium (material) property and the self-trapping mechanism, spatial solitons are broadly divided into three classes : (i) Kerr and non-Kerr like solitons, (ii) photorefractive solitons, and (iii) quadratic solitons. The physics of Kerr and non-Kerr solitons lies in the intensity dependent instantaneous change in the refractive index of the medium which can have any of the several possible origins, like, the electronic origin or the thermal origin or via carrier generation etc. On the other hand, the physics of photorefractive solitons lies in the electro-optic effect and the phemomenon of self-trapping and requires a DC electric field. Quadratic solitons, unlike the previous two types of solitons, have their physics in the phase-matched parametric interaction between different frequency components of the optical field.

4.2 Kerr and non-Kerr like spatial solitons

As stated above the physics of such solitons lies in the intensity dependent instantaneous change in the refractive index of the medium. The evolution of such a soliton in a bulk medium is governed by the NLSE which is derived from the Maxwell's equations in the paraxial and slowly varying envelope approximation (SVEA)[5]. It has the following form :

$$2i\beta_0\frac{\partial A}{\partial z}+\frac{\partial^2 A}{\partial x^2}+\frac{\partial^2 A}{\partial y^2}+2\beta_0 k_0 n_{nl}(|A|^2)A=0, \tag{20}$$

where

$$\beta_0 = n_0 k_0 = \frac{2\pi n_0}{\lambda_0} \tag{21}$$

is the propagation constant. n_0, k_0, and λ_0 are the linear refractive index, the wave number and the carrier wavelength, respectively. A(x,y,z) is the complex envelope amplitude of the electric field of the beam and $n_{nl}(I),\ I=|A|^2$ is the nonlinear refractive index change. The beam is supposed to propagate in the z-direction and diffract or self-focus in the x and y directions, i.e., transverse to the direction of propagation.

For a Kerr nonlinear medium with Kerr coefficient n_2, we have $n_{nl}(I)=n_2 I$ and the above equation takes the familiar dimensionless form :

$$i\frac{\partial u}{\partial Z}+\frac{1}{2}\left(\frac{\partial^2 u}{\partial X^2}+\frac{\partial^2 u}{\partial Y^2}\right)\pm|u|^2u=0 \tag{22}$$

where plus is taken for a self-focusing medium ($n_2 > 0$) , while minus for a self-defocusing medium ($n_2 < 0$) and the following nondimensionalisation has been carried out:

$$X=\frac{x}{w_0},\ Y=\frac{y}{w_0},\ Z=\frac{z}{L_d},\ u=(k_0|n_2|L_d)^{1/2}A. \tag{23}$$

Here, W_0 is the input beam width and $L_d=\beta_0 W_0^2$ is the characteristic diffraction length of the beam. The above equation is known as the (2+1) dimensional NLSE.

4.2.1 One dimensional spatial solitons

If the medium is in the form of a planar waveguide, the above equation takes the form :

$$i\frac{\partial u}{\partial Z}+\frac{1}{2}\frac{\partial^2 u}{\partial X^2}\pm|u|^2u=0, \tag{24}$$

where it has been assumed that the beam is confined in y-direction, due to waveguiding, and can diffract only in the x-direction in the absence of nonlinearity. Eq.(24) is a (1+1) dimensional NLSE. It is easily integrated by ISTM.
Bright soliton solution For a self-focusing medium one gets the fundamental bright soliton solution in the following form

$$u(Z,X) = a\ sech[a(X - vZ)]\ e^{i[Xv + \frac{(X^2 - v^2)}{2}Z + \phi]}, \tag{25}$$

where a and v are related to the amplitude and the velocity of the soliton. When $v \neq 0$ the soliton travels at an angle to the z-axis and, hence, v is the measure of the angular displacement of the soliton beam. This aspect of soliton propagation renders a special property to such solitons : the ability to steer light. By changing the angle of propagation one is able to steer light.The above solution represents a two parameter family of solutions. The bright spatial Kerr-solitons have been observed in[27] CS_2, glass[28], semiconductor[29] and polymer[30] waveguides.
Dark soliton solutions If the waveguide is made of a self-defocusing material one obtains what is called a dark soliton, satisfying the boundary condition $u \longrightarrow a_0$ as $|X| \longrightarrow \infty$. The fundamental dark soliton is given by

$$u(Z,X) = a_0\,(B\ tanh[a_0B(X - Aa_0Z)] + iA)\, e^{(-ia_0^2Z)}. \tag{26}$$

Here the parameters A and B satisfy $A^2 + B^2 = 1$. If we introduce $A = cos\ \phi$ and $B = sin\ \phi$, then the above solution can be written as

$$u(Z,X) = a_0\,(sin\phi\ tanh[a_0tan\phi(Xcosec\phi - a_0Z)] + icos\ \phi])\, e^{(-ia_0^2Z)}, \tag{27}$$

which has only one free parameter ϕ, which equals half the total phase shift across the dark soliton. The intensity of the soliton can be written as :

$$|u|^2 = a_0^2\left(1 - cos^2\phi\ sech^2[a_0\ cos\ \phi\ (X - a_0\ sin\ \phi\ Z)]\right). \tag{28}$$

The above soliton solution represents a localized dark hole on a CW background whose dip at the center is governed by $cos^2\phi$. The velocity of a dark soliton depends on the phase shift ϕ, and hence on the amplitude. The quantity $a_0\ sin\ \phi$ represents the relative velocity of the dark soliton with respect to the CW background.

Note that unlike the bright soliton the phase does not remain the same across the entire wave form. Since the hyperbolic tangent is an odd function of its argument, the dark soliton undergoes a π phase shift at $X = 0$ where its intensity also goes to zero. Such a soliton is called black and is stationary with respect to the background. For non-zero ϕ the intensity is also non-zero at the dip center and such a soliton is called a gray soliton. Dark spatial solitons have also been generated and observed experimentally[31,32].

4.2.2 Two dimensional spatial solitons

In two spatial dimensions the NLSE takes the form

$$i\ \frac{\partial u}{\partial Z} + \vec{\nabla}_{\perp}^2 u + F(|u|^2)u = 0 \tag{29}$$

where $F(|u|^2)$ is the nonlinear function whose form depends on the medium properties. Here

$$\vec{\nabla}_{\perp}^2 = \frac{\partial^2}{\partial X^2} + \frac{\partial^2}{\partial Y^2} = \frac{\partial^2}{\partial r^2} + \frac{1}{r}\frac{\partial}{\partial r} + \frac{1}{r^2}\frac{\partial^2}{\partial \theta^2}. \tag{30}$$

is the transverse Laplace operator. Note that, because of the stability properties of the soliton solution, usually the saturating form of nonlinearity is considered. For this case

$$F(|A|^2) = \frac{|A|^2}{1 + \alpha_s |A|^2} \tag{31}$$

where α_s is the saturation parameter. The resulting dimensionless NLSE is given by

$$i\frac{\partial u}{\partial Z} + \vec{\nabla}_{\perp}^2 u + \frac{|u|^2}{1+|u|^2}u = 0, \tag{32}$$

which goes into the well known cubic-quintic NLSE for the case of non-resonant saturation. The stationary and cylindrically symmetric solution is looked for in the form

$$u = U(r)\ e^{i\beta Z} \tag{33}$$

where β is the propagation constant shift. The function U sartisfies

$$\frac{\partial^2 U}{\partial r^2} + \frac{1}{r}\frac{\partial U}{\partial r} + \frac{|U|^2}{1+|U|^2}U = \beta U \tag{34}$$

The soliton solutions with maximum amplitude at $r = 0$ and satisfying the localization condition at $r \longrightarrow \infty$, have been found numerically. The set of solutions consists of nodeless as well as nodal solutions. Note that at a node the optical field vanishes. Because of this the soliton solutions are labelled with an integer n which stands for the number of nodes of a given solution. The solution without any nodes, i.e., with n=0, is called the fundamental two dimensional spatial soliton.

Clearly, the stability of spatial solitons is of utmost importance for their practical applications. It has been established that for Kerr nonlinearity bright spatial solitons are stable only in (1+1) dimension[26]. Stability analysis shows that two dimensional fundamental (non-nodal) solitons are always stable in a nonlinear medium with saturating nonlinearity. The higher-order nodal solutions are unstable and break up during propagation[33–36]. The growth rate of instability varies from medium to medium and is quite considerably suppressed for media with saturable nonlibnearity.

In a bulk medium, a two dimensional spatial soliton acquires the form of a planar soliton stripe and is susceptiple to transverse instability. This instability occurs when transverse modulations locally decrease the soliton energy. It has been shown that two dimensional bright solitons are unstable in a Kerr medium. Linear stability analysis shows that, in general, depending on the higher-order dispersive and dissipative effects, different regimes of soliton dynamics can result :

Collapse regime : It is characterized by the formation of a chain of two dimensional singularities along the front of the soliton.
Transition to a chain of two-dimensional solitons.
Soliton break-up into localized states which eventually disperse.
Periodic growth and damping of transverse modulations along the front of the soliton.

The bright solitons of the corresponding NLSE, with anomalous dispersion, show the first scenario, where as bright solitons of the NLSE with normal dispersion show the third scenario. The dark solitons, on the other hand show either the second or the third scenario which depends on their parameters. Spatial solitons have been observed in bulk media[37-39] also.

4.2.3 Solitons with angular momentum

In addition to the above mentioned soliton solutions, another class of higher order localized solutions have also been studied. They are ring shaped with zero intensity at the center and outside the ring. They have a spiral phase structure with a singularity at the origin, carry a finite angular momentum and resemble an optical vortex. Therefore, they are called vortex solitons.

The simplest of the above solutions is the scalar spinning soliton which evolves according to equation (34) with $\vec{\nabla}_{\perp}^2$ given by Eq.(30). The vortex solution of Eq.(34) is looked for in the following form

$$u(X,Y,Z) = U(X,Y)\ e^{i(\beta Z+\phi(X,Y))} \tag{35}$$

where U, ϕ and β are the amplitude, phase and the propagation constant, respectively. Substituting for $u(X,Y,Z)$ from (35) into (34), and separating the real and the imaginary parts, we obtain

$$\vec{\nabla}_{\perp}^2 U - \beta U - (\vec{\nabla}\phi)^2 U + F(U^2)U = 0, \tag{36}$$

$$\vec{\nabla}_{\perp}^2 \phi + 2\vec{\nabla}_{\perp} U \cdot \vec{\nabla}_{\perp}\phi = 0. \tag{37}$$

The vortex type bright soliton solution of the above equation corresponds to a bright spatial soliton whose optical field depends not only on the radial coordinate r but also on co-ordinate θ. This property results in a finite angular momentum possesed by such solutions. The vortex solutions correspond to the solutions of Eq.(36) and Eq.(37) with a radially symmetric amplitude $U(r)$ such that $U = 0$ at the center of the beam, i.e., at $r = 0$. They have a rotating spiral phase ϕ varying linearly with θ, that is, $\phi = m\ \theta$, where m is an integer. The integer m represents a phase twist around the intensity ring and is called the topological charge which is conserved[5]. Solitons having fractional topological charge also exist. Such structures with integral or fractional angular momentum are quasi-stable when they have the form of a necklace beam or a necklace-ring vector soliton beam. They suffer from the so-called azimuthal instability which breaks them up into a series of fundamental solitons which fly off the ring.

Since the angular momentum must be conserved during this decay process the solitons fly off along the tangent to the ring. This process has been observed experimentally.

The linear momentum of the vortex solution is given by :

$$\vec{p_l} = Im\left(\int u^*\vec{\nabla}_\perp u\right)d^3X = \int U^2\vec{\nabla}_\perp\phi d^3X, \tag{38}$$

and the angular momentum is given by

$$\vec{p_a} = \int U^2(r)\frac{\partial\phi}{\partial\theta}d^3X, \qquad r = \sqrt{X^2+Y^2}. \tag{39}$$

The spin of the soliton is defined as :

$$s = \frac{p_a}{p_0}, \tag{40}$$

where

$$p_0 = \int |u|^2 d^3X \tag{41}$$

is the soliton power. If we use $\phi(X,Y,) = m\theta$, then $s = m$, and the spin equals just the topological charge of the soliton. s is equal to zero for the fundamental soliton.

4.3 Photorefractive spatial solitons

Photorefractive (PR) solitons[40,41] are formed due to strong beam nonlinearity that takes place in a PR sample of an electro-optic crystal having a small concentration of deep donor and acceptor impurities. Physically, donors in such a sample, are ionized by a light beam of suitable wavelength. As a result, the corresponding electrons move into the conduction band of the crystal, and are free to move under the combined action of diffusion and of an electrostatic field, before they recombine with the donors which were ionized earlier, or with the available acceptors. Since the donors remain in a fixed position after ionization, a space charge results that generates an electric field. This electric field changes the refractive index of the medium via electro-optic effect. The resulting variation in the refractive index acts back on the beam and the propagation characteristics of the beam are modified as the beam gets self-trapped. PR solitons result as a consequence of a complete balance between diffractive divergence and nonlinear compression due to self-trapping. PR solitons are continuous wave visible micro-meter sized beams that propagate without diffractive divergence. In contrast to the instantaneous Kerr nonlinearity, the nonlinear refractive index change Δn in a PR material depends on the light intensity, I, in a much more complicated manner. For instance, Δn at point $\vec{r}$ and time t depends on the value $I(\vec{r}', t')$ of the light intensity at the point $\vec{r}'$, in the vicinity of the point $\vec{r}$, at time $t' < t$. That is the dependence of $\Delta n(I)$ on intensity is both spatially and temporally non-local[41]. As a result the self-trapping mechanism is not instantaneous and requires the space-charge field in the material to reach the seady-state self-trapping distribution. This process takes a finite time to occur during which the energy is supplied to the system. Consequently, PR solitons need relatively low powers for their experimental observation. They can occur as bright, black and gray spatial solitons in a photorefractive medium[42]. They can also occur in a photorefractive medium with externally applied field[43] and have been experimentally seen in several experiments[44]. They are tipped to play important role in signal processing and data storage systems.

4.4 Quadratic spatial solitons

Quadratic solitons are formed as a consequence of mutual trapping and phase locking of multifrequency waves mediated by a second-order nonlinearity[45]. The formation of solitons in the process of second harmonic generation, or optical parametric generation, in which a fundamental frequency wave and its harmonics generate each other inside a quadratic crystal, is the classic example of such solitons. The resulting soliton contains both the fundamental and the harmonic fields. As a result, the quadratic solitons, unlike Kerr or photorefractive solitons, are essentially multicoloured.

When light propagates in a nonlinear quadratic crystal, once again two characteristic length scales : the diffraction length that determines the diffractive beam divergence and the nonlinear length that determines the strength of frequency conversion process, play vital roles in the resulting physics. If, by choosing the parameters correctly, the conditions are set appropriate to have these length scales comparable to each other, a quadratic soliton forms. It has been observed that after the soliton is formed the energy exchange between the waves ceases and the envelopes of all the waves are phase-locked.

So far as interaction of spatial solitons in (1+1) dimension is concerned, they, like the temporal solitons, attract each other when their relative phase, θ, is zero and repel when their relative phase equals π. For the intermediate values of θ, given by $0 < \theta < \pi$, the interaction is inelastic and the solitons exchange energy. Two dimensional solitons also show similar behaviour. In a bulk medium, however, some new features are seen. It turns out that, due to their interaction, the solitons can change their direction of propagation, fuse together and even spiral around each other.

4.5 Applications of spatial solitons

Spatial solitons, owing to their localization against diffractive divergence in a nonlinear medium, are considered to be information carrying units. (3+1)-dimensional light bullets[46] which are localized both in space and time are tipped to be ideal for use, in an all-optical communication system, as bits of information. They are also very useful for all-optical switching process and for producing clean optical beams in optical parametric devices, such as optical parametric generators and optical parametric amplifiers. Collision properties of spatial solitons can be used for performing logic operations and optical computing.

It is well established that spatial solitons, apart from changing the refractive index of the medium in which they propagate, can also attract and guide beams. Because of this it is usually said that spatial solitons induce optical waveguide that can guide another beam of a different frequency or polarization via the induced cross-phase modulation. Several such colliding waveguides are expected to create a self-configurable waveguide circuitry which will have considerable advantage over the conventional waveguides and devices. The corner stone of such an integrated optical circuit is the basic property of soliton beams, owing to which they can pass through each other without disturbing each other considerably. It opens up the possibility of designing dense re-configurable self-induced optical interconnects.

References

1. A. Hasegawa and F. Tappert, "Transmission of stationary nonlinear optical pulses in dispersive dielectric fibers : I. Anomalous Dispersion", *Appl. Phys. Lett.*, **23**, 142 (1973).
2. L.F. Mollenauer, R.H. Stolen and J.P. Gordon, "Experimental observation of picosecond pulse narrowing and solitons in optical fibers", *Phys. Rev. Lett.*, **45**, 1095(1980).
3. G.P. Agrawal, Nonlinear Fiber Optics, Academic Press, (1995).
4. G.P. Agrawal, Fiber-Optic Communication Systems, John Wiley and Sons, (1992).
5. Y.S. Kivshar and G.P. Agrawal, Optical Solitons, Academic Press, (2003).
6. H.A. Haus, "Fiber optical solitons their properties and uses", *Proc. of the IEEE*, **81**, 970 (1993).
7. A. Hasegawa and Y. Kodama, Solitons in optical communications, Clarendon Press, Oxford, UK (1995)
8. Ajit Kumar, "Soliton dynamics in a monomode optical fiber", *Phys. Rep.*, **187**, 95 (1990)
9. Ajit Kumar, "Invariant of motion method for nonlinear pulse propagation in optical fibers", *Phys. Rev. A*, **44**, 2130 (1990).
10. V.E. Zakharov and A.B. Shabat, "Exact theory of two-dimensional self-modulation of waves in nonlinear media", *Sov. Phys. JETP*, **34**, 62 (1972).
11. J. Satsuma and N. Yajima, "Initial value problem of one-dimensional self-modulation of nonlinear waves in dispersive media", *Prog. Theor. Phys.*, **55**, 284 (1974).
12. W.J. Tomlinson, R.J. Hawkins, A.W. Weiner, J.P. Heritage and R.N. Thurston, "Dark optical solitons with finite-width background pulses", *J. Opt. Soc. Am. B*, **6**, 329 (1989).
13. Ph. Emplit, J.P. Hamade, F. Reynaud, G. Froehly and A. Barthelemy, " Picosecond steps and dark pulses through nonlinear single mode fiber ", *Opt. Commun.*, **62**, 374 (1987).
14. A.W. Weiner, J.P. Heritage, R.J. Hawkins R.N. Thurston, E.M. Krischner, D.E. Learid and W.J. Tomlinson, "Experimental observation of fundamental dark solitons in optical fibers", *Phys. Rev. Lett.*, **61**, 2445 (1988).
15. A. Hasegawa, "Amplification and reshaping of optical solitons in a glass fiber-IV : Use of Raman Process", *Opt. Lett.*, **8**, 650 (1983).
16. L.F. Mollenauer, R.H. Stolen and M.N. Islam, "Experimental demonstra-

tion of soliton propagation in long fibers: loss compensated by Raman gain", *Opt. Lett.*, **10**, 229, 1985.
17. L.F. Mollenauer and K. Smith, "Demonstration of soliton transmission over more than 4000 km in fiber with loss periodically compensated by Raman gain" *Opt. Lett.*, **13**, 675 (1988).
18. P.L. Chu and C. Desem, "Mutual interaction of solitons with unequal amplitudes in optical fibers", *Electron. Lett.*, **21**, 228 (1985).
19. J.P. Gordon, "Interaction forces among solitons in optical fibers", *Opt. Lett.*, **8**, 596 (1983).
20. J.P. Gordon and H. Haus, "Random walk of coherently amplified solitons in optical fiber transmission", *Opt. Lett.*,**11**, 665 (1986).
21. Y. Kodama and A. Hasegawa, "Generation of asymptotically stable solitons and suppression of the Gordon-Haus effect", *Opt. Lett.*, **17**, 31 (1992).
22. A. Mecozzi, J.D. Moores, H.A. Haus and Y.Lai, "Soliton transmission control", *Opt. Lett.*, **16**, 1841 (1991).
23. D. Marcuse, "Simulations to demonstrate the reduction of the Gordon-Haus effect", *Opt. Lett.*, **17**, 34 (1992).
24. M. Nakazawa, K. Suzuku, E. Yamada, H. Kubota, Y. Kimura and M. Takaya, "Experimental demonstration of soliton data transmission over unlimited distances with soliton control in time and frequency domain", *Electron. Lett.*, **29**, 729 (1993).
25. L.F. Mollenauer, E. Lichtman, M.J. Neubelt, and G.T. Harvey, "Demonstration, using sliding frequency guiding filters of error free soliton transmission over more than 20 Mm at 10 Gb/sec single channel and over more than 13 Mm at 20 Gb/sec in a two channel WDM", *Electron. Lett.*, **29**, 910 (1993).
26. G.I. Stegeman and M. Segev, "Optical spatial sdolitons and their interactions : Universality and diversity", *Science*, **286**, 1518 (1999).
27. A. Barthelemy, F. Manuef and C. Froehly, "Propagation soliton et autoconfinement de faiseaux laser par non linearite optique de Kerr", *Opt. Commun.*, **55**, 201 (1985).
28. J.S. Atchison, A.M. Weiner,Y.Silverberg, M.K. Oliver, J.L. Jackel, D.E. Leaid, E.M. Vogel and P.W.E. Smith, "Observation of spatial optical soliton in a nonlinear glass waveguide", *Opt. Lett.*, **15**, 471 (1990) .
29. J.S. Atchison, K. Al-Hemyari, C.N. Ironside, R.S. Grant and W. Sibbett, "Observation of spatial solitons in AlGaAS waveguides" *Electron. Lett.*, **28**, 1879 (1992).
30. U. Bartuch, U. Peschel, Th. Gabler, R. Waldhaus and H.H. Horhold, " Experimental investigation and numerical simulations of spatial solitons in planar polymer waveguides", *Opt. Commun.*, **134**, 49 (1997).
31. D.R. Andersen, D.E. Hooton, G.A. Swartzlander,Jr., and A.E. Kaplan, "Direct measurement of the transverse velocity of dark spatial solitons",*Opt. Lett.*, **15**, 783 (1990).
32. G.R. Allan, S.R. Skinner, D.R. Andersen, and A.L. Smirl, "Observation of dark spatial solitons in semiconductors using picosecond pulses", *Opt. Lett.*, **16**, 156 (1991).

33. J.M. Soto-Crespo, D.R. Heatly, E.M. Wright and N.N. Akhmediev, " Stability of higher-order bound states in a saturable self-focusing medium", *Phys. Rev. A*, **44**, 636 (1991). (Beark-up of higher order Spatial solitons)
34. D. Edmundson, "Unstable higher modes of a three dimensional nonlinear Schroedinger equation", *Phys. Rev. E*, **55**, 7636 (1997). (Beark-up of higher order Spatial solitons)
35. J.Yang, "Internal oscillations and instability characteristics of (2+1) dimensional solitons in a saturable nonlinear medium", *Phys. Rev E*, **66**, 026601 (2002). (Beark-up of higher order Spatial solitons)
36. N.N. Akhmediev, "Spatial solitons in Kerr and Kerr-like media", *Opt. Quant. Electron.*, **30**, 535 (1998).
37. J.E. Bjorkholm and A. Ashkin, "cw self-focusing and self-trapping of light in sodium vapour", *Phys, Rev. lett.*,**32**, 129 (1974).
38. G. Khitrova, H.M. Gibbs, Y. Kawamura, H. Iwamura, T.Ikagami, J.E. Sipe and L.Ming, "Spatial solitons in a self-focusing gain medium", *Phys. Rev. Lett.*, **70**, 920 (1993).
39. W. Torruellas, B. Lawrence and G.I. Stegeman, "Self-focusing and 2D spatial solitons in PTS", *Electron. Lett.*, **32**, 2092 (1996).
40. M. Segev, B. Crosignani, A. Yariv and B. Fischer, "Spatial solitons in photorefractive media", *Phys. Rev. Lett.*, **68**, 923 (1992).
41. B. Crosignani aqnd G. Salamo, "Photorefractive Solitons", *Opt. Photon. News*, **13**, (2002) 38.
42. D.N. Christodoulides and M.I. Carvalho, "Bright, dark and gray spatial soliton states in photorefractive media", *J. Opt. Soc. Am.*, **12**, 1628 (1995).
43.M. Segev, G.C. Valley, B. Crosignani, P. Diporto, and A. Yariv, "Steady state spatial screening solitons in photorefractive materials with external applied field", *Phys. Rev. Lett.*, **73**, 3211 (1994)..
44. G. Duree, Jr., J.L. Shultz, G.J. Salamo, M. Segev, A. Yariv, B. Crosignani. P. Diporto, E.J. Sharp and R. Neurgaonkar, "Observation of self-trapping of an optical beam due to photorefractive effect", *Phys. Rev. lett.*, **71**, 533 (1993).
45. L. Torner and A.P. Sukhorukov, "Quadratic Solitons", *Opt. Photon. News*, **13**, (2002) 42.
46. Y. Silberberg, "Collapse of optical pulses", *Opt. Lett.*, **15**, (1990) 1282.

5

Cerenkov Nonlinear Interactions in Optical Waveguides

K. Thyagarajan

1. Introduction

At the large power densities provided by lasers, matter behaves in a nonlinear fashion and we come across new optical phenomena such as second harmonic generation (SHG), sum and difference frequency generation etc.[1,2] In SHG, an incident light beam at frequency ω interacts with the medium and generates a new light wave at frequency 2ω. In sum and difference frequency generation, two incident beams at frequencies ω_1 and ω_2 mix with each other producing sum $(\omega_1 + \omega_2)$ and difference $(\omega_1 - \omega_2)$ frequencies at the output. Higher order nonlinear effects such as self phase modulation, four wave mixing etc. can also be routinely observed today and are the limiting factors in fiber optic communication systems.[3,4] A large number of applications require compact laser sources in the 200 nm to 4000 nm range. These applications include high-density data storage, printing, optical communications, displays, spectroscopy etc. Nonlinear optics provides us with possibilities of achieving compact, tunable and efficient solid state lasers to cater to such applications.

Nonlinear optical interactions depend on the strength of the electric field of the light wave and hence become prominent when the optical power densities are high and interaction takes place over long lengths. Optical waveguides in which the light beam is confined to a small cross sectional area are ideal for achieving highly efficient nonlinear interactions. Since the efficiency of nonlinear effects depends on the intensity of the light beam, in order to achieve high efficiencies in the case of bulk, the beam has to be focussed. Strong focussing leads to a very small effective length of interaction due to inherent diffraction in the beam. In contrast, in optical waveguides, the beam can be made to have extremely small cross sectional areas (25 μm^2) over very long interaction lengths : 50 mm in the case of integrated optic waveguides (see Fig. 1) to kilometers in length in the case of optical fibers. This leads to very much increased efficiencies even

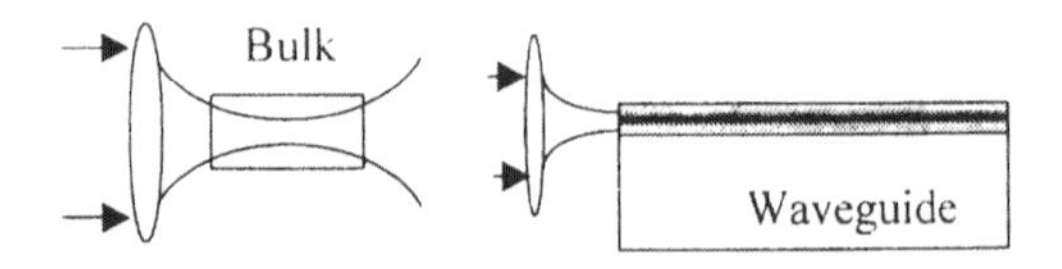

Figure 1 Nonlinear interactions in bulk are limited by diffraction effects.

at moderate fundamental *powers* (~ few tens of mW), thus opening the possibility of realizing efficient nonlinear devices based on high power semiconductor laser sources.

In this chapter we will discuss the second order nonlinear effects and bring out the importance of phase matching in such interactions. The technique of Cerenkov phase matching can be combined with quasi phase matching to achieve interesting possibilities; we discuss some of our results in this area. Finally we will discuss a relatively new class of waveguides namely periodic segmented waveguides.

2. Nonlinear polarization

In a linear medium the electric polarization $\mathcal{P}$ is related to the electric field $\mathcal{E}$ of the light wave by the following equation:

$$\mathcal{P} = \varepsilon_0 \chi \mathcal{E} \tag{1}$$

where χ represents the linear electric susceptibility of the medium. For strong electric fields the polarization generated in the medium is no more related linearly to the applied electric field and we get

$$\mathcal{P} = \varepsilon_0 \chi \mathcal{E} + 2\varepsilon_0 d \mathcal{E}^2 + \varepsilon_0 \chi^{(3)} \mathcal{E}^3 ... \tag{2}$$

where d and $\chi^{(3)}$ are constants. The terms proportional to the square, cube etc. of the electric field correspond to the various nonlinear terms and lead to various nonlinear optical effects. The term proportional to the square of the electric field gives rise to second order nonlinear effects such as second harmonic generation (SHG), sum frequency generation (SFG), difference frequency generation (DFG), parametric amplification and oscillation etc. For media possessing inversion symmetry it can be shown that $d = 0$ and such media do not exhibit second order nonlinear effects.

The term proportional to cube of the electric field gives rise to effects such as self phase modulation (SPM), cross phase modulation (XPM), four wave mixing (FWM) etc. This term is present in all media and is very important for optical fiber communication systems[3, 4].

In this chapter we will restrict ourselves to the second order nonlinear effects in waveguides.

3. Second order nonlinear effects

In order to appreciate the effect of the second order nonlinear effect, consider the incidence of a plane light wave propagating along the z-direction in a medium with non-zero value of d. We write for the electric field of the incident light wave as

$$\mathcal{E}_1 = E_1 \cos(\omega t - k_1 z) \tag{3}$$

In such a case the nonlinear polarization generated in the medium is given by

$$\begin{aligned} \mathcal{P}_{NL} &= 2\varepsilon_0 \, d \, E_1^2 \cos^2(\omega t - k_1 z) \\ &= \varepsilon_0 \, d \, E_1^2 + \varepsilon_0 \, d \, E_1^2 \cos[2(\omega t - k_1 z)] \end{aligned} \tag{4}$$

The above equation shows that in the presence of the nonlinear term, apart from oscillation at frequency ω due to the linear term in the polarization, the polarization has now a component at a frequency 2ω. Since polarization is nothing but dipole moment per unit volume, this implies that there are dipoles oscillating with a component at 2ω. When dipoles oscillate at 2ω then they radiate electromagnetic waves at this frequency. The electromagnetic wave at this frequency will build if the radiation from all the dipoles add constructively.

The electromagnetic wave at frequency 2ω will propagate according to the equation:

$$\mathcal{E}_2 = E_2 \cos(2\omega t - k_2 z) \quad (5)$$

where k_2 represents the propagation constant of the medium at the second harmonic frequency.

Equations (4) and (5) show that the nonlinear polarization is traveling at a velocity ω/k_1 while the electromagnetic wave that it is trying to generate has a velocity $2\omega/k_2$. Since polarization is the source of electromagnetic radiation, maximum radiation at the second harmonic will be generated if the two velocities are equal, i. e.,

$$k_2 = 2k_1 \quad (6)$$

This condition is referred to as the phase matching condition and is very important to be satisfied for the nonlinear interaction to be efficient.

By expressing the propagation constants in terms of refractive indices Eq. (6) gives

$$n(2\omega) = n(\omega) \quad (7)$$

i.e., the refractive index of the medium at the two frequencies must be the same. This is usually not possible due to the presence of dispersion in the medium. This condition however can be satisfied by using the birefringence of the medium or by using the technique of quasi phase matching (QPM).

3.1 Second harmonic generation

The amplitude E_2 of the electric field of the second harmonic wave can be shown to satisfy the following equation[2]:

$$\frac{dE_2}{dz} = -i\frac{\omega d}{cn_2} E_1^2 e^{i\Delta k z} \quad (8)$$

where

$$\Delta k = k_2 - 2k_1 \quad (9)$$

is the phase mismatch. Assuming low conversion efficiency, Eq. (8) can be readily integrated to give the following expression for the efficiency of second harmonic generation

$$\eta = \frac{P_2(z)}{P_1(0)} = \frac{2\mu_0 \omega^2}{c\, n_1^2\, n_2} d^2 z^2 \frac{P_1}{A} \operatorname{sinc}^2\left(\frac{\Delta k\, z}{2}\right) \quad (10)$$

where P_1 is the power of the incident wave at ω, P_2 is the power in the second harmonic and A represents the area of cross section of the beam and $n_1 = n(\omega)$ and $n_2 = n(2\omega)$. The above equation shows that for maximum efficiency the phase mismatch has to be zero, i.e. $\Delta k = 0$ which is nothing but Eq. (6), the phase matching condition. We also note that when phase matching takes place the efficiency increases

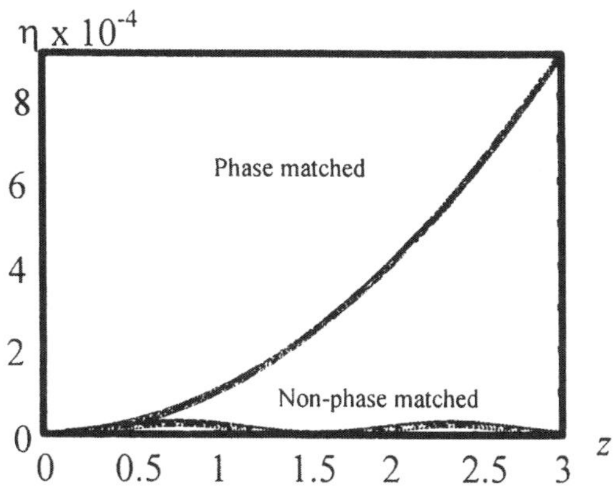

Figure 2 Variation of efficiency of SHG with interaction length in the case of perfect phase matching and non-phase matched operation.

quadratically with the length of interaction (see Fig. 2). Figure 3 shows the vector diagram corresponding to Eq. (6).

Figure 4 shows a schematic of the nonlinear polarization wave and the electromagnetic wave at frequency 2ω when $\Delta k \neq 0$. As the two waves propagate along the *z*-direction, they accumulate a phase difference of π between them at which point the nonlinear polarization instead of feeding energy into the second harmonic, draws energy from it and feeds back to the fundamental wave at ω. Thus when $\Delta k \neq 0$, the efficiency of second harmonic generation varies periodically along the *z*-direction as shown in Fig. 2. The maximum efficiency in such a case will be obtained when the argument of the sinc term in Eq. (10) becomes π/2 which gives the following value for the length of interaction:

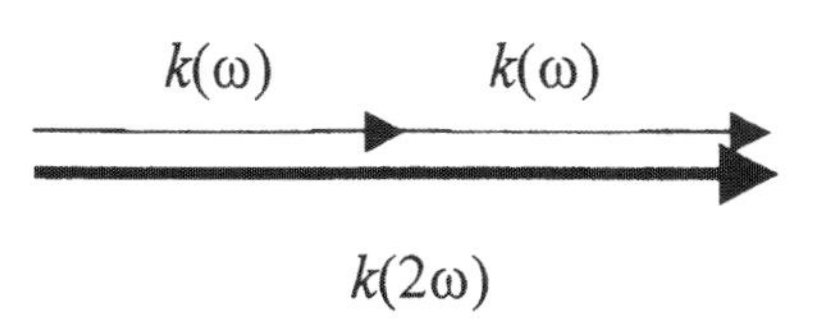

Figure 3 Vector diagram for SHG.

$$L_c = \frac{\pi}{\Delta k} \tag{11}$$

The length L_c is also referred to as the coherence length and represents the distance where the nonlinear polarization and the electromagnetic wave that it is trying to generate have an accumulated phase difference of π. This is also the minimum length

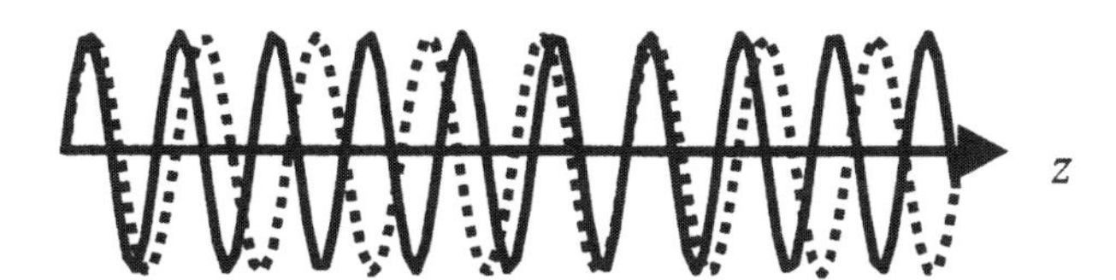

Figure 4 Schematic of the nonlinear polarization wave (dashed curve) and the electromagnetic wave (solid curve) at frequency 2ω.

of the medium required to achieve the maximum efficiency. Figure 5 shows the variation of the efficiency with $\Delta k\, z/2$ showing that larger lengths of interaction can tolerate only smaller values of phase mismatch.

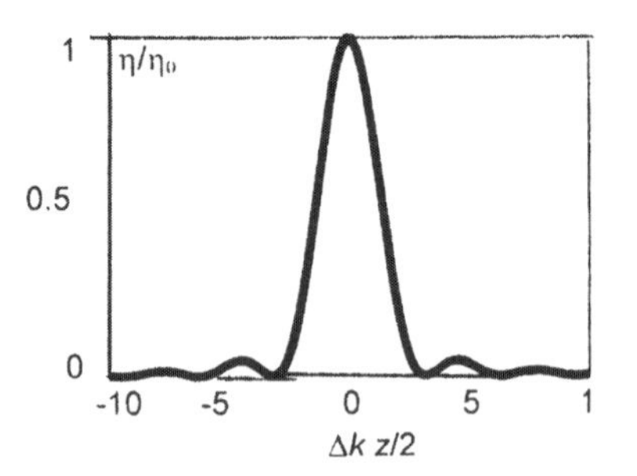

Figure 5 Variation of normalized SHG efficiency with $\Delta k\, z/2$.

3.2 Phase matching

Conventionally, phase matching is achieved using the birefringence in the material. Since birefringent media are characterized by two refractive indices,[2] by choosing the fundamental and the second harmonic with appropriate polarizations one can achieve phase matching.

Recently the technique of *quasi phase matching* (QPM) has become a convenient and practical way to achieve phase matching at any desired wavelength in any material including isotropic materials. QPM relies on the fact that the phase mismatch in the interacting beams can be compensated by readjusting the phase of interaction through periodically modulating the nonlinear characteristics of the medium[5]. Whenever the

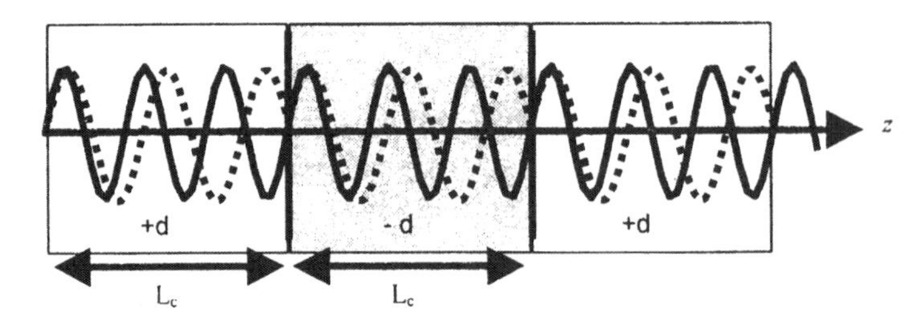

Figure 6 Schematic of the nonlinear polarization wave and the electromagnetic wave at 2ω for the case of reversal of nonlinear coefficient every coherence length L_c.

phase difference between the nonlinear polarization and the generated electromagnetic wave reaches the value π, then the sign of the nonlinear coefficient is reversed bringing the nonlinear polarization and the electromagnetic wave back into phase (see Fig. 6). This way the nonlinear polarization continuously feeds energy into the electromagnetic wave, which then continues to build (see Fig. 7). In a ferroelectric material such as $LiNbO_3$, the signs of the nonlinear coefficients are linked to the direction of the spontaneous polarization. Thus a periodic reversal of the domains of the crystal can indeed be used for achieving QPM and current devices based on $LiNbO_3$ are based on such

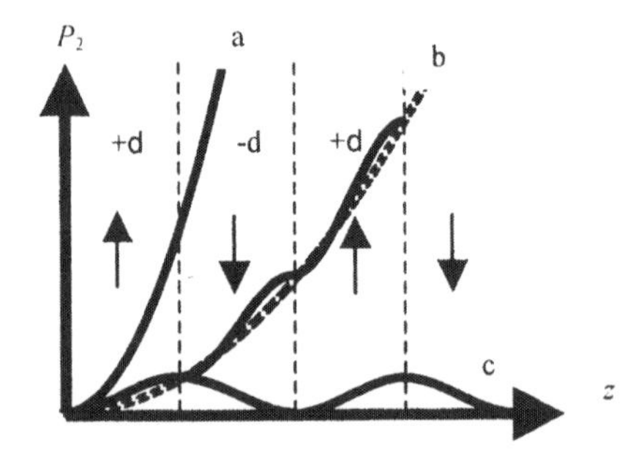

Figure 7 Variation of SHG efficiency with length for a: Perfect phase matching, b: Quasi phase matching, c: Non phase matched.

substrates.

In order to understand QPM, we consider the case when the nonlinear coefficient d varies sinusoidally along the z-direction:

$$d(z) = d_0 \sin(Kz) \tag{12}$$

where d_0 is a constant and K represents the spatial frequency of the variation. If we use this value of d in Eq.(8) it follows that by choosing a value of $K = \Delta k$, we can make the phase term on the right hand side of Eq. (8) disappear and this would lead to a continuous growth of the second harmonic rather than an oscillatory variation. Thus in this case the argument of the sinc function in Eq. (10) will become $(\Delta k - K)z/2$, where the non-phase matched term has been neglected. This implies that if the spatial period Λ is chosen to have a value

$$\Lambda = \frac{2\pi}{K} = \frac{2\pi}{\Delta k} = 2L_c \tag{13}$$

then we would obtain continuous growth (rather than an oscillatory behavior) of the second harmonic. This is what is precisely shown in Fig. 7.

If we now assume that the nonlinear coefficient d is reversed every coherence length L_c (instead of having a sinusoidal variation as given by Eq. (12)), then we can pick the term of the type given by Eq. (12) from the Fourier expansion and this term will ensure that the second harmonic will grow with length. Of course the value of d_0 will correspond to the amplitude of the corresponding Fourier term in the expansion and one would have to maximize this term for maximum efficiency. Figure 8 shows the vector diagram corresponding to second harmonic generation under QPM interaction

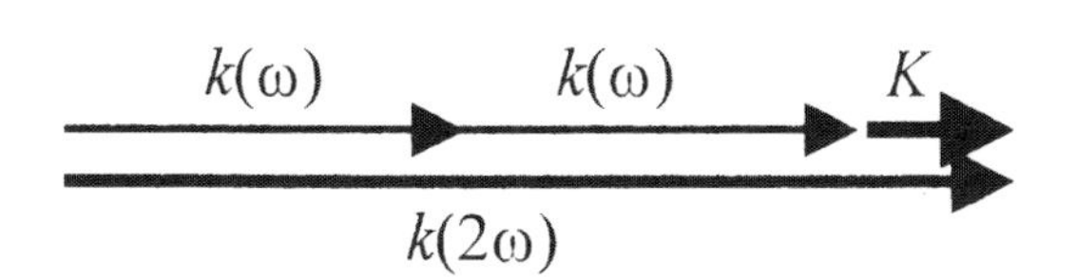

Figure 8 Vector diagram for QPM SHG.

4. Nonlinear interactions in waveguides

When nonlinear interactions take place in bulk, then in order to increase the intensity the beam needs to be focused. The focused beam diffracts after reaching a minimum spot size and this leads to reduced intensities in the beam. Optical waveguides overcome diffraction by waveguiding mechanism and thus light beams can be confined to small areas of cross section over much longer interaction lengths (see Fig. 1). Also since the beam areas can be quite small, the powers required to generate high intensities are also small. Thus nonlinear interactions in optical waveguides are very attractive.

In the case of nonlinear interactions in waveguides, instead of plane waves we need to consider propagation in terms of modes of the waveguide. Thus the fundamental wave at ω and the second harmonic wave at 2ω would be propagating in the form of modes with effective indices $n_{eff}(\omega)$ and $n_{eff}(2\omega)$. Phase matching would require these two to be equal. Since many different modes can propagate in waveguides, one can in principle choose the interaction to take place between different modes of the waveguide and hence satisfy the phase matching condition. At the same time since nonlinear interaction would depend on the overlap between the interacting modes by choosing different modes for interaction, this overlap may reduce drastically resulting in significantly reduced efficiency.

On the basis of increased overlap, the most efficient nonlinear interaction would be one in which all the interacting waves are guided modes. However, a novel configuration referred to as the *Cerenkov configuration* (in which the nonlinearly generated wave radiates freely into the substrate) has been extensively studied. Since waveguide radiation modes form a continuum, phase matching is automatically satisfied in such a configuration, unlike the case of guided to guided mode conversion. Coupled with QPM techniques, this leads to a number of advantages such as increased gain bandwidth for signal amplification in a parametric amplifier as well as increased tolerance towards waveguide dimensions and pump wavelength variations in a parametric interaction process.

The price we pay for increased phase matching tolerance is the fall in the efficiency of interaction. This happens due to a poor overlap between the guided mode at the fundamental frequency and the radiation modes at the second harmonic.

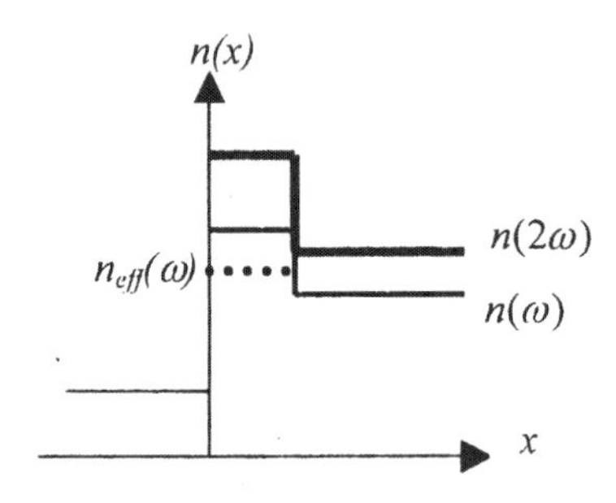

Figure 9 The mode at the fundamental frequency ω is guided and it generates a polarization at 2ω which radiates into the substrate.

5. Cerenkov phase matching

Figure 9 shows the refractive index profile of the waveguide at ω and 2ω. If the effective index of the mode at the fundamental frequency is less than the substrate refractive index at the second harmonic then the nonlinear polarization at the second harmonic can generate electromagnetic radiation at 2ω, which is radiating into the substrate as a radiation mode (see Fig.10). Figure 11 shows the corresponding phase matching condition between the fundamental guided mode at ω and the radiating second harmonic at 2 ω. Due to phase matching, the radiation at 2 ω will appear at the angle θ shown in the figure. Depending on the value of β(ω), the angle θ can adjust itself leading to automatic phase matching.

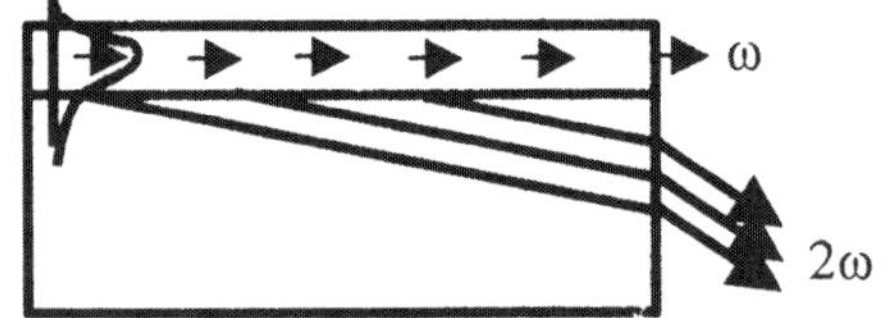

Figure 10 In Cerenkov phase matching the second harmonic propagates as a radiation mode into the substrate at an angle determined by the phase matching condition.

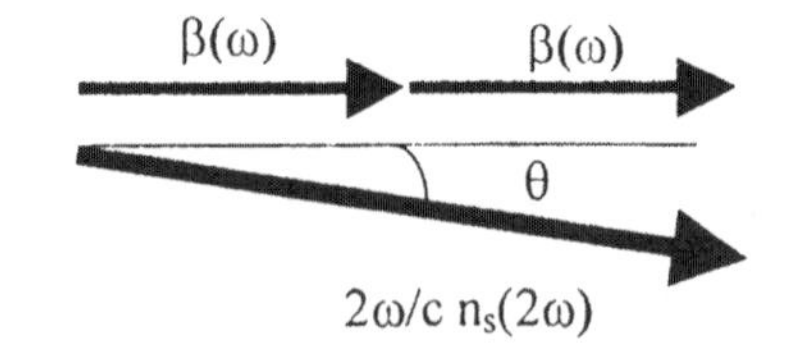

Figure 11 Vector diagram corresponding to Cerenkov phase matching.

The major problem here would be the overlap of the modal field of the guided mode at ω and the radiation mode at 2ω. Hence this configuration has much lower efficiency as compared to the all guided case. One can combine both the Cerenkov geometry as well as quasi phase matching to have additional benefits. By choosing a periodic nonlinear grating, the angle θ can be controlled. Thus by decreasing the angle θ for phase matching, it is possible to increase the overlap between the interacting modes and at the same time retaining the advantages of Cerenkov geometry. Such a configuration is referred to as QPM Cerenkov[6].

Let us consider a waveguide with a refractive index profile shown in Fig. 12. In the continuum of radiation modes with effective indices below that of the substrate refractive index n_s but above that of the intermediate layer of refractive index n_i, there are special modes referred to as *quasi modes* for which the field confinement can be quite significant[2, 7]. Thus if the guided mode at ω (with its effective index greater than n_s at that frequency) is phase matched to the quasi mode at 2ω, then one can achieve both automatic phase matching as well as a good efficiency. By adjusting the thickness and refractive indices of the intermediate leaky layer, one can indeed adjust the efficiency and bandwidth to lie anywhere between all guided interaction and pure Cerenkov interaction[8]. Figure 13 shows a typical variation of SHG efficiency with grating period as the thickness of the intermediate layer changes. The structure will tend to a purely guiding structure in the limit of t tending to infinity while it becomes a Cerenkov structure for t tending to zero. The former has very high efficiency but low tolerance towards various parameters while the latter has high tolerance but lower efficiency. Thus by choosing appropriate value of t one can trade bandwidth for efficiency. Parametric interactions in leaky waveguides have also been investigated with interesting properties[9].

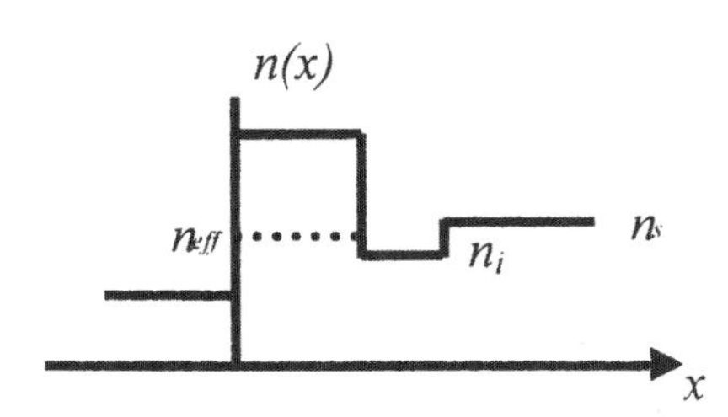

Figure 12 All values of effective indices below n_s are allowed. Some of these modes are *quasi modes* having a significant fraction of power within the guiding region

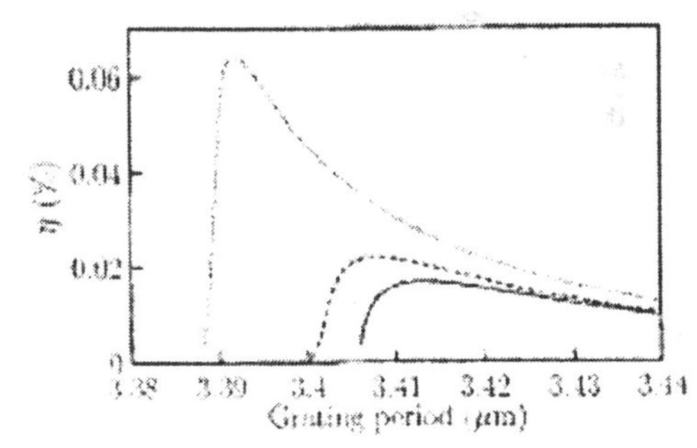

Figure 13 Variation of SHG efficiency with grating period for different thicknesses of the intermediate layer. Solid curve: $t = 0$, dashed curve $t = 0.01$ μm, dotted curve $t = 0.04$ μm (Adapted from Ref. [8]).

6. Optical parametric amplification

One of the most important nonlinear interactions is *optical parametric amplification (OPA)*. In the optical parametric amplification process, a high power pump wave at frequency ω_p mixes with a signal wave at frequency ω_s ($< \omega_p$) to generate a new idler wave at frequency ω_i ($=\omega_p - \omega_s$). During this process, the signal wave at frequency ω_s gets amplified. Thus this interaction process behaves as an amplifier for frequency ω_s. By properly tuning the temperature or the propagation direction in the crystal, we can tune the signal frequency that can get amplified. If the OPA is placed inside an optical resonator, one can create oscillation at the frequency ω_s and this leads to what is referred to as an *optical parametric oscillator* (OPO)[1]. This process can thus provide us with solid state miniature optical sources that are tunable over a wide wavelength range, especially in the near infrared and such sources are becoming very interesting from the practical point of view.

The three frequencies are related by the following equation:

$$\omega_p = \omega_s + \omega_i \tag{14}$$

The phase matching condition for this interaction is given by

$$\beta_p = \beta_s + \beta_i \tag{15}$$

where β_p, β_s and β_i are the propagation constants of the interacting modes at the pump, signal and the idler frequencies respectively. If QPM techniques are used then Eq. (15) gets modified to

$$\beta_p = \beta_s + \beta_i + K \tag{16}$$

where $K=2\pi/\Lambda$ is the spatial frequency of the nonlinear grating with Λ being the corresponding period.

All guided QPM is one of the most widely used techniques for phase matching in the parametric process. In this the interacting pump, signal and idler waves are all guided modes of the waveguide. Since the interaction is through a periodic variation in nonlinear coefficient, as λ_p changes or as the waveguide parameters change, the interaction can get out of phase matching leading to very much reduced efficiencies. Also for a given period of domain inversion and the pump wavelength, the bandwidth for amplification is rather restricted. Recently a novel configuration referred to as the Cerenkov-idler configuration (in contrast to the all guided configuration) in which the pump and signal are guided waves while the idler is a radiating wave has been proposed[10, 11]. Phase matching in the presence of periodic domain reversals takes place between the guided pump and signal waves and the radiating idler wave (see Fig. 14). Since radiation modes form a continuum, as the signal wavelength changes, the idler wave adjusts itself to keep the phase matching intact. This leads to increased bandwidth for signal amplification as well as increased tolerance towards pump wavelength variations. Figure 15 shows the signal gain variation with wavelength corresponding to the all guided case and the Cerenkov idler case. Thus the Cerenkov idler configuration provides much greater signal gain bandwidth compared to the all guided configuration. Of course since the overlap between the interacting waves has reduced, this would lead to reduced efficiencies. The tolerance towards pump wavelength variation for this case is also large.

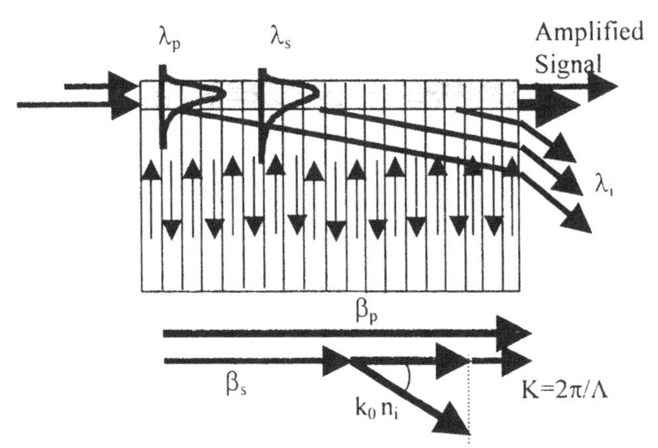

Figure 14 The pump and signal are guided modes while the idler radiates into the substrate. The lower figure is the vector diagram.

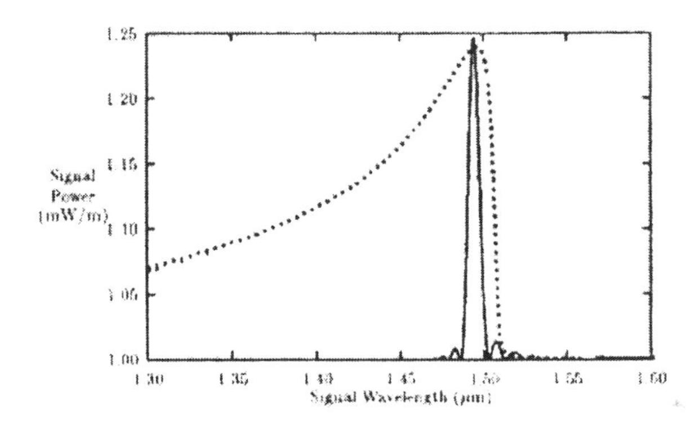

Figure 15 Variation of signal power with signal wavelength for the case of all-guided (solid) and cerenkov (dashed) parametric amplifier configuration (Adapted from Ref. [11]).

7. Periodic Segmented Waveguides (PSW)

Periodic segmented waveguides consist of a periodic array of high index regions in a lower index surrounding (see Fig. 16). Such waveguides were first fabricated for applications to nonlinear interactions. PSWs possesses an interesting characteristic namely the effective index of the mode can be changed by simply changing the duty cycle of segmentation. Thus we can achieve tapering in a channel waveguide structure by simply changing the duty cycle of segmentation along the length of the waveguide. Such waveguides have been shown to possess very low loss. In order to optimize their performance for various applications, it is important to understand the propagation characteristics of such waveguides. In this respect, a parabolic index segmented waveguide has been proposed and analysed which provide us with analytical expressions for the mode effective index and mode spot size[12, 13] An interesting feature of such waveguides is their instability for certain parameter range. A ray optical analysis of a PSW has also been performed[14]. This gives an easy physical understanding of waveguidance in such a waveguide. Ray analysis also predicts unstable regions of operation of the waveguide. PSWs can also be made waveguiding with other guiding mechanisms[15, 16].

We can consider the PSW as a periodic waveguide and thus describe their properties using a coupled mode analysis. The results of this method for both parabolic index as well as step index PSWs matches very well with the finite difference beam propagation method[17]. Figure 17 shows the propagation of a beam in the stable and unstable regions of propagation through a periodic segmented waveguide.

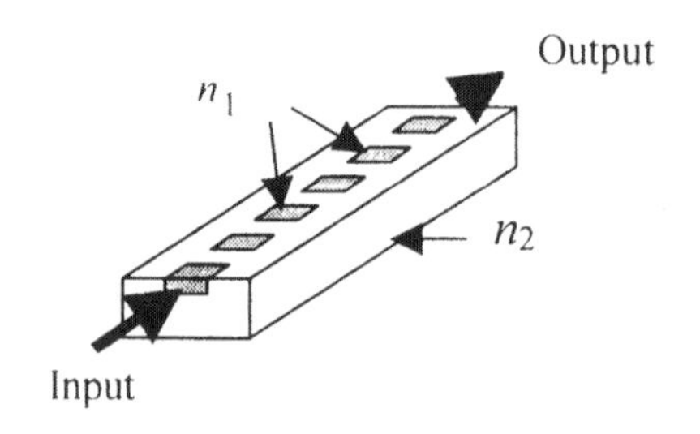

Figure 16 A periodic segmented waveguide.

The analyses of PSWs should help in optimization of the structure of such waveguides and this should in turn lead to their increased application in various devices in integrated optics.

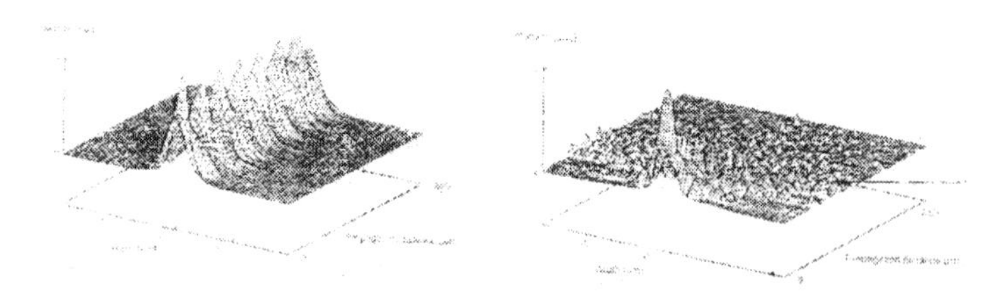

Figure 17 Modal evolution in the stable and unstable regions of operation of a PSW (After Ref. [17]).

8. Summary

Waveguides offer an ideal geometry for nonlinear interactions. One can achieve good conversion efficiencies with low pump powers. Although all guided interaction has the highest efficiency, the Cerenkov configuration has its own advantages. Indeed by using leaky waveguides, one can trade efficiency with interaction bandwidth. The new class of waveguides, namely periodic segmented waveguides have very interesting properties and are finding applications in linear as well as nonlinear waveguide devices.

Acknowledgements: The work in the area of nonlinear interactions in integrated optical waveguides was partially supported by a collaborative project sponsored by the Indo French Center for the Promotion of Advanced Research (IFCPAR), New Delhi between IIT Delhi, India and the University of Nice Sophia Antipolis, Nice France. The author gratefully acknowledges the support of IFCPAR/CEFIPRA, which made this collaboration between the Indian and French Scientists possible.

References

1. A. Yariv, *Optical Electronics*, Harcourt Brace, Fort Worth, 1971.

2. A.K. Ghatak and K. Thyagarajan, *Optical Electronics*, Cambridge University Press, UK, 1989.
3. A.K. Ghatak and K. Thyagarajan, *Introduction to Fiber Optics*, Cambridge University Press, UK, 1998.
4. G. P. Agarwal, *Nonlinear fiber optics*, Academic Press, NY, (1995)
5. M.M. Fejer, G.A. Magel, D.H. Jundt and R.L. Byer, Quasi phase matched second harmonic generation: tuning and tolerances, *IEEE J. Quant. Electron.* **28**, 2631 (1992)
6. K. Thyagarajan, V. Mahalakshmi and M.R. Shenoy, Performance comparison of different configurations for SHG in planar waveguides, , *Int. J. Optoelectronics*, **8,** 319-332 (1993).
7. A.K. Ghatak, Leaky modes in optical waveguides, *Opt. Quantum Electron.*, **17**, 311 (1985).
8. Minu Vaya, K. Thyagarajan, and Arun Kumar, "Leaky waveguide configuration for quasi-phase-matched second harmonic generation," *J. Opt. Soc. Am. B*,. vol. **15**, pp.1322-1328, 1998.
9. Minu Vaya, K. Thyagarajan and Arun Kumar, Parametric amplification studies in 4-layered qpm leaky waveguides, *Fiber and Integrated Optics*, **20** (2001) 367-376.
10. K. Thyagarajan, Vipul Rastogi, M.R. Shenoy, D.B. Ostrowsky, M.De Micheli, and P. Baldi, "Modeling of parametric amplification in the Cerenkov-idler configuration in planar waveguides", *Opt. Lett.*, vol. **21**, pp. 1631-1633, 1996
11. Vipul Rastogi, K.Thyagarajan, M.R. Shenoy, P. Baldi, M. De Micheli, and D. B. Ostrowsky, "Modeling of large-bandwidth parametric amplification in the Cerenkov-idler configuration in planar waveguides", *J. Opt. Soc. Am. B*, Special issue on Nonlinear Guided Waves: Physics and Applications, vol. **14**, pp. 3191-3196, 1997.
12. K. Thyagarajan, V. Mahalakshmi & M.R. Shenoy, Propagation characteristics of planar segmented waveguides with parabolic index segments, *Optics Letts.*, **19,** 2113-2215 (1994).
13. K. Thyagarajan, V. Mahalakshmi and M.R. Shenoy, Equivalent waveguide model for parabolic index planar segmented waveguides, *Optics Comm.* **121,** 27 (1995).
14. Vipul Rastogi, A.K. Ghatak, D.B. Ostrowsky, K. Thyagarajan, M.R. Shenoy, "Ray analysis of parabolic index segmented planar waveguides," *Applied Optics*, **37**, pp. 4851-4856, 1998.
15. Vipul Rastogi, V. Mahalakshmi, M.R. Shenoy, and K. Thyagarajan, "Propagation characteristics of a novel complementary-structure planar segmented waveguide"', *Opt. Commun.*, **148**, pp. 230-235, 1998.
16. K. Thyagarajan, Vipul Rastogi, V. Mahalakshmi and M.R. Shenoy, Propagation characteristics of novel gain guided segmented planar waveguides, *Optics Comm.* **121** (1995) 19.
17. Pierre Aschieri, "*Etude numérique et expérimentale des guides d'ondes optiques segmentés*" Université de Nice-Sophia Antipolis, 1999.

6

Photonic Crystal Fibers

Anurag Sharma

1. Introduction

Ever since the first low-loss fiber was fabricated in 1970 at Corning, efforts to improve fibers have been made and the fiber has evolved from the first step-index multimode fiber to graded-index multimode fiber, single mode fiber, zero-dispersion fiber, dispersion shifted fiber, dispersion flattened fiber, non-zero dispersion fiber and dispersion compensating fiber. In all these cases, the basic structure of the fiber – high refractive index region surrounded by a uniform region of slightly lower index – has been preserved. However, in the last decade a new concept based on microstructured material has emerged. These fibers are referred to as *photonic crystal fibers*.

The idea of photonic crystal is based on inhibiting the propagation of light of certain frequencies using periodic variation of refractive index much the same way in which electron flow is inhibited by periodic potential in crystalline solids.[1-4] Such inhibition of propagation in one dimension (1-*D*) has been used for long in the form of Bragg mirrors (periodic dielectric stacks), in which normally incident light of certain range of wavelengths (or frequencies) is completely reflected or, in other words, is not allowed to propagate within the dielectric stack. This gives rise to *photonic band gap* (PBG) exactly the same way as electronic bandgaps in crystalline solids. However, the use of 2-*D* and 3-*D* periodic structure, or a photonic crystal, with a PBG for propagation in more than one direction was suggested[1,2] only in 1987. Since then a variety of applications have been suggested including waveguidance, and photonic crystal fibers[5] have been studied, techniques for their fabrication have been established, and some of these are now commercially available.

2. Photonic Crystals

It is well known that a stack of alternate quarter wave thick layers of high, n_1, and low index, n_2, material reflects strongly the light in a wavelength-band, $\Delta\lambda$, around a designated wavelength λ_0. For propagation within the stack, therefore, the light in this band is inhibited. Thus, the stack is said to have a photonic bandgap of $\Delta\lambda$ around λ_0. This bandgap is larger for higher index contrast (larger difference between n_1 and n_2) as shown in Fig.1. Now if we consider propagation in the *x-z* plane and designate the component of the propagation vector along the *z*-axis as k_z, then a plot of frequency, ω, versus k_z gives the dispersion curve. In a uniform

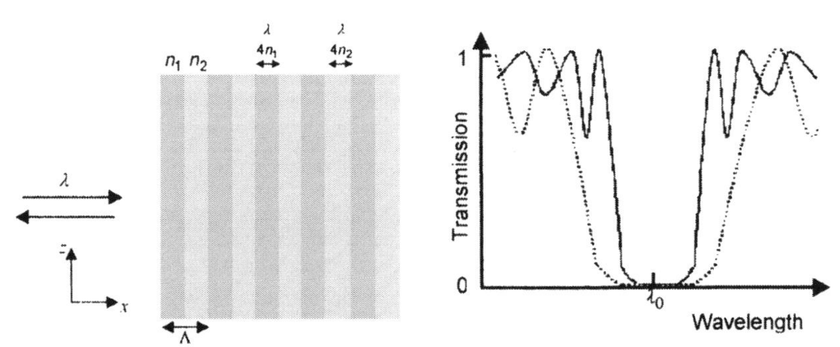

Figure 1 Multilayer stack of high and low index materials with periodicity Λ and its spectral transmission. Zero transmission represents non-propagating waves through the stack and is the photonic bandgap. For higher index contrast, the width of the bandgap is larger (the broken curve).

medium of index *n*, the curve would be straight line of slope $1/n$ (ignoring material dispersion; otherwise it would depart slightly from the straight line). In the case of normal incidence on a 1-*D* periodic stack, $k_z = 0$ and hence, the curve would be a vertical line at $k_z = 0$ with some gaps corresponding to the bandgap $\Delta\lambda$. For oblique incidence, the position and the width of the bandgap would vary giving rise to regions of allowed propagation and forbidden propagation on $\omega - k_z$ plane. Further, due to the planar geometry, one can treat the TE and TM polarizations separately. Figure 2 shows that for planar (2-*D*) propagation in a multilayer planar stack (1-*D* photonic crystal), although bandgaps exist for both polarizations over some range of frequencies for some range of directions of propagation, it is not possible to find a range of frequencies for which both TE and TM waves *do not* propagate through the crystal for all possible directions in the plane. Thus, complete photonic bandgaps do not exist in the 1-*D* crystals. This happens due to absence of any periodicity in the *z*-direction. Therefore, it is necessary to have periodicity along the *z*-direction also to have a possibility of a complete bandgap. Such structures would be termed as 2-*D* photonic crystals. While for a 1-*D* propagation in a 1-*D*

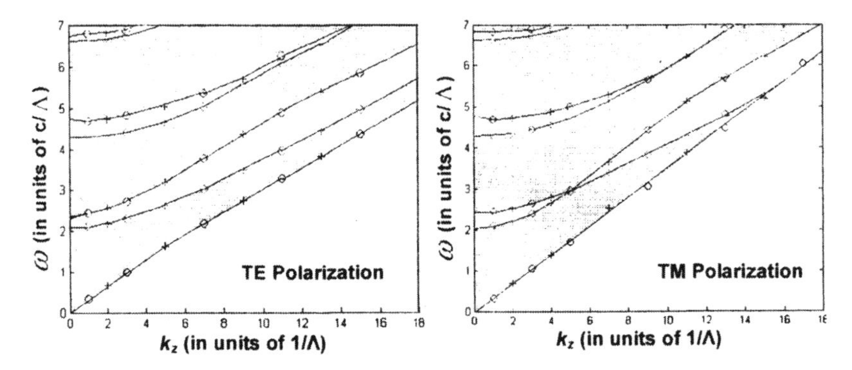

Figure 2 Dispersion plots for TE and TM waves for 2-*D* propagation in a multilayer stack. Shaded regions correspond to allowed propagation and white regions to non-propagating waves (*courtesy Pushkar Singh*).

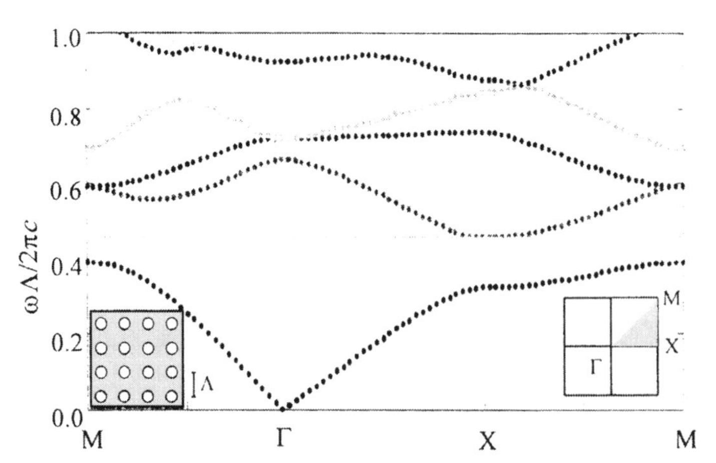

Figure 3 Photonics band structure of a triangular lattice (see inset on the left) of air holes in a dielectric material (n^2=5) with 18.9% area of air (*courtesy V. Karthik and A. Arvind*)

crystal, any contrast of the indices would result in a bandgap, for planar propagation in a 2-*D* crystal a high contrast is needed. This makes realization of such crystals more difficult. Further, the type of lattice (two dimensional periodic distribution of index) also affects the position and width of the bandgaps. As an example Fig.3 shows the bandgaps obtained for *E*-polarization of the field for a square lattice of air holes in a material with $n^2 = 5$. The filing fraction, the fractional area of air holes, is 0.189. The figure shows the existence of a bandgap. However, if one considers the *H*-polarization, a similar band structure would be obtained, but the bandgap would be located in a different frequency range. Thus, such a structure *does not* have a (universal) bandgap. However, another lattice arrangement may have such bandgaps. Figure 4a shows a 2-*D* photonic crystal with a triangular lattice (each lattice point is surrounded by six lattice points on the corners of a regular hexagon); typically it may represent a lattice of air holes in a block of pure silica. However, for silica/air structure with realizable periodicity of about $2\,\mu$m, it is not possible to obtain bandgaps in the wavelength region of interest (0.5-1.6 μm), although at much

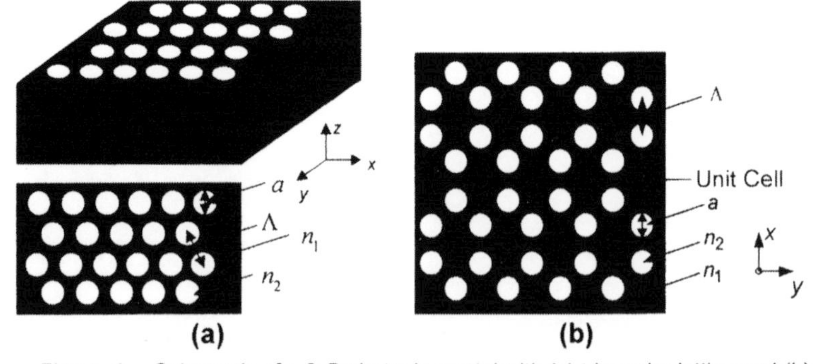

Figure 4 Schematic of a 2-*D* photonic crystal with (a) triangular lattice and (b) honeycomb lattice. It could be a block of silica (n_1) with air holes (n_2) drilled along the *z*-axis.

higher wavelengths (~ 3-4 μm) a bandgap exists. It has been shown[6,7] that a honeycomb lattice (Fig.4b) can be designed and realized to obtain bandgaps in a silica/air structure at wavelengths of interest. Once again if we consider 3-*D* propagation in a 2-*D* crystal, we do not have complete bandgaps, but bandgaps do occur for some range of directions for some range of frequencies. One can have a 3-*D* photonic crystal, which inhibits propagation in all direction within its bandgap. Such crystals have many applications, but are of no concern to waveguides or fibers and hence, we will not discuss these any further.

3. Waveguidance through Photonic Crystals

Let us first consider propagation in a uniform medium with index n_2. The dispersion in such a medium would give a straight line at a slope proportional to $1/n_2$ as shown in Fig.5a (material dispersion has been ignored). All waves corresponding to points above this line (darker area would be propagating waves while those below (lighter area) would be non-propagating evanescent waves. Thus, the latter can be termed as the forbidden region. Now, if we create a narrow region of higher index n_1 and width d (Fig.5b), the dispersion curve in this medium would be a straight line below the n_2 line, with propagating waves above the line and non-propagating waves below the line. Evidently, it is possible to have waves that would propagate in medium n_1, but not in medium n_2. Such waves would correspond to points in the shaded region shown on the dispersion plot in Fig.5b. Thus, if we consider waves of a particular frequency (shown by the line CBA), waves in medium n_1 corresponding to points A and B would be confined to this medium as these can not propagate in medium n_2, but the waves corresponding to point C would propagate in both media. However, not all points on the segment of the line in the shaded region would correspond to propagating or guided modes, since an additional condition of resonance between waves reflected from the upper interface and the lower interface must be satisfied. This requirement leads to only few points on this line segment which correspond to guided modes; points A and B could be such points with A being a lower order (and more confined) mode. If the allowed region for guided moded (shaded region) is narrow enough, then only one mode may be allowed to give single mode guidance. Such waveguides are called single mode waveguides and are of great importance to communication, signal processing and other applications. It is evident from the figure that as the frequency increases, the width of the allowed region increases, and more and more modes are possible.

Now let us consider that the upper and lower surrounding media are replaced by 1-*D* photonic crystal and the layer between them now has an index n_0 which could be lower than either of the medium in the periodic layers; in fact, it could be air. The photonic crystal layer could have a forbidden region above the n_0 line, say, for the TE polarization. This is shown schematically by the shaded in the dispersion plot in Fig.5c. The waves that are allowed to propagate within the photonic crystal layer correspond to the darker region, but not to the shaded region. Thus, if we consider waves of a particular frequency shown by the line PQ on the dispersion plot, point P corresponds to those waves, which would propagate in the n_0 layer as well as in the photonic crystal layer and hence are not confined within the n_0 layer. On the other

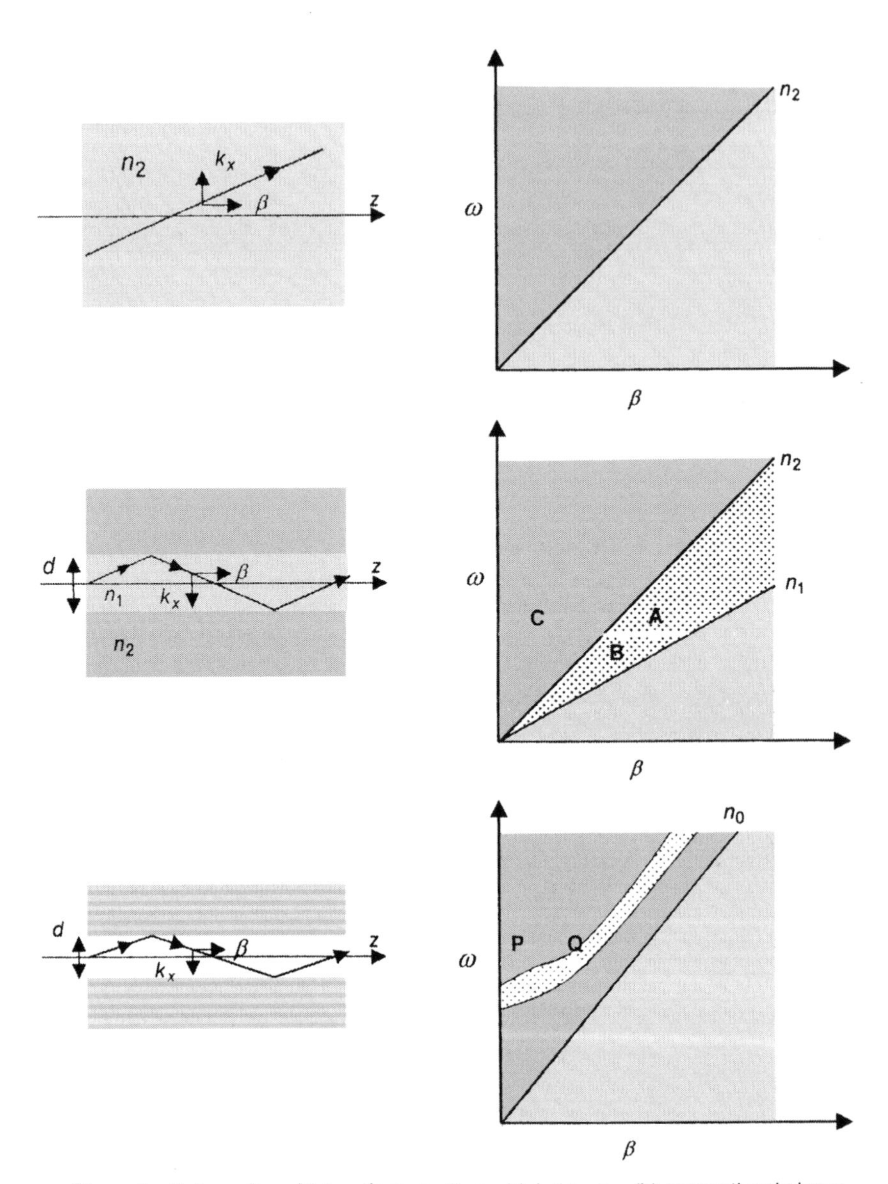

Figure 5 Schematics of (a) uniform medium with index n_2, (b) conventional planar waveguide and (c) a photonic crystal planar waveguide and the corresponding dispersion plots. The index variation in (b) and (c) is along the *x*-axis and the guidance is along the z-axis. The z-component of the propagation vector is represented by β.

hand, the waves at an angle corresponding to point Q would propagate in the n_0 layer, but would not propagate in the photonic crystal layer as it lies in the bandgap. However, for this wave to become a guided mode, the width *d* of the n_0 layer should be such that the transverse resonance condition is satisfied. One notable difference between the conventional waveguidance or *index guidance* and the photonic bandgap (PBG) guidance is that in the latter case, waves are guided in a medium of lower index than that of either of the media used in the multilayer stack whereas in the former case, it is necessary to have a core of higher index. PBG guidance thus makes it possible to have a waveguide with core of vacuum or air. This is of tremendous significance, as we shall see later.

We must add here that if the waves do not correspond to photonic bandgap and if the core is of higher index, waves are index-guided through the core of a photonic crystal waveguide. In such cases, the advantage of having cladding of photonic crystal material is that the effective refractive index of cladding has very interesting spectral characteristics, as we shall see in the next section.

The waveguides considered above are planar waveguides since the waves are confined only in one direction (x-direction) and waves would still diffract along the y-direction. For practical waveguides, it is necessary to have confinement in both transverse directions. In a fiber, it could be provided either by a 2-D lattice such as the ones shown in Fig.4, or, by radially periodic layers. We discuss both types in the following sections.

4. Index Guided PCF

First photonic crystal fibers were made of pure silica with air hole forming the triangular lattice[8]. Due to the presence air holes, these were referred to as *holey fibers* in the early literature, but now PCF is commonly used. A missing hole at the center provided the high index region for the core. Thus, these fibers can be considered[9] to have a core of silica with index, n_{co}, while the cladding index, n_{cl}, is between the index of air and that of silica, depending on the relative distribution of power of the wave in the air and silica regions of the lattice. In fact, it is the effective index of the fundamental mode supported by the infinite lattice; this mode is called the space-filling mode (SFM). Since, as the wavelength decreases, more and more power is concentrated in the high index regions, the cladding index increases and the effective relative index difference between core and cladding indices reduces. This helps in keeping the normalized frequency, $V = (2\pi\rho/\lambda)\sqrt{n_{co}^2 - n_{cl}^2}$, relatively insensitive to wavelength (here, ρ is the effective

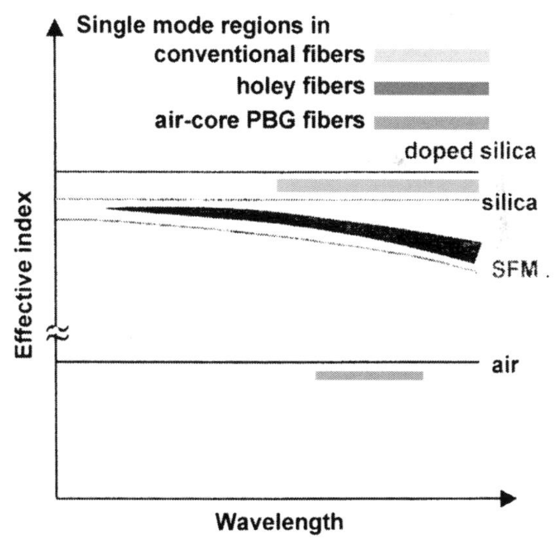

Figure 6 Schematic plot of the effective index ($\beta c/\omega$) as a function of wavelength ($2\pi c/\omega$) showing the regions of single mode propagation for different types of fibers material dispersion has been ignored).

radius of the core, generally taken as 0.65Λ, where Λ is the pitch of the lattice). This makes the wavelength range for single mode operation very large. In fact, it has been shown that if the ratio of the air hole radius to the period of the crystal, a/Λ, is kept small (< 0.3), the fiber would be *endlessly* single moded[10]. The range of single mode regions in various types of fibers is shown schematically in Fig.6. Another consequence of strong dependence of effective index on wavelength is that holey fibers can be designed to have very large negative chromatic dispersion. This makes them very useful for dispersion compensation in communication systems (see Chapter 1). Further, by making the size of the holes large, a larger core-cladding index difference could be achieved. This makes it possible to have a very small core and hence, very small effective mode area giving a very high nonlinear phase shift.[11] The first holey fiber was made[8] in 1996 at the University of Bath, UK. It was made by an intricate process using tubes of pure silica. Salient steps in the fabrication process are shown in Fig.7. An SEM image of the end of a holey fiber is shown in Fig.8a. The technology of fabrication has since been refined and these fibers are now available commercially.[12,13]

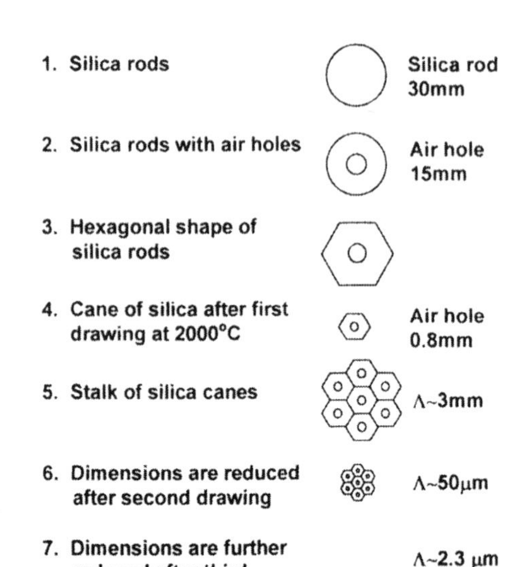

Figure 7 Main steps involved in fabrication of silica/air-hole fibers.

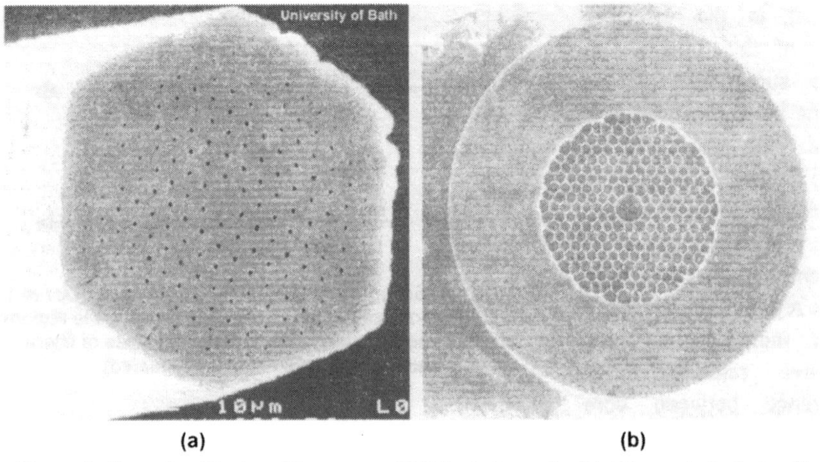

Figure 8 Scanning Electron Microscope (SEM) photograph of (a) the end of a holey fiber and (b) the end of an air-core PBG fiber (*reproduced with permission, University of Bath*).

5. PBG Fibers

Another type of photonic crystal fibers is the one in which the waves are guided along a low index core in otherwise periodic lattice in two transverse dimensions. In fact, fibers with a hole at the centre, *i.e.*, with air core, have been fabricated. The first of such fibers was also made[14] at the University of Bath, UK in 1998. Figure 8b shows an SEM micrograph of such a fiber. In these fibers, the effective index of the guided mode is less than the index of the core and hence, in air-core fibers the effective index is less than unity (see Fig.6). The advantage of hollow core is that very high powers can be transmitted through such fibers due to the absence of any absorptive or nonlinear effects to hinder such transmission. On the other hand, the hollow core could be filled with a gas or a liquid to have its interaction with the confined light over long lengths. This would not only be useful in enhancing nonlinear effects in such materials, but also in gas sensing and monitoring. Since single transverse mode is generally formed in such fibers, which could have vacuum in the core, these offer possibility of guiding atoms by intense light beams. A recent review discusses the development such fibers and their applications.[15]

6. Radially Periodic Fibers

Another type of fiber based on photonic bandgap principle is the one in which alternate layers of high and low index materials concentrically surround a core of low index material or air. These can be considered as true cylindrical analogues of planar stacks used as Bragg mirrors, though in the cylindrical case, the periodicity is not uniform and the thickness of successive layers is determined by zeroes of Bessel's functions. However, the departure from uniform thickness is not much, particularly for layers away from the cores and even equal thickness layers show useful bandgap characteristics. In the first paper[16], these were called *Bragg fibers*. The term photonic crystal fiber came later. Such fibers with hollow core were first made[17] at MIT, USA. Since this group approached this idea through their studies on omnidirectional reflectors, they termed these fibers as *omniguides*. In addition to hollow core fibers, coaxial fibers have also been proposed and studied[18] with the aim of propagating a TEM mode just like the one in a metal coaxial waveguide. The advantage of such fibers is that the fundamental mode is axially symmetric eliminating the problem of rotation of polarization on propagation, and there is possibility of zero dispersion at a specific wavelength.[18] Figure 9 shows schematically the structure of hollow core and coaxial Bragg fibers. Recently, these omnidirectional dielectric fibers have been fabricate[19] using polymer (PES, polyether sulfone) and As_2Se_3, which have photonic bandgaps in the mid-IR and have outer

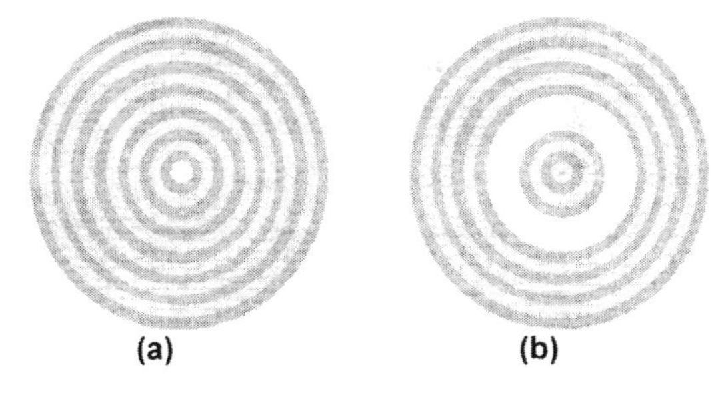

Figure 9 Cross-sections of (a) a hollow core radial fiber and (b) a coaxial fiber.

diameters ranging between 175-500 μm. These fibers could be woven into fabrics for spectral identification and could be used as flexible radiation barriers. Other applications of these hollow fibers include infrared laser delivery for medical use and very high power laser (CO_2 laser) delivery for industrial applications. These fibers are now being commercially produced.[20]

References:

1. E. Yablolonvitch, "Inhibited spontaneous emission in solid state physics and electronics," *Phys. Rev. Lett.*, **58**, 2059 (1987).
2. S. John, "Strong localization of photons in certain disordered dielectric superlattices," *Phys. Rev. Lett.*, **58**, 2486-2489 (1987).
3. J.D. Joannopoulos, R.D. Meade and J.N. Winn, *Photonic Crystals: Molding the Flow of Light*, Princeton University Press, Princeton, USA (1995).
4. K. Sakoda, *Optical Properties of Photonic Crystals*, Springer, Berlin, Germany (2001).
5. A. Bjarklev, J. Broeng, and A.S. Bjarklev, *Photonics Crystal Fibres*, Kluwer Academic Publishers, Boston, USA (2003).
6. A. Barra, D. Cassagne and C. Jouanin, "Existence of two-dimensional absolute photonic band gap in the visible," *Appl. Phys. Lett.* **72**, 627 (1998).
7. J. Broeng, S.E. Barkou, A. Bjarklev, J.C. Knight, T.A. Birks and P.St.J. Russell, "Highly increased photonic band gaps in silica/air structures," *Opt. Commun.* **156**, 240 (1998).
8. J.C. Knight, T.A. Birks, P.St.J. Russell and D.M. Atkin, "All-silica single-mode optical fiber with photonic crystal cladding," *Opt. Lett.* **21**, 1547-1549 (1996); errata: **22**, 484-485 (1997).
9. J.C. Knight, T.A. Birks, P.St.J. Russell and J.-P. de Sandro, "Properties of photoic crystal fiber and the effective index model," *J. Opt. Soc. Am A* **15**, 748 (1998).
10. T.A. Birks, J.C. Knight and P.St.J. Russell, "Endlessly single-mode photonic crystal fiber," *Opt. Lett.*, **22**, 961-963 (1997).
11. N.G.R. Broderick, T.M. Monro, P.J. Bennett and D.J. Richardson, "Nonlinearity in holey optical fibers: measurement and future opportunities," *Opt Lett.* 24, 1395-1397 (1999).
12. http://www.crystal-fibre.com/.
13. http://www.blazephotonics.com/.
14. J.C. Knight, T.A. Birks and P.St.J. Russell, "Photonic bandgap guidance in optical fibers," *Science* **282**, 1476-1549 (1998).
15. P.St.J. Russell, "Photonic crystal fibers," *Science*, **299**, 358-362 (2003).
16. P. Yeh, A. Yariv and E. Marom, "Theory of Bragg fiber," *J. Opt. Soc. Am.* **68**, 1196-1201 (1978).

17. Y. Fink, D.J. Ripin, S. Fan, C. Chen J.D. Joannopoulos and E.L. Thomas, "Guiding optical light in air using an all-dielectric structure," *J. Lightwave Technology* **17**, 2039-2041 (1999).
18. M. Ibanescu, Y. Fink, S. Fan, E.L. Thomas and J.D. Joannopoulos, "An all-dielectric coaxial waveguide," *Science*, **289**, 415-419 (2000).
19. S.D. Hart, G.R. Maskaly, B. Temelkuran, P.H. Prideaux, J.D. Joannopoulos and Y. Fink, "External reflection from omnidirectional dielectric mirror fibers," *Science* **296**, 510-513 (2002).
20. http://www.omni-guide.com/.

7

Side-Polished Fiber Coupler Half Block and Devices

R. K. Varshney

1. Introduction

With increased penetration of optical fibers into the subscriber loop and their applications in various kinds of sensors and other optical processing applications, there is a growing demand for fiber optic components capable of performing various functions such as modulation, splitting, filtering etc. Normally, these functions can be performed by taking light out of the fiber and using bulk optical components, and then coupling the light back into the fiber. This will involve interruption of light as it propagates in the fiber and would lead to high optical insertion loss, problem of stability of components and larger size. These problems can be overcome by using in-line fiber optic components in which the processing is performed without taking light out of the fiber and which are completely compatible with the transmission medium, namely, the optical fiber. A large number of in-line fiber optic devices have been fabricated based on the interaction of the evanescent field existing outside the core region of the fiber. This evanescent field of the fiber can be accessed by the side-polishing technique, in which an appropriate portion of the cladding of the fiber is removed from one side of the fiber fixed in a quartz block. Such a block is known as '*side-polished fiber coupler half block*' or simply a '*side-polished fiber block*'. A schematic diagram of a side-polished fiber block is shown in Fig. 1. Such side polished fiber blocks are basic building blocks of many in-line fiber optic components/devices such as tunable directional couplers, switches, modulators, polarizers, mode combiners, wavelength filters, etc.[1-13]. In this chapter, we have briefly discussed the side-polished fiber block and some of the e devices such as gain flattening of EDFA, mode splitters, temperature sensors, tunable directional couplers, polarizers, filters, etc. based on it.

2. Side-polished fiber coupler half block

To access the evanescent field, an unjacketed portion of the fiber is glued into a groove, cut in the shape of an arc of radius ~ 25 cm in a quartz block. The entire assembly is then grounded and polished to remove the fiber cladding from one side till the required extent of evanescent field exposure is achieved. Depth of polishing (minimum remaining cladding thickness over the core) is estimated by the liquid drop method, which involves periodic measurement of the power loss of the structure when a liquid of refractive index greater than the fiber effective index is applied to the upper surface of the device (see Fig.1). The interaction length (polished region of the fiber) depends on the radius of curvature of the groove and the depth of polishing. A typical inter- action length is ~ 2-3 mm and can be achieved by making a groove with radius of curvature ~ 25 cm and depth of polishing ~ 2 μm. The main steps involved in fabrication of such side polished fiber blocks are briefly described below.

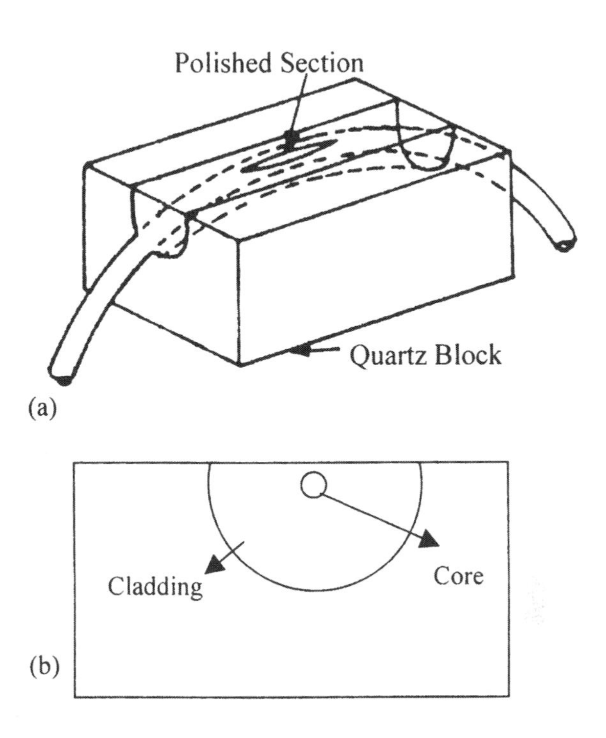

Figure 1 Schematic diagram of a side polished fiber coupler half block: (a) three dimensional view (b) side view.

2.1. Fabrication and characterization of a curved slot/groove in a quartz block

Since the fiber is very flexible and fragile, it can't be polished directly. Therefore, it has to be fixed in a hard material block like a quartz block. In order to fix the fiber in the quartz block, a groove of the appropriate radius of curvature has to be made. The groove can be made by moving a diamond particle coated wire over the block with appropriate pressure on the wire. The radius of curvature of the groove, which is a very important parameter for estimating the '*depth of polishing*', can be obtained by measuring the depth of the groove along its length. A typical groove profile is shown in Fig. 2. The radius of curvature (R), of the groove can be calculated from the following simple geometrical relation[14]

$$R^2 = (L/2)^2 + (R + d_1 - d_2)^2 \tag{1}$$

where L is the length of the block, d_1 and d_2 are depths of groove at the centre and end of the block, respectively. For a fiber block with typical parameters $d_1 = 0.205$ mm, $d_2 = 0.585$ mm and $L = 30.0$ mm, the radius of curvature of the groove is ~ 29.8 cm.

2.2. Fixing and polishing of the fiber

The unjacketed portion of the fiber is first fixed in the groove by using a suitable epoxy and heating it to an appropriate temperature. Now, the silica block mounted on

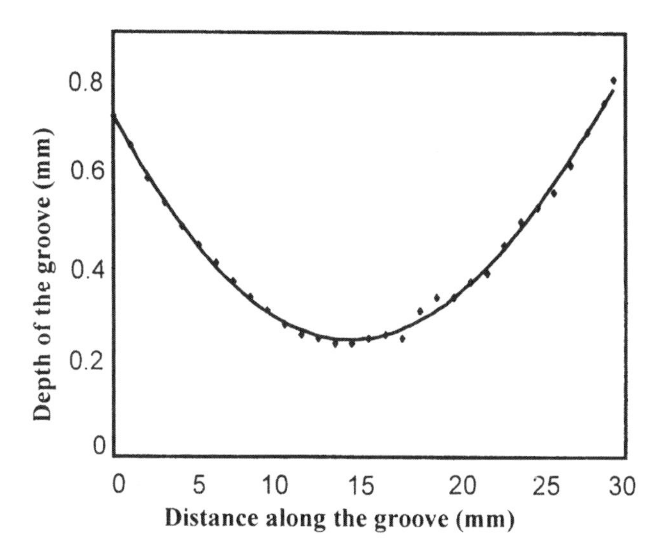

Figure 2 Depth profile of the groove made in the quartz block.

the zig is polished by using the appropriate emery solutions. As soon as cladding of the fiber starts getting polished and depleted, an ellipse can be seen in the polished region whose dimension increases with the depth of polishing. By measuring the length of the major axis of this observed ellipse, and knowing the radius of curvature of the groove, one can estimate the depth of polishing. A schematic diagram used to calculate the depth of polishing of the fiber is shown in Fig. 3. From this diagram, one can easily obtain the following relation[14] for the depth of polishing d.

$$d \cong \frac{l^2}{2(R+D)} \tag{2}$$

where D *and* l are the diameter of the fiber cladding and length of semi-major axis of the ellipse, respectively.

2.3. Characterization of side polished fiber block

A more accurate estimation of the depth of polishing can be obtained by measuring the throughput power loss of the structure when a liquid of refractive index greater than the fiber effective index is applied to the polished region of the fiber[4, 14]. This technique is commonly known as the index matching liquid technique. In this technique, laser light is coupled at the input end of the side polished fiber block and power coming out from other end of the fiber is measured with and without the oil overlay. A typical variation of the throughput power loss (= -10 log (P_{air}/P_{oil}); where P_{oil} and P_{air} are the power measured with and without the oil overlay, respectively) as a function of the oil index is shown in Fig. 4[4]. This figure clearly indicates that there is no power loss when the refractive index of the oil overlay is less than the cladding index of the fiber, and this loss increases rapidly as the refractive index of the oil approaches the effective index of the mode of the fiber. The maximum power loss occurs when the refractive index of the oil is equal to the mode index of the fiber. The figure also indicates that the throughput power loss increases with the increase in the depth of polishing (i.e., decrease of the remaining cladding thickness (d)).

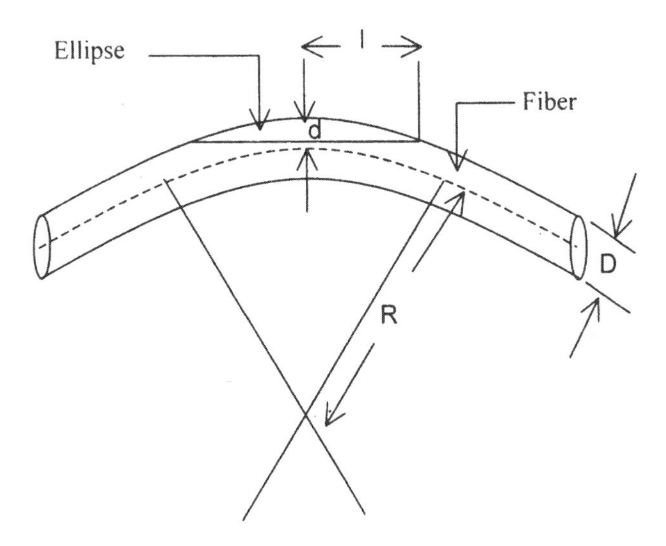

Figure 3 Diagram used to calculate the depth of polishing.

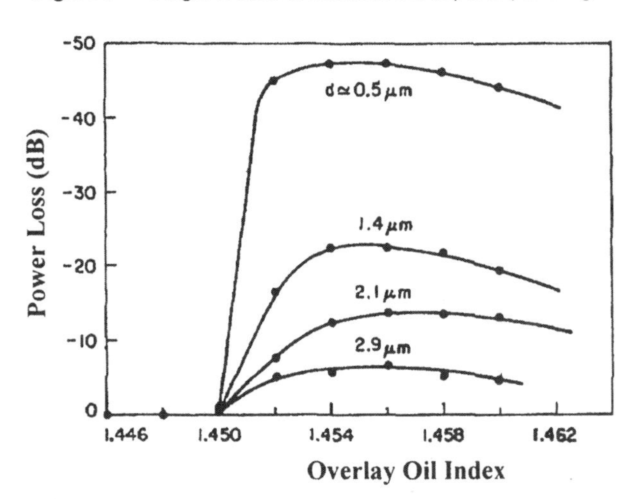

Figure 4 Variation of the throughput power loss as a function of the overlay oil index (adapted from Ref.[4]; © Optical Society of America).

3. Gain-flattening of Erbium Doped Fiber Amplifier

The Erbium Doped Fiber Amplifier (EDFA) has become a very important component of any long haul dense wavelength division multiplexed (DWDM) optical fiber communication system, that can enhance tremendously the information carrying capacity of the existing system. A typical amplified spontaneous emission

(ASE) spectrum of an EDFA is shown in Fig.5.[15, 16] It has a peak around 1530 nm and a broad band with smaller gain for other wavelengths. This non-uniformity in the ASE spectrum of the EDFA creates problems in the optical signal to noise ratio between different channels[16]. Therefore, gain-flattening of EDFAs is very essential. Recently, it has been proposed and shown that gain-flattening of EDFAs can be achieved by using a side polished single mode fiber block, loaded with an appropriate multimode overlay waveguide (MMOW).[17] A schematic diagram of this side polished fiber based SPF-MMOW device is shown in Fig.6. It consists of a planar multimode overlay waveguide of appropriate thickness and refractive index, evanescently coupled to a side polished single mode fiber. The MMOW can be formed by introducing a liquid of appropriate refractive index into the narrow gap created by placing another quartz block on top of the side polished fiber block with suitable spacers at the ends, as shown in Fig. 6. A schematic of the refractive index profile of the side polished fiber and MMOW are shown in Fig.7. Here, n_c and n_{cl} are core and cladding indices of the fiber, n_o and n_s ($<n_{cl}$) are film and substrate indices of the waveguide, and dashed lines correspond to effective indices of the modes. The device essentially functions as an asymmetric directional coupler with wavelength dependent resonant coupling between the mode of the side polished fiber and one or more modes of the MMOW. Since, interaction between the fiber and MMOW takes place along one vertical direction (say x-direct-ion), in order to study various characteristics of such a device

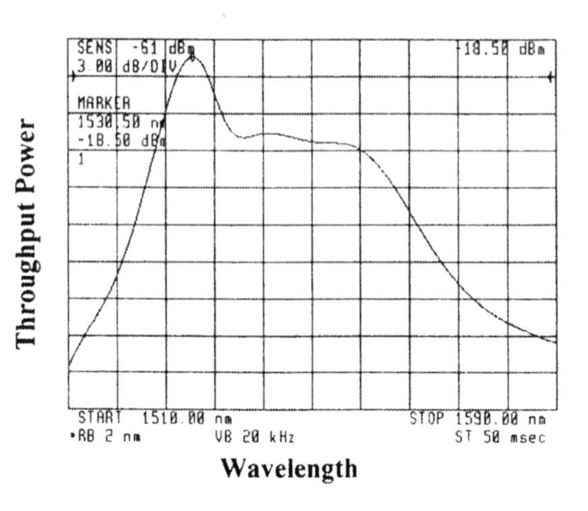

Figure 5 Measured ASE spectrum of the EDFA.

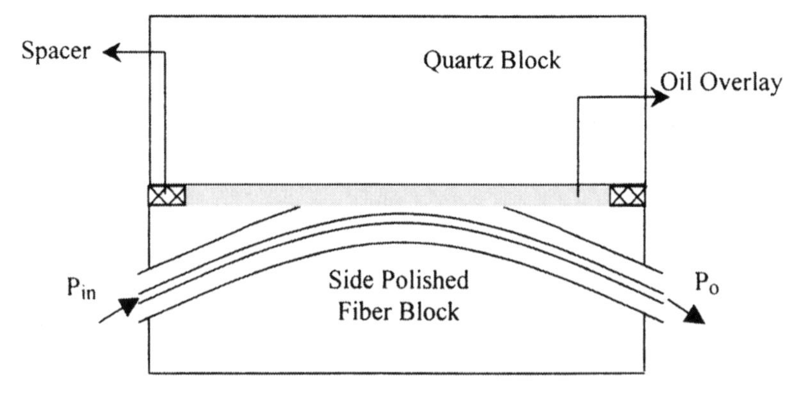

Figure 6 Schematic of the SPF-MMOW device.

we can replace fiber by an equivalent planar waveguide (EPW).[18] Thus, the SPF-MMOW device can be treated as a multilayer planar waveguide configuration along x-direction, which is schematically shown in Fig. 8. The equivalent planar waveg-uide is obtained in the form of a step-index slab waveguide by *matching* the modal field in the direction of interaction and the propagation constant of the side polished fiber with that of the fundamental mode of the equivalent planar waveguide. Using the empirical relations, the index distribution of such an x-slab equivalent planar waveguide with core index (n_1), cladding index (n_2) and width $(2\sigma a)$ can be written as[18]

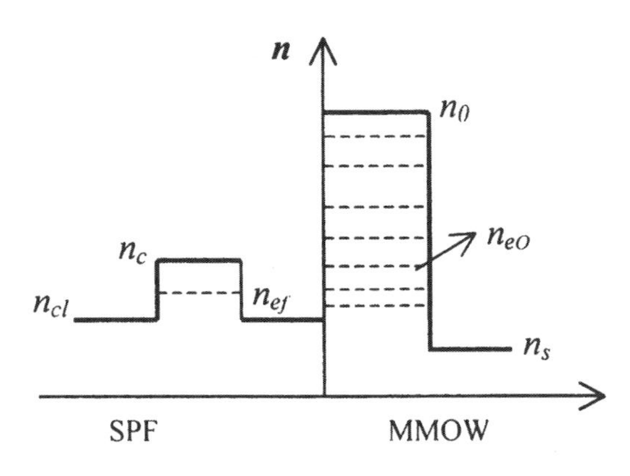

Figure 7 Refractive index profile of SPF and MMOW; dashed lines correspond to effective indices of their modes.

$$
\begin{aligned}
n^2(x) &= n_1^2 = n_c^2 - \frac{(U^2 - p^2)}{(k_0 a)^2}, \qquad |x| < \sigma a \\
&= n_2^2 = n_1^2 - \frac{p^2 \sec^2(p\sigma)}{(k_0 a)^2}, \qquad |x| > \sigma a
\end{aligned}
\tag{3}
$$

where $k_o(= 2\pi / \lambda)$ is the free space wave number of the propagating light, a is the core radius of the fiber,

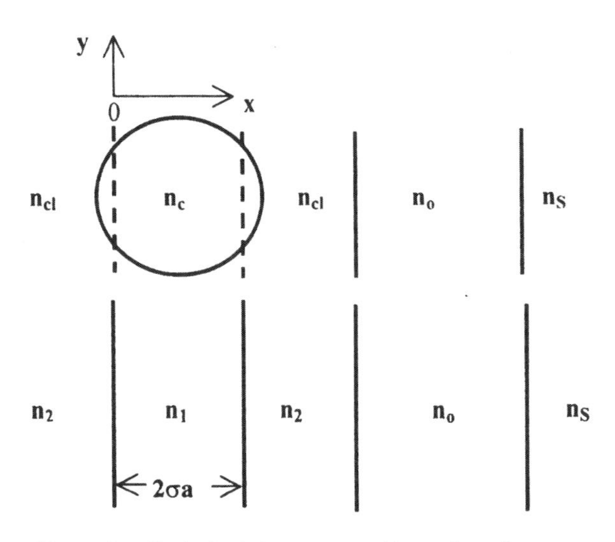

Figure 8 Equivalent planar waveguide configuration.

$$p^3 = -1.3528 + 1.6880V - 0.1894V^2 \quad (4)$$

$$\sigma = 0.8404 + 0.0251V - 0.0046V^2 \quad (5)$$

$$V = k_0 a\sqrt{n_c^2 - n_{cl}^2} \quad (6)$$

and for $1.5 \le V \le 2.5$,

$$U = \sqrt{V^2 - (1.1428V - 0.996)^2} \quad (7)$$

Once an equivalent planar waveguide is obtained, the composite structure can be analyzed using any method that deals with multilayer structures (e.g., Matrix method[19, 20], Runge-Kutta method[21], etc.). The throughput power of the SPF-MMOW device can be obtained as follows:

In order to study the normal modes of such a multilayered structure by using the matrix method, let us assume that there are N regions (in the present case shown in Fig. 8, N=5) and the refractive index (n_i) in each region (i^{th} region, i = 1, 2, 3, ...N) is real and constant. The modal field (ψ) in any region can be obtained by solving the following wave equation:

$$\frac{\partial^2\psi}{\partial x^2} + (k_0^2 n_i^2 - \beta^2)\psi = 0 \quad (8)$$

where, β is the propagation constant of the mode. The solution of Eq. (8) in the ith region can be written as[19]

$$\begin{aligned}\psi_i &= A_i \cos(\gamma_i(\xi - \sigma_{i-1}) + B_i \sin(\gamma_i(\xi - \sigma_{i-1}) \quad \text{for } \kappa_i^2 > 0\\ &= A_i \cosh(\gamma_i(\xi - \sigma_{i-1}) + B_i \sinh(\gamma_i(\xi - \sigma_{i-1}) \quad \text{for } \kappa_i^2 < 0\end{aligned} \quad (9)$$

where, i = 1, 2, 3, ...N

$$\begin{aligned}&\kappa_i^2 = V^2(F_i - b); \qquad \gamma_i = |\kappa_i|\\ &\xi = x/h; \ \sigma_i = d_i/h \text{ with } \sigma_0 = 0, \ d_1 = 0\\ &V = k_0 h\sqrt{(n_m^2 - n_{cl}^2)}; \ b = \frac{(\beta/k_0)^2 - n_{cl}^2}{(n_m^2 - n_{cl}^2)}\\ &F_i = \frac{n_i^2 - n_{cl}^2}{(n_m^2 - n_{cl}^2)}\end{aligned} \quad (10)$$

and d_i represent the position coordinate of the right hand boundary of the ith region, n_m is the maximum refractive index in the structure and h is the thickness of the film, and $n_{cl} > n_s$. The relation between field coefficients (A and B) in two successive regions (i^{th} say and $(i+1)^{th}$ regions) can be obtained by applying proper boundaries conditions and can be written as[20]:

$$\begin{pmatrix} A_{i+1} \\ B_{i+1} \end{pmatrix} = S_i \begin{pmatrix} A_i \\ B_i \end{pmatrix} \quad (11)$$

where, S_i, which is known as the transfer matrix of the ith region, is given by

$$S_i = \begin{cases} \begin{pmatrix} \cos\Delta_i & \sin\Delta_i \\ -\dfrac{\gamma_i}{\gamma_{i+1}}\alpha_i \sin\Delta_i & \dfrac{\gamma_i}{\gamma_{i+1}}\alpha_i \cos\Delta_i \end{pmatrix} & \text{for } F_i > b \\ \begin{pmatrix} \cosh\Delta_i & \sinh\Delta_i \\ \dfrac{\gamma_i}{\gamma_{i+1}}\alpha_i h \sinh\Delta_i & \dfrac{\gamma_i}{\gamma_{i+1}}\alpha_i \cosh\Delta_i \end{pmatrix} & \text{for } F_i < b \end{cases} \tag{12}$$

and

$$\begin{aligned} \alpha_i &= 1 \quad \text{for TE mode} \\ &= \frac{n_{i+1}^2}{n_i^2} \quad \text{for TM mode} \end{aligned} ; \qquad \Delta_i = \gamma_i(\sigma_i - \sigma_{i-1}) \tag{13}$$

By successive application of Eq. (11), we can write

$$\begin{pmatrix} A_{i+1} \\ B_{i+1} \end{pmatrix} = S_i S_{i-1} S_{i-2} \ldots\ldots S_2 S_1 \begin{pmatrix} A_1 \\ B_1 \end{pmatrix} \tag{14}$$

Thus, $$\begin{pmatrix} A_N \\ B_N \end{pmatrix} = G \begin{pmatrix} A_1 \\ B_1 \end{pmatrix} \tag{15}$$

where the G matrix is given by

$$G = S_{N-1} S_{N-2} \ldots\ldots S_2 S_1 \tag{16}$$

Since for a guided mode the modal field $\psi = 0$ at $x = \pm\infty$, we get[20]:

$$A_1 = B_1 \qquad \& \quad A_N + B_N = 0 \tag{17}$$

Thus, the propagation constant and the modal field, ψ, of any guided mode of a multilayered structure can be obtained by solving Eq. (17) and using Eqs. (9)-(13). Physically, power P_{in} launched into the fiber [see, Fig. 6] excites all the normal modes as it enters the coupling region of the structure (i.e., $z = 0$). At the output end of the interaction length ($z = L$), the optical power would be redistributed among the various modes of the interacting waveguides depending on the phase accumulated along the z-direction. The normalized throughput power P_f (= P_O / P_{in}) coming out from the side polished fiber can be evaluated by determining the overlap between the mode of the equivalent planar waveguide and the normal modes of the multilayered structure through simple overlap integrals, given by[7]

$$P_f = \frac{P_0}{P_{in}} = \sum_{i=1}^{N_m} M_{i1}^2 + 2\sum_{j=1}^{N_m - 1} \sum_{k=j+1}^{N_m} M_{j1} M_{k1} \cos((\beta_j - \beta_k)L) \tag{18}$$

where, N_m corresponds to total number of guided modes of the multilayered structure and M_{j1} represents the strength of interaction between the equivalent planar waveguide mode (with modal field E_1) and the normal j^{th} mode of the multilayered structure, and is given by

$$M_{j1} = \left(\int_{-\infty}^{\infty} \psi_j E_1 \, dx \right)^2 \Bigg/ \left\{ \left(\int_{-\infty}^{\infty} |\psi_j|^2 dx \right) \left(\int_{-\infty}^{\infty} |E_1|^2 dx \right) \right\} \tag{19}$$

A typical spectral response of the throughput power of a side polished fiber loaded with an overlay waveguide is shown in Fig. 9. The wavelength corresponding to the dip in the throughput power in the spectral response is known as resonance wavelength (λ_r). At this wavelength, the effective index (n_{ef}) of the mode of the side polished fiber matches with the effective index (n_{ef}) of one of the modes of the overlay waveguide. For a thick overlay waveguide supporting a large number of guided modes, the resonance wavelength (λ_r), is given approximately by the following relation:

$$\frac{2\pi d\sqrt{(n_o^2 - n_{ef}^2)}}{\lambda_r} = m\pi + \phi_S + \phi_2 \tag{20}$$

where, m is the mode order and d is the thickness of the overlay waveguide. Further, ϕ_s and ϕ_2 are the phase shifts of the overlay waveguide mode at the upper and lower boundaries of the waveguide (see, Fig.8) respectively, and can be written as

$$\phi_i = \tan^{-1}\left(\eta_i \frac{\sqrt{n_{ef}^2 - n_i^2}}{\sqrt{n_o^2 - n_{ef}^2}}\right) \tag{21}$$

where $\eta_i = 1$(or $(n_o / n_i)^2$) for TE (TM) mode of the waveguide and i corresponds to S and 2. Using Eqs. (20) & (21), one can obtain the change ($\Delta\lambda_r$) in wavelength required to change coupling from the mth mode to the $(m+1)^{th}$ mode of the overlay waveguide as

$$\Delta\lambda_r = \lambda_r^2 / [\lambda + 2d\sqrt{(n_0^2 - n_{ef}^2)}] \tag{22}$$

This change in wavelength, $\Delta\lambda_r$, which corresponds to the wavelength separation between two successive dips is commonly known as channel separation. The full width at half maximum (FWHM) of the dip (power fall) is called the linewidth. These parameters, known as characteristics parameters of the device, critically depend on the parameters of the fiber and waveguide, and can be obtained by using the equations given above. By choosing an appropriate value of the refractive index

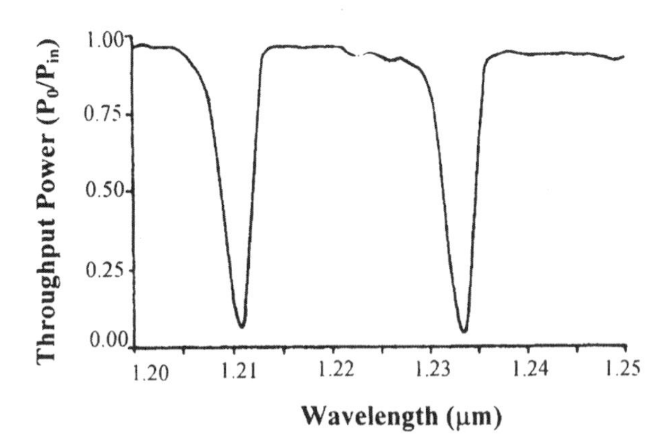

Figure 9 Spectral response of the throughput power of a SPF-MMOW device.

and thickness of the MMOW, one can obtain the transmission characteristics shown in Fig. 10. Thus, the SPF-MMOW can work as a gain-flattening device for EDFAs if the transmission characteristic is tuned to match the unwanted gain peak spectrum of the EDFA. Measured ASE spectrum of an EDFA after passing through the SPF-MMOW device is shown in Fig. 11, which clearly demonstrates the gain-flattening function of the SPF-MMOW device[15, 17].

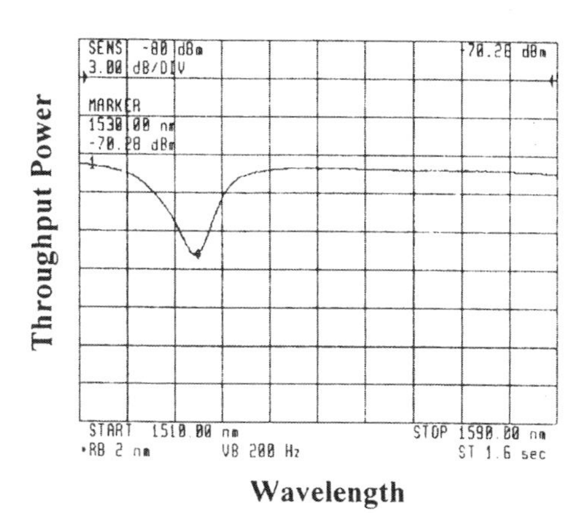

Figure 10 Spectral response of the throughput power of the SPF-MMOW device.

4. Mode splitter/filter

Dual mode fibers have been used in realizing many fiber optic devices such as sensors, intermodal couplers, frequency shifters, and dispersion compensators etc.[22-27] Such devices can be used effectively only if one is able to couple in/filter out either of the modes selectively. Such a mode splitter/ filter is similar to a normal directional coupler and can be fabricated by using two identical dual mode fiber based side polished fiber blocks. A schematic diagram of the mode splitter made by using two side polished fiber blocks is shown in Fig. 12.

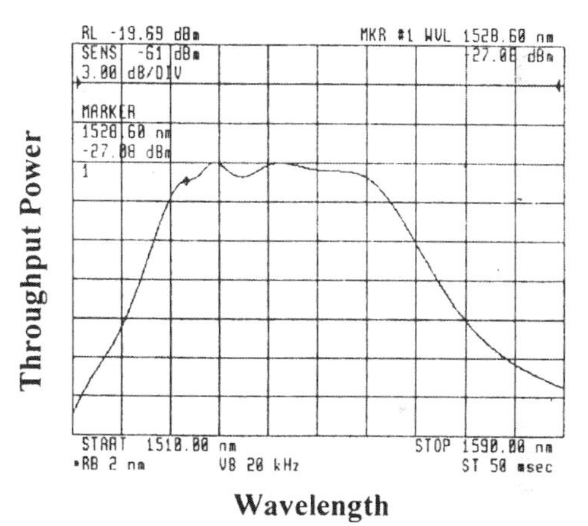

Figure 11 ASE spectrum of the EDFA after passing through the SPF-MMOW device.

To study such a coupler, it is assumed that two identical circular-core dual-mode fibers each of radius 'a' interact with each other over a distance 'L'. In order to keep the mathematics simple, it is also assumed that the centre-to-centre distance ('d') between the two fibers remains same throughout the interaction region. The schematic diagram of a dual mode fiber coupler is shown in Fig. 13. Each fiber supports three modes, namely, LP_{01}, LP_{01}^{even} and LP_{01}^{odd} modes (neglecting the polarization degeneracy of the modes). Further, one can neglect the coupling between the LP_{01} mode of one fiber and the LP_{11} mode of the other fiber, which would be negligibly small as compared to the coupling between similar modes. By using coupled mode theory, the coupling coefficients corresponding to the above three modes are given as[28, 29]:

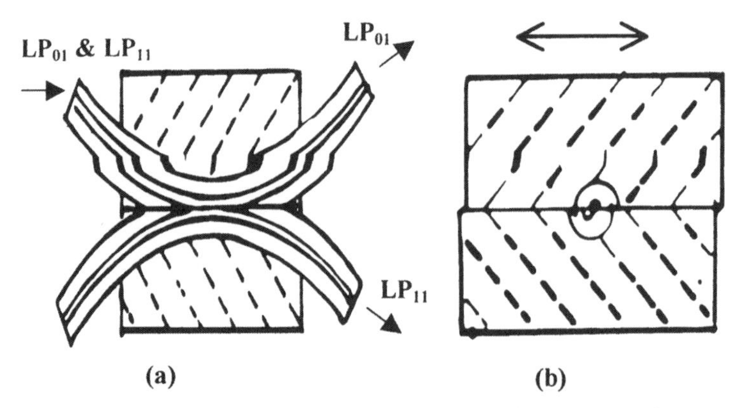

Figure 12 **(a)** Mode splitter consisting of two dual moded SPF blocks. **(b)** Offset between the two SPF blocks.

(i) For LP_{01} to LP_{01} mode coupling,

$$C_{01} = \frac{\sqrt{2\Delta}}{a}\frac{U^2}{V^3}\frac{K_0(Wd/a)}{K_1^2(W)} \tag{23}$$

(ii) For $LP_{01}{}^{odd}$ to $LP_{01}{}^{odd}$ mode coupling,

$$C_{11}^{odd} = \frac{\sqrt{2\Delta}}{a}\frac{U^2}{V^3}\frac{K_2(Wd/a)+K_0(Wd/a)}{K_0(W)K_2(W)} \tag{24}$$

(iii) For $LP_{01}{}^{even}$ to $LP_{01}{}^{even}$ mode coupling,

$$C_{11}^{even} = \frac{\sqrt{2\Delta}}{a}\frac{U^2}{V^3}\frac{K_2(Wd/a)-K_0(Wd/a)}{K_0(W)K_2(W)} \tag{25}$$

where $U = a\sqrt{(k_0^2 n_1^2 - \beta^2)}$, $W = a\sqrt{(\beta^2 - k_0^2 n_2^2)}$,

$V = \sqrt{U^2 + W^2} = k_0 a\sqrt{(n_1^2 - n_2^2)}$, $\Delta = \dfrac{n_1^2 - n_2^2}{2n_1^2}$

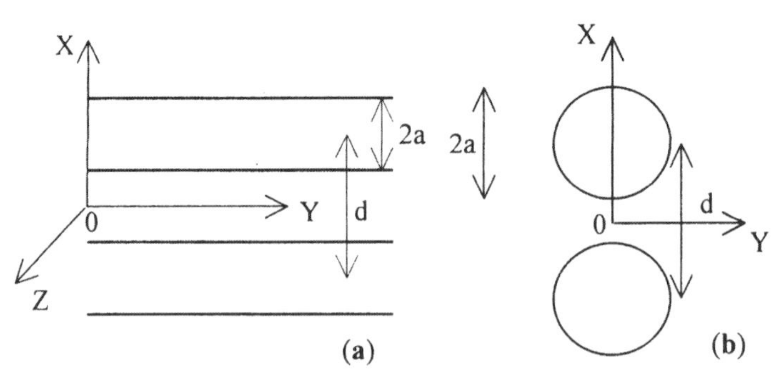

Figure 13 Schematic of a dual mode fiber coupler **(a)** transverse view **(b)** cross-sectional view.

β and k_0 represent the propagation constant of the mode and free space wavenumber, respectively, and n_1, n_2 represent the core and cladding refractive indices. Further, $K_n(x)$ represents the modified Bessel function of order *n*. The corresponding coupling lengths can be obtained using the following equations:

$$(i)\ L_{c0} = \frac{\pi}{2C_{01}} \tag{26}$$

$$(ii)\ L_{cl}^{odd} = \frac{\pi}{2C_{11}^{odd}} \tag{27}$$

$$(iii)\ L_{cl}^{even} = \frac{\pi}{2C_{11}^{even}} \tag{28}$$

The modal powers corresponding to the LP_{01} and LP_{11} modes will be separated at the output ports if the interaction length is such that either of the following conditions is satisfied,

$$L = 2m_1 L_{c0} = (2m_2 + 1) L_{cl}^{even} = (2m_3 + 1) L_{cl}^{odd} \tag{29a}$$

or,

$$L = (2m_1 + 1) L_{c0} = 2m_2 L_{cl}^{even} = 2m_3 L_{cl}^{odd} \tag{29b}$$

where m_1, m_2 and m_3 are integers such that $m_3 < m_2$. It is obvious from the above analysis that for a mode splitter to work, the three different conditions given by Eqs. (29-a or 29-b) should be satisfied simultaneously.

A mode splitter was fabricated by using two side polished dual mode fiber blocks as shown in Fig. 12. The fiber is dual moded at the He-Ne red laser (wavelength λ = 0.6328 μm). Light from a He-Ne red laser was coupled at one of the input ports and the outputs from the two fibers were observed on a screen. By displacing the upper polished block laterally with respect to the lower one, the effective interaction length was changed and at one position of the two blocks, the two modes were found to

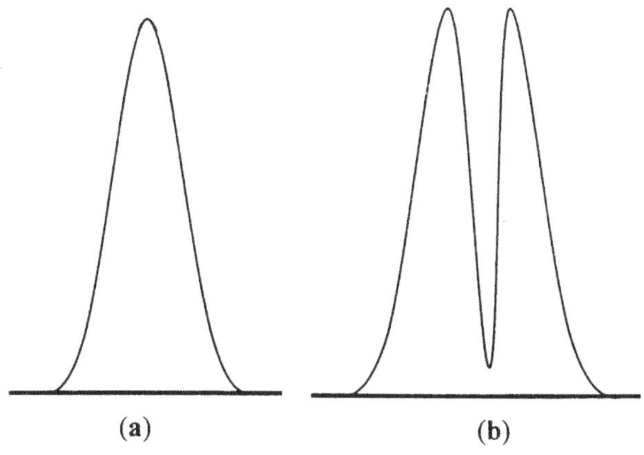

Figure 14 Measured intensity profile of the power coming out from the output ports of a mode splitter.

split at the output ports. The far-field patterns corresponding to the power coming out from both the output ports were scanned by using a moving-detector connected to a x-y recorder. The measured intensity variations corresponding to the two ports are shown in Figs. 14(a) & 14(b)[30]. The intensity pattern shown in Fig. 14(a) corresponds to LP_{01} mode and that shown in Fig. 14(b) corresponds to almost LP_{11} mode. A slight mismatch between two peaks and nonzero value of the central power in Fig. 14(b) corresponds to the presence of a very small amount of the fundamental mode power, which can be removed by fine-tuning.

5. Tunable directional coupler

Some side polished fiber based devices like directional couplers, polarizers, wavelength filters etc. have already been discussed in details[1-3]. However, for completeness, we are discussing these devices briefly. Directional coupler is usually a 4 port device, which is used to split the optical power from one input fiber to two output fibers or to combine two optical powers. A tunable coupler can be made by using two single moded side polished fiber blocks as shown in Fig.15. We may mention that although the geometry of the directional coupler is the same as for the mode splitter shown in Fig. 12 (a), however, here both blocks use a single mode fiber. Depending on the values of the throughput power (P_1) and cross coupled power (P_2), a directional coupler is known as 3 dB coupler ($P_1 = P_2 = P_{in}/2$), parallel switch ($P_1 = P_{in}$ and $P_2 = 0$) or a cross switch ($P_2 = P_{in}$ and $P_1 = 0$). In a directional coupler made of two identical fibers, P_1 and P_2 are given by[1]

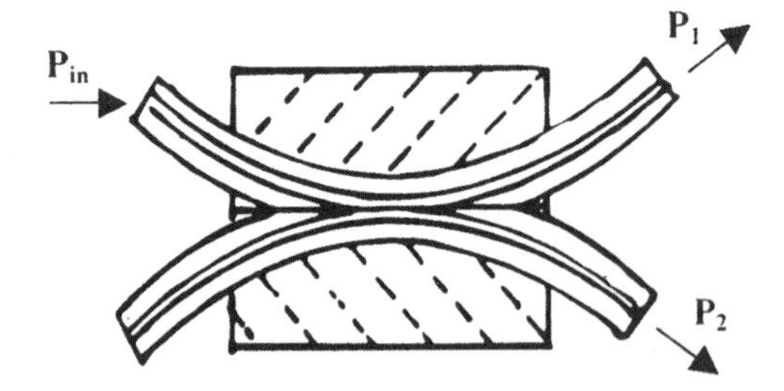

Figure 15 Directional coupler consisting of two SPF blocks.

$$\frac{P_1}{P_{in}} = \cos^2(\kappa L) \quad \& \quad \frac{P_2}{P_{in}} = \sin^2(\kappa L) \tag{30}$$

where L is the interaction length over which coupling/interaction between the two fibers takes place, and κ is the coupling coefficient which is proportional to the overlap integral between the modal fields of the two interacting fibers[1]. Such couplers are also characterized in terms of coupling length $L_c (= \pi/2\kappa)$, which is the minimum coupling length of the coupler at which input power is completely transferred to the second fiber ($P_2 = P_{in}$ and $P_1 = 0$). These parameters (L & κ), and hence P_1 & P_2, can be changed by moving one block with respect to other (as shown in Fig. 12(b)). Thus, any desired power splitting can be achieved by a directional coupler made of two single moded side polished fiber blocks and such couplers are known as tunable directional couplers. Variation of the normalized throughput power (P_1/P_{in}) and cross-coupled power (P_2/P_{in}) of a directional coupler with fiber-offset is shown in Fig. 16. Dots and triangles represent experimental results corresponding to powers (P_1/P_{in}) and (P_2/P_{in}) respectively, while solid line corresponds to a theoretical fit[31]. The device behaves as a cross switch when both fibers are perfectly aligned (i.e., offset is zero). This figure clearly indicates that any desired power ratio (P_1 / P_2) can be achieved by

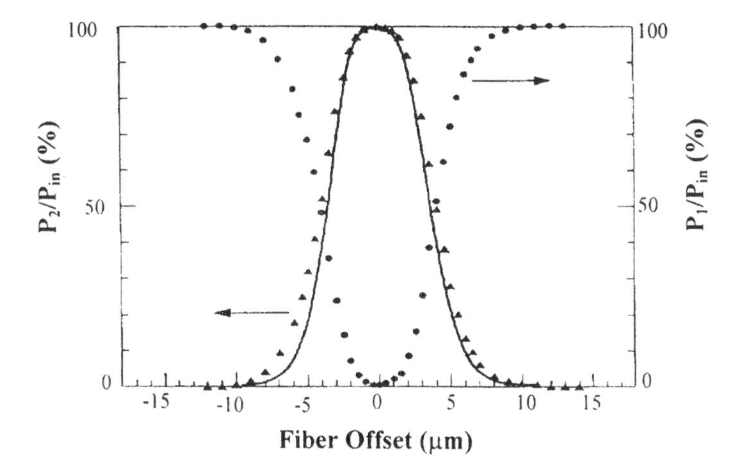

Figure 16 Variation of the output power of the coupler with fiber offset. Dots and triangles represent experimental results corresponding to power P_1/P_{in} and P_2/P_{in} respectively, while solid line corresponds to theoretical fit (adapted from Ref. [31]; © IEEE)

such a coupler by adjusting the offset between two side polished fiber blocks.

6. Polarizers

Fiber optic polarizers are often required for coherent optical communication systems, fiber optic sensors etc. Polarizers can be made by depositing an appropriate material (birefringent crystal or metal) over a SPF block[1-3, 8, 9, 32] described below briefly:

a) ***Birefringent crystal based polarizer***: Such polarizers have been made by using birefringent crystals (like Potassium Pentaborate, lquid crystal) over the SPF block so that the refractive index of the crystal is lower than fiber cladding index for TM polarized light (electric field perpendicular to the polished surface), while for TE polarized light (electric field parallel to the polished surface) the refractive index of the crystal is equal or greater than effective index of the fiber. In this case, loss of the throughput power is negligible for TM polarized light, however, for TE polarized light this loss is very large. Extinction ratio, which is defined as *-10 log (Power of unwanted polarized light/ Power of desired polarized light)*, of such polarizers can be as high as 60 dB[8].

b) ***Thin metal clad polarizer***: Such polarizers are formed by depositing a thin metal film of thickness less than the penetration depth of the mode, and then a thick dielectric layer over the SPF block [1-3, 8,9]. These polarizers are based on the resonance coupling between the SPF mode and the surface plasmon modes supported by the metal dielectric interface. A polarization extinction ratio ~ 60 dB can be obtained in such polrizers[8]. Polarization dependent synchronous coupling between the SPF mode and the mode of the high index overlay waveguide (without metal) is also reported to realize fiber optic polarizers with large polarization extinction ratio ~ 80 dB[9].

7. Channel Drop Filters

Channel drop filters are used to filter out certain wavelengths from the input signal. Such devices can be made by depositing a thin, high refractive index ($n_0 > n_1$ fiber core index) layer over an SPF block, and then, a thick dielectric layer of lower refractive index n_3 ($\leq$ cladding index (n_2) of the fiber). This waveguide behaves as a multimode overlay waveguide (MMOW). Such a SPF-MMOW device configuration is shown schematically in Fig. 6. This device essentially functions as an asymmetric directional coupler, with a channel dropping filter attributed to the wavelength-dependent resonant coupling between the mode of the SPF and one or more modes of the MMOW. The modes of the MMOW have effective indices between n_0 and n_2. Thus, for a SPF-MMOW device, there would be resonance wavelengths at which the effective indices of the fiber mode and one of the modes of the MMOW will become identical. This is known as phase matching of modes. At such wavelengths, the power propagating in the optical fiber can efficiently couple to the overlay waveguide, and there will be a drop in transmitted power of the fiber. At the other wavelengths when the modes are not phase matched, there will be very little coupling of power from the fiber to the overlay waveguide, and the fiber will transmit almost the entire power. A typical spectral response of the throughput power of a SPF-MMOW device is shown in Fig. 9. Hence, resonance wavelengths are filtered out from the output signal of the device[5, 11-13]. Such a SPF-MMOW device has been used as a wavelength filter with line width < 3nm[5]. Tunability of the device can also be achieved by choosing a suitable electro-optic or thermo-optic material for the MMOW.

8. SPF-MMOW device as a Sensor (Refractometer/ Temperature Sensor)

A large number of sensing applications in refractometers, pH sensors, water detectors, temperature sensors etc. have been demonstrated by using the properties of light coupling between a side polished single mode fiber and a plannar multimode overlay waveguide[33-38]. As discussed above, the SPF-MMOW device has a resonant coupling between the side polished single mode fiber and the planar multimode overlay waveguide at certain resonance wavelengths (λ_r), which depend on the parameters of the side polished single mode fiber and multimode overlay waveguide (see, Eqs. (20) – (22)). Hence, variation of the refractive index leads to periodic (resonance like) drops in throughput power as an individual higher-order mode of the overlay waveguide is tuned in and out of resonance with the fiber mode. Values of the resonance wavelengths change on varying the refractive index of the multimode overlay waveguide. A thin layer of the liquid, whose refractive index has to be measured, is introduced over the side polished single mode fiber by using appropriate spacers between the polished surface and a quartz block (as shown in Fig. 6). The refractive index of the liquid can be measured by measuring the shift ($\Delta\lambda_r$) in the resonance wavelength. Thus, the SPF-MMOW device can work as a refractometer. Since, the refractive index of a liquid varies on changing the temperature of the liquid, temperature can also be measured by measuring the shift in the resonance wavelength[38]. In order to study the temperature sensing of the fabricated SPF-MMOW device, the liquid is heated from room temperature upto 70 ^0C. To measure the temperature of the liquid, a digital thermometer is placed in contact with the liquid. Since, the refractive index of the liquid varies inversely with temperature, an increase in temperature induces a decrease in refractive index, which in turn leads to a shift in the resonant positions ($\Delta\lambda_r$) in the spectral output of the fiber. On the spectrum analyzer attention was focused on one particular resonant dip, whose position (λ_r) is noted down at an interval of 1.0 ^0C during the process of cooling. The shift in terms of resonant wavelength ($\Delta\lambda_r$) with temperature is shown in Fig. 17. Circular points shown in the figure corresponds to measured data. This figure indicates that change in λ_r (i.e., $\Delta\lambda_r$) with temperature is almost linear and its slope ($d\Delta\lambda_r / dT$) is $\approx$ 6 $nm/^0C$[38]. Since a wavelength shift of $0.1nm$ in $\Delta\lambda_r$ can be easily measured by the system, we can measure temperature change as low as 0.02 ^0C in the temperature range of 23-60 ^0C.

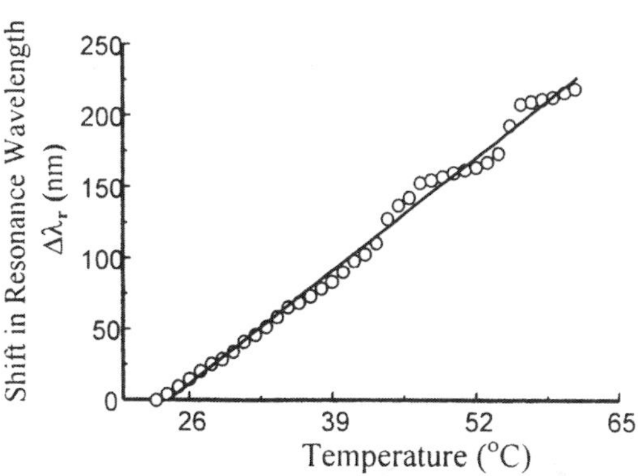

Figure 17 Shift in resonance wavelength with temperature.

9. Summary

In this chapter, we have discussed fabrication and characterization of side-polished fiber blocks. Many fiber-optic devices, such as mode splitters, tunable directional couplers, temperature sensors and those for gain flattening of EDFA, which are based on interaction between the single mode fiber and the multimode planar overlay waveguide have also been discussed.

Acknowledgements: Author is thankful to Professors Arun Kumar, I.C. Goyal, K, Thyagarajan, B.P. Pal and Anurag Sharma, and to DR. M.R. Shenoy for their encouragement and interest in the work.

References

1. Ajoy Ghatak and K. Thygarajan, *Introduction to Fiber Optics* (Cambridge University Press, Cambridge, 1998).
2. B. P. Pal (Ed), *Fundamental of Fiber Optics in Telecommunication and Sensor Systems* (Wiley Eastern Limited, New Delhi, 1994).
3. W. Johnstone, "In line Fiber Optics Components", Ch7, in *Recent Trends in Fiber Optics and Optical Communication* by A.K. Ghatak, B. Culshaw, V. Nagarajan and B. Khurana (Ed.) (Viva, New Delhi, 1995).
4. M. J. F.Digonnet, J.R. Feth, L.F.Stokes and H.J. Shaw, "Measurement of the core proximity in polished fiber substrates and couplers", *Optics Lett.* **10**, 463-465 (1985).

5. W. Johnstone, G. Thursby, D. Moodie, R.K. Varshney and B. Culshaw, "Fiber optic wavelength channel selector with high resolution", *Elect. Lett.* **28**, 1364-1365 (1992).
6. B.P. Pal, G. Raizada and R.K. Varshney, "Modelling a fiber half block with multimode overlay waveguide", *J. Optical Commui.* **17**, 179-183 (1996).
7. G. Raizada, B.P. Pal and R.K. Varshney, "Estimating performance of fiber optic modulators/switches with multimode electro-optic overlay/interlay waveguide", *Optical Fiber Techn.* **2**, 89-97(1996).
8. S.P. Ma and S.M. Tseng, "High performance side polished fibers and applications as liquid crystal clad fiber polarizers", *J. Lightwave Tech.* **15**, 1554-1558 (1997).
9. A. Kumar, R. Jindal, R. K. Varshney and R. Kashyap, "Fiber optical polariser using resonant tunneling", *Optical Fiber Technology* **3**, 339-346 (1997).
10. A. Kumar, R. K. Varshney and G.C. Mishra,, "Fabrication of a single to dual mode fiber coupler for realising two mode fiber couplers", *SPIE Proc. on selected papers from International conference on 'Fiber Optics & Photonics'98'* **3666**, 62-66 (1998).
11. K. R. Sohn and J. W. Song, "Thermo-optically tunable side polished fiber comb filter and its application", *IEEE Photonics tech. lett.* **14**, 1575-1577 (2002).
12. K. T. Kim, S. Hwangbo, G.I. Kweon and S. R. Choi, "Wide tunable filter on side polished polarization maintaining fiber coupled with thermo-optic polymer overlay", *Elect. lett.* **40**, 1330-1331 (2004).
13. N. K. Chen, S. Chi and S. M. Tseng, "Wideband tunable fiber short-pass filter based on side polished fiber with dispersive polymer overlay", *Opt. lett.* **29**, 2219-2221 (2004).
14. Y. Kumar, *Studies on side polished fiber blocks and their application as fiber polarizers*, (M. Sc., IIT Delhi, 2003).
15. V.K. Singh, *Gain-flattening of Erbium doped fiber amplifier using side polished fiber half block technology*, (M. Tech. (Opto-electronics), IIT Delhi, 2002).
16. Desurvire, E., *Erbium Doped Fiber amplifiers: Principles and applications*, (Jhon Wiley and Sons, 1994).
17. R.K. Varshney, V.K. Singh, K. Pande, M.R. Shenoy, B.P. Pal and K. Thyagarajan, "EDFA gain-flattening using side-polished fiber half-coupler with an overlay waveguide", *8th Opto-electoronics and Communications Conference (OECC), Shanghai (China)* **23**, 243-244 (Oct. 13-16, 2003).
18. A. Sharma, J. Kompella and P. K. Mishra, "Analysis of fiber directional coupler half blocks using a new simple model for single mode fiber", *J. Lightwave Technology* **8**, 143 (1990).
19. A.K. Ghatak, K. Thygarajan and M.R. Shenoy, "Numerical analysis of planar optical waveguides using matrix method", *J. Lightwave Technology* **5**, 660 (1987).
20. A.K. Ghatak, I.C. Goyal and R. K. Varshney, *Fiber Optica* (Viva Publications, New Delhi, 1999).
21. I.C. Goyal, R. K. Varshney and A.K. Ghatak, "Design of a small residual dispersion fiber (SRDF) and a corresponding dispersion compensating fiber (DCF) for DWDM systems", *Optical Engin.* **42**, 977-980 (2003).

22. A.M. Vengsarkar, J.A. Greene, and K.A. Murphy, "Photo-induced refractive-index changes in two-mode elliptical-core fibers: sensing applications", *Opt. Lett.* **16**, 1541 (1991).
23. K. Bohnert and P. Peruignot, "Inherent temperature compensation of a dual-mode fiber voltage sensor with coherence-tuned interrogation", *J. Lightwave Technol.* **16**, 598 (1998).
24. C.D. Poole, C.D. Townsend, and K.T. Nelson, "Helical-grating two-mode fiber spatial-mode coupler", *J. Lightwave Technol.* **9**, 598 (1991).
25. H.G. Park and B.Y. Kim, "Intermodal coupling using permanently photo-induced grating in two-mode optical fiber", *Electron. Lett.* **25**, 797 (1989).
26. B.Y. Kim, J.N. Blake. H.E. Engan, and H.J. Shaw, "All-fiber acoustic-optic frequency shifter", *Opt. Lett.* **11**, 389 (1986).
27. A.J. Antos and D.K. Smith, "Design and characterization of dispersion compensating fiber based on the LP_{01} mode." *J. Lightwave Technol.* **12**, 1739 (1994).
28. K. Ogawa, "Simplified theory of the multimode fiber coupler", *J. Bell Sys. Tech.* **56**, 729 (1977).
29. Arun Kumar, R. K. Varshney and Siny Antony C., "Coupling characteristics of fiber optic couplers consisting of two circular core dual mode fibers", *J. Optical Soc. India*, In-press.
30. G. C. Mishra, *Fabrication and characterization of a mode splitter and a mode combiner for realizing two mode fiber sensors*, (M. Tech. (Opto-electronics), IIT Delhi, 1997).
31. M. J. F.Digonnet and H.J. Shaw, "Analysis of a tunable single mode optical fiber coupler", *J. Quant Elect.* **QE-18**, 746-754 (1982).
32. K. Thygarajan, S. Diggavi and A. K. Ghatak, W. Johnstone, G. Stewart and B. Culshaw, "Thin metal clad waveguide polarizers: analysis and comparison with experiment", *Optics Letters* **15**, 1041(1990).
33. W. Johnstone, G. Fawcett and L.W.K. Yim, "Inline fiber optic refractometry using index sensitive resonance position in single mode fiber to planner waveguide coupler", *IEE Proc., Optoelectron.* **141**, 299-302 (1994).
34. G. Raizada and B.P. Pal, "Refractometers and tunable components based on side polished fibers with multimode overlay waveguides: role of the superstrate" *Optics Letters* **21**, 399-401 (1996).
35. D. Flannery, S.W. James, R.P.Tatam and G.J. Ashwell, "pH sensors using Langmuir Blodgett overlay on polished optical fibers", *Optics Lett.* **15**, 567-569 (1997).
36. K.R. Sohn, K. T. Kim and S. W. Kang, "Optical fiber sensor for water detection using side polished fiber with a planar glass overlay waveguide", *Sensor Act.-A*, **101**, 137-142 (2002).
37. W. G. Jung, S. W. Kim, K. T. Kim, E. S. Kim and S. W. Kang, "High sensitivity temperature sensor using a side polished single mode fiber covered with the polymer planar waveguide", *IEEE Photonics Tech. Lett.* **13**, 1209-1211 (2001).
38. R. K. Varshney, H. S. Pattanaik and B. P. Pal, "Temperature sensor based on side polished fiber half-coupler", *International Conference on Laser Applications and Optical Metrology, IIT Delhi*, 460-463 (Dec. 1- 4, 2003).

8

Fused Fiber Coupler-Based Components for Optical Fiber Communication

M.R. Shenoy and B. P. Pal

1. Introduction

With the availability of low loss silica fibers at the 1310 nm and 1550 nm wavelength windows, optical fiber communication technology has witnessed a tremendous growth as an area of R & D with a broad range of applications. Optical fibers have already penetrated the local loops and subscriber premises in the access networks. CATV and other local networks like computer communication and data transmission. Most of these applications require a host of *branching components,* which have two or more ports that distribute optical signal among fibers. Some of these components are passive, data-format transparent, and are able to combine/split optical power or wavelengths in some pre-determined ratio (coupling ratio), regardless of the information content of the signal. Very often it is desirable to have the component in an all-fiber form. All-fiber components are also, alternatively, referred to in the literature as *in-line fiber components*. By far, the most important and widely used all-fiber component is a single-mode fiber directional coupler.

In general, a fiber directional coupler may have more than one input/output terminals in the form of fibers. In such fiber-based in-line branching components, the necessary functions of manipulation/processing of optical signal is performed whilst the signal is still guided by the fiber. It has the advantage that the component(s) can be readily spliced to the signal-carrying fiber and it can be readily installed into an optical fiber circuit with standard fiber-handling tools.[1] Fiber-based components, by virtue of their all-fiber nature, are directly compatible with the fiber networks and systems; thus, fiber circuits (which consist of fiber components interconnected by lengths of fiber) can be developed, which have certain advantages over the competing technologies of *integrated optics* and *bulk optics*. Many of the in-line fiber components are designed around three foundation technologies – side-polished fiber half-coupler,[2] fused biconical tapered (FBT) fiber coupler[3], and monolithic fiber coupler technology.[4] Amongst these, the FBT fiber coupler technology is a versatile, inexpensive and widely accepted technology for realising a large number of branching components.[5] The basic device in the FBT coupler technology is a 2×2 *fused fiber coupler*, which can be used to form power splitter/combiners, wavelength division multiplexers and demultiplexers, band-pass/ band-stop filters, broadband and wavelength selective couplers, polarization splitters, signal taps, etc. Here we briefly describe the basic structure, fabrication and applications of some of the devices based on the FBT fiber coupler technology that are relevant to optical fiber communication. In particular, two specific devices namely all-fiber Mach-Zehnder interferometer (MZI) based *wavelength interleaver* and *fiber loop reflector* have been discussed in detail. These are among the recent developments in all-fiber components that use FBT fiber couplers.

2. Fabrication of FBT fiber couplers

FBT fiber coupler components are usually realized in the form of 2×2, 1×3, 3×3, and 1×2 port devices. A 2×2 coupler is the basic building block for a host of higher order arrays and other components. The fabrication process involves (see Fig.1) stretching a pair of single-mode fibers together, which are held in intimate contact across a short unjacketed length, in a high temperature flame (or electrical furnace). This process of heating and stretching results in narrowing of the two fibers into a single biconical tapered junction.[6] The tapered region essentially transforms to a multimode near-rectangular waveguide with its core comprising of the original fiber cladding material, surrounded by air as the cladding.[7] Light from the input fiber excites predominantly the two lowest order *supermodes* of the composite guide. Beating between these supermodes with propagation decides the power distribution, and eventual coupling of light into the two output fibers.[7] Since the basic process in this fabrication technique consists of fusion, pulling and simultaneously tapering, the process is also referred to as *fuse-pull-taper* method. Since the input and output tapers are identical, either side may be used as the 'input end', for which these components are also called *bi-directional* couplers. During the 'fusion and elongation', light is injected into one of the fiber input ports, and the light exiting from the two fiber output ports are monitored (see Fig.1). On attaining the targeted distribution of light between the outputs, the heat source is withdrawn and the pulling is stopped. The device is then appropriately packaged and subjected to

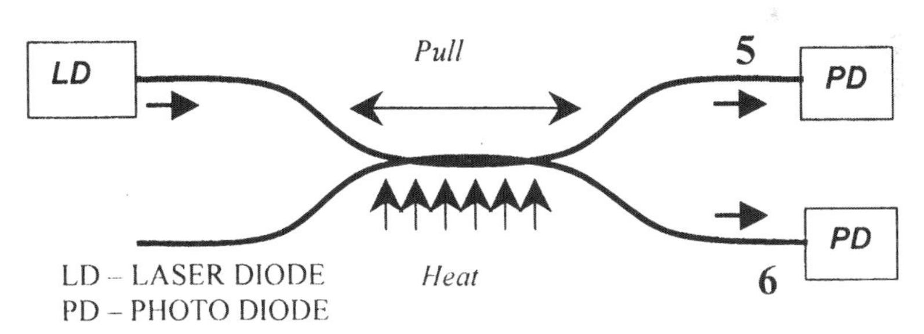

Figure 1 Schematic of the basic fabrication technique of FBT fiber coupler: the fuse-pull-taper method.

several environmental tests [*Generic requirements, Bellcore specifications* (1998)], before being labeled as a commercial product.

The coupling behaviour of a fused single-mode coupler depends on the interaction length and the operating wavelength. This fact has been readily exploited to tailor the characteristics of an FBT fiber coupler for realizing various application-specific devices. Such design tailoring requires a thorough understanding of the wavelength dependence of the splitting ratio, and its optimization by controlling the factors that influence it.[8] The rate of change of spectral splitting ratio i.e., how fast or slow the power transfer varies with wavelength, determines the wavelength period of power-transfer oscillations. On the other hand, at a given monitoring wavelength of the input source, the light power distribution at the output ports vary during the fuse-pull-taper process due to the increasing interaction between the two fibers. A plot of the variation of throughput power as a function of the pulled length of the fiber, called *pulling signature*, indicates the extent of pulling that is necessary to realize a specific coupler-based component. Knowledge of the pulling signature, for a given set of fabrication parameters, provides the necessary guidelines for controlling the taper shape, and subsequent signature of the desired wavelength periodicity.[9] Control and optimization of spectral splitting ratio of a fused coupler can yield a number of in-line components such as power splitter, wavelength division multiplexer, wavelength flattened coupler, bandpass and bandstop filters. As mentioned above, fiber loop reflectors, and wavelength interleavers form another set of important devices that are realizable through the FBT coupler technology. Functional principles of operation of some of these devices are briefly discussed in the following sections.

3. FBT coupler based beam splitter/combiner

This is the most common class of FBT coupler-based branching components, which are widely used in optical networks. A splitter/combiner is a symmetric coupler in which the constituent fibers are identical so as to achieve, if required, a 100% power transfer. The commonly used components of this species are 3-dB couplers, tap/access couplers, tree couplers, etc.

- **3-dB coupler**

A 3-dB coupler is a 2×2 symmetric fiber coupler, which equally splits input optical power at a designed wavelength between the throughput (T)- and the coupled (C) ports; the effective interaction length of the device is $L_C/2$, where L_C is the coupling length.[10] The 50:50 splitting of input power at a given wavelength (say, 1310 or 1550 nm) is ensured by real-time monitoring of powers from the T- and C-ports during the fabrication. These couplers find extensive use in optical telecommunication networks, test equipments, optical fiber sensors (e.g. interferometric fiber sensors[11-12]) and CATV networks. These are reciprocal devices, and could be used to either split or combine signals.

- **Tap/access coupler**

A tap coupler is essentially a beam splitter with a desired splitting ratio, usually in the range of 5 to 50%. Such a branching component permits passive add/drop of a signal at an optical network node. These are widely used as power-taps for monitoring stability of an optical source, optimizing power budgets in optical networks, accessing Rayleigh back-scattered signal from the test fiber in an OTDR, monitoring signal level status in fiber transmission links and Erbium doped fiber amplifiers (EDFAs)[10].

- **Tree coupler**

This is a bi-directional branching component, configurable through integration of a number of FBT couplers, which distributes the optical signal from one fiber to several others ($1\times n$ ports) or combines signals from several fibers into one (see Fig.2). Typically, it splits input power equally to all output ports. The number of

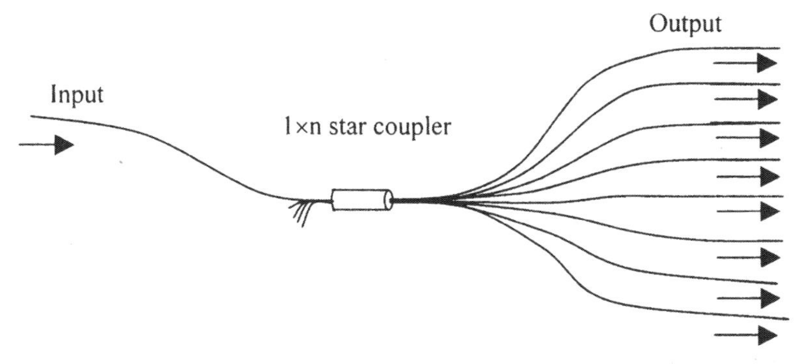

Figure 2 Schematic of fused fiber coupler-based $1\times n$ star coupler having one input and several outputs.

output ports could range from 2 to 32. However, this configuration in an all-fiber form is ideal for a lower order array (for $n \leq 8$). A tree coupler is also realized by fabricating 3-dB couplers on each of the T- and C- arms of the previous 3-dB coupler or through splicing of 3-dB couplers in tandem (see Fig.3).

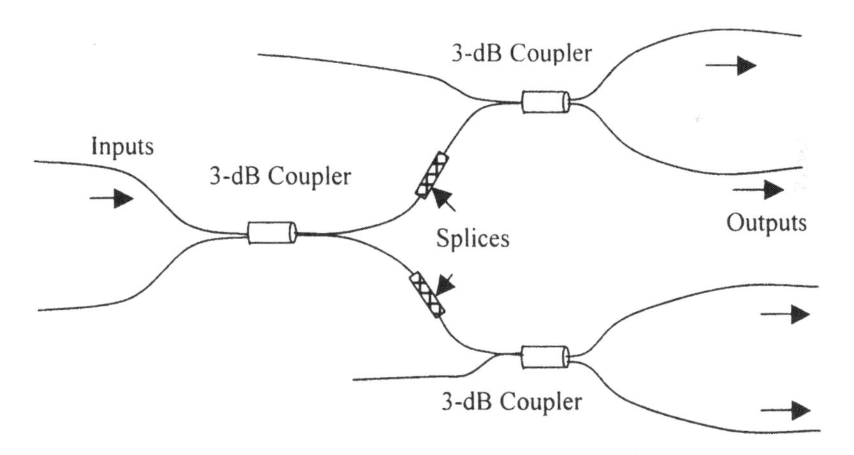

Figure 3 Schematic of a 1×4 power splitter configured through three 3-dB couplers.

4. Wavelength division multiplexer (WDM)

A WDM is also a symmetric 2×2 FBT fiber coupler in which two inputs at two different wavelengths, say λ_1 and λ_2, from the two input ports, are combined at one output port. This is a reciprocal device (see Fig.4). Such a design owes its origin to the fact that, for a given coupler, the coupling coefficients and the effective lengths of interaction at two different wavelengths, say 1310 and 1550 nm, are different. Therefore, the splitting ratios at these wavelengths are different. For a coupler to function as a WDM at these two operating wavelengths, the fabrication process has to be tailored such that the splitting ratio of the device is a maximum at one of these wavelengths and a minimum at the other.[13] This implies that all the input power at λ_1 will emerge at one output port, and all the input power at λ_2 will emerge at the second output port. A WDM, which is used for telecommunication applications, is the 1310/1550 nm WDM, which is also referred to as 'classical WDM' in order to distinguish it from dense-WDMs (DWDMs).[14] The wavelength 1310 nm is usually used to transmit telephones/data signals while 1550 nm is used to transmit TV/video signals.

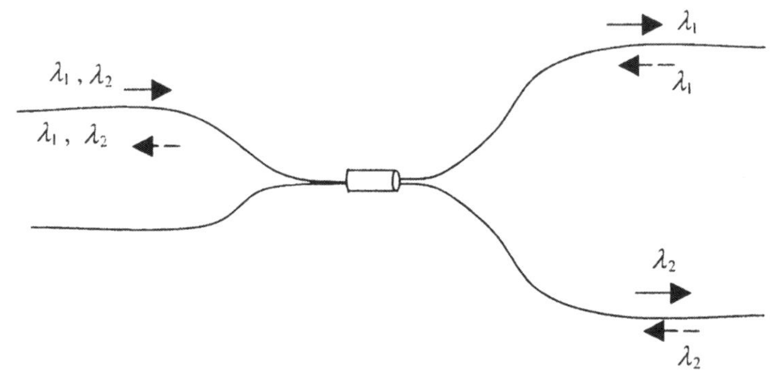

Figure 4 Schematic of a WDM at wavelengths λ_1 and λ_2 designed around 2×2 fused coupler.

The two specific wavelengths at which a WDM is required to operate are called the *channels*. When configured around a 2×2 FBT fiber coupler, the separation between the two wavelengths, known as *channel spacing*, requires the coupler to exhibit a specific wavelength response. Based on the channel spacing, classical WDM are classified as *narrow channel spacing* (< 100 nm) and *wide channel spacing* devices. Wide channel-spacing WDMs are usually characterized by a larger bandwidth owing to a relatively large wavelength-period in their spectral response. A very important wide channel-spacing WDM is the 980/1550 nm WDM, which is used in EDFAs, where the pump and signal at these two wavelengths are multiplexed in the amplifying (Erbium doped) fiber.[15] An alternative WDM to combine pump and signal in EDFAs is the 1480/1550 nm WDM, which is an example of a narrow channel WDM.[16] In general, for the coupler to function as a WDM of narrow channel-spacing, one is required to go through a relatively larger number of power-coupling cycles. Consequently, these couplers may exhibit a slightly higher excess loss and stronger polarization dependence.[17-18]

5. Wavelength Interleaver

All-fiber interferometric devices in which FBT coupler is the basic component[19] have been developed for a variety of sensing applications.[11] They could also be used as wavelength filters.[19] In this section we will discuss the concept of wavelength interleaving based on all-fiber Mach-Zehnder interferometer in detail. We begin with a brief description on the relevance of wavelength interleaving in DWDM systems.

A key figure of merit for a DWDM optical transmission system is known as the *spectral efficiency*, which is defined as the ratio of the bit rate per channel to the channel spacing,; and it is expressed in bits/s/Hz. Due to the well-known difficulties encountered in attaining high-speed electronic components at speeds beyond 40 Gbit/s, a decrease in the channel spacing has evolved as the best near-term option to achieve high spectral efficiency in a DWDM link. In contrast to classical WDM, DWDM involves simultaneous transmission of at least 4 wavelengths in the EDFA bands (e.g., C-band: 1530-1565 nm, L-band: 1570-1610 nm); typically, number of wavelengths could be 8, 16, 32, 64, and so on. In order to avoid haphazard growth of DWDM fiber links, International Telecommunication Union (ITU) has set certain wavelength standards, referred to as ITU wavelength grids, for DWDM optical transmission systems. As per ITU standards, the reference frequency (ν_0) is chosen to be that corresponding to the Krypton line i.e. 193.1 THz. (≡ 1552.52 nm in wavelength), and the chosen channel spacings are expected to follow the relation: $\nu_0 \pm \Delta\nu\,(\text{THz.}) = 193.1 \pm 0.1I$, I is an integer. The recommended standard channel spacings are 200 GHz (≡ 1.6 nm), 100 GHz (≡ 0.8 nm), and 50 GHz (≡ 0.4 nm). As the demand for more and more bandwidth grows, the system designers are expected to pack-in more and more wavelength channels within the amplifier band. This would necessarily lead to smaller channel spacings, which would entail tighter specifications for wavelength selective components like MUX/DMUX and wavelength filters.

A *wavelength interleaver* is a device that combines two input stream of wavelength-channels having a constant spacing in the frequency domain, $\Delta\nu$ (e.g. as per ITU grid), into a single dense stream of channels with separation $\Delta\nu/2$ at the output (see Fig.5). The device being reciprocal[20], it functions as a *wavelength slicer* when operated in reverse direction, thereby greatly relaxing the tight tolerance otherwise required on detection of associated DWDM signals, since the adjacent

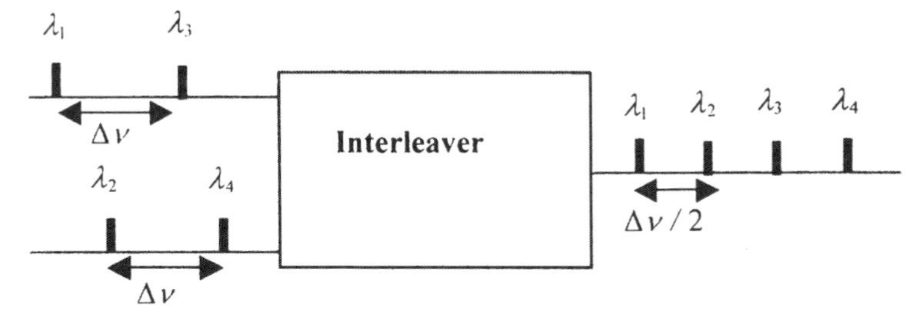

Figure 5 Schematic representation of a wavelength interleaver which interleaves the input channels with spacing in frequency domain $\Delta\nu$ into a densely packed output having inter channel spacing $\Delta\nu/2$; here $\lambda_1, \lambda_2, \lambda_3, \lambda_4$...... represent the centre wavelengths corresponding to channels 1, 2, 3, 4 and so on.

channels would then be spaced apart by double the original frequency spacing. Out of the few alternative routes followed to realize a wavelength interleaver, the one based on unbalanced MZI is perhaps the most popular,[21-23] and it can be realized either in an integrated optic form[24] or in an all-fiber form.[22-23]

5.1 All-fiber MZI based wavelength interleaver

In an all-fiber form, a single-stage unbalanced MZI can be formed by joining two 2×2 fiber couplers in such a way that the lengths of the two arms are slightly unequal by an amount ΔL.[23] A schematic of an all-fiber MZI is shown in Fig.6. The differential path length (ΔL) is equivalent to a delay line in one of the paths. The corresponding differential phase delay $\Delta\varphi$ suffered by any wavelength λ (while traversing ΔL) is given by

$$\Delta\varphi = 2\pi n_{eff}\,\Delta L / \lambda \tag{1}$$

where n_{eff} is the effective index ($= \beta / k_0$) of the guided mode propagating through the fiber. The transmission characteristics of an interleaver based on Mach-Zehnder interferometer can easily be described in terms of transfer matrices of individual couplers and delay lines. From the coupled mode equations for the electric field amplitudes, the transfer matrix for the coupler of splitting ratio $\sin^2(\kappa z)$ has the form:[19]

$$M_{coupler} = \begin{pmatrix} c & -js \\ -js & c \end{pmatrix} \tag{2}$$

where $c = \cos(\kappa z)$, and $s = \sin(\kappa z)$; κ is the coupling coefficient.[10] The transfer matrix corresponding to the differential delay between the two arms is given by

$$M_{\Delta L} = \begin{pmatrix} e^{j\Delta\varphi} & 0 \\ 0 & 1 \end{pmatrix} \tag{3}$$

Thus, the transfer function of an unbalanced MZI (shown in Fig.6) is given by

$$M_{MZI} = M_{coupler2}\, M_{\Delta L}\, M_{coupler1} \tag{4}$$

If E_1 and E_2 are input fields to the MZI at Ports 1 and 2, respectively, then the output fields E_T and E_C at Ports 3 and 4 can be expressed using Eq.(4) as:

$$\begin{pmatrix} E_T \\ E_C \end{pmatrix} = M_{MZI} \begin{pmatrix} E_1 \\ E_2 \end{pmatrix} \tag{5}$$

Thus, if $P_1(\lambda_1)$ and $P_2(\lambda_2)$ represent powers at the two input ports corresponding to wavelengths λ_1 and λ_2, respectively, then the power at the output ports are given by

$$P_T = P_1(\lambda_1)\sin^2\frac{\Delta\varphi(\lambda_1)}{2} + P_2(\lambda_2)\cos^2\frac{\Delta\varphi(\lambda_2)}{2} \tag{6}$$

$$P_C = P_1(\lambda_1)\cos^2\frac{\Delta\varphi(\lambda_1)}{2} + P_2(\lambda_2)\sin^2\frac{\Delta\varphi(\lambda_2)}{2} \tag{7}$$

Here, we have assumed that the splitting ratios of the couplers 1 and 2 in the MZI configuration (see Fig.6) are 50:50. In order to achieve wavelength-interleaving using an unbalanced MZI configuration, Eqs.(6) and (7) have to satisfy following condition:

$$\Delta\varphi(\lambda_1) = (2n+1)\pi \text{ and } \Delta\varphi(\lambda_2) = 2n\pi \tag{8}$$

or

$$\Delta\varphi(\lambda_1) = 2n\pi \text{ and } \Delta\varphi(\lambda_2) = (2n+1)\pi \tag{9}$$

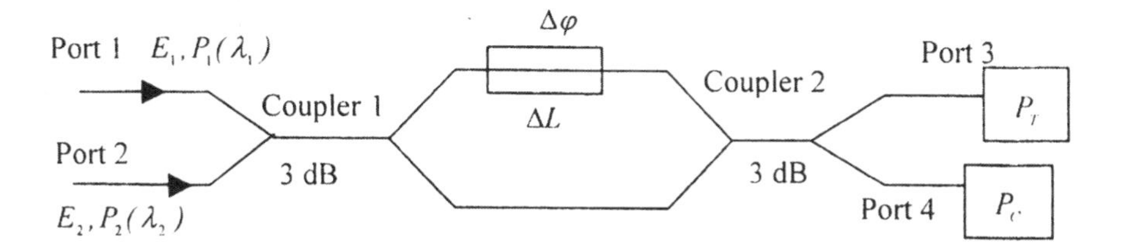

Figure 6 Schematic of an all-fiber Mach-Zehnder interferometer based interleaver through concatenation of two 3 dB couplers.

where n is an integer. From Eqs.(8) and (9), we can observe that $|\Delta\varphi(\lambda_1) - \Delta\varphi(\lambda_2)| = \pi$. Thus, substituting the values of $\Delta\varphi(\lambda_1)$ and $\Delta\varphi(\lambda_2)$ from Eq.(1), we get

$$\Delta L = \frac{\lambda_1 \lambda_2}{2 n_{eff} \Delta\lambda}. \tag{10}$$

For a typical single-mode fiber with $V = 2.1$, $n_{eff} = 1.445$ and $\Delta\lambda = 0.4$ nm ($\Delta\nu = 50$ GHz) around 1550nm, the required path difference ΔL between the two arms of the MZI is ≈ 2.1 mm. If the DWDM signal channels are input at Port 1, the signal wavelengths that suffer a differential phase delay of $(2n+1)\pi$, will exit through Port 3, and the wavelengths that suffer a differential phase delay of $2n\pi$ will exit from Port 4. These wavelengths correspond to the peaks in the spectral response. Such a configuration could separate/interleave the wavelength channels with a sinusoidal response function. The wavelength separation between the consecutive peaks at any output port is known as the *free spectral range* (*FSR*) of the configuration. The spectral response of such a configuration with an *FSR* of 25 GHz is shown in Fig.7. If we change ΔL by an amount $\lambda / 2n_{eff}$ (which introduces an additional phase change of π), the spectral response at the output ports gets reversed, as can be observed from Fig. 8.

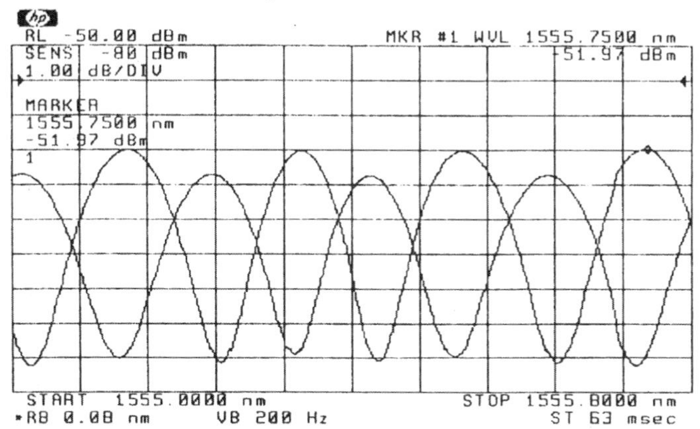

Figure 7 Measured spectral response of the all-fiber wavelength interleaver, tuned for an FSR of 25 GHz, based on single stage MZI; the marker at one of the peaks corresponds to $\lambda = 1555.75$ nm.

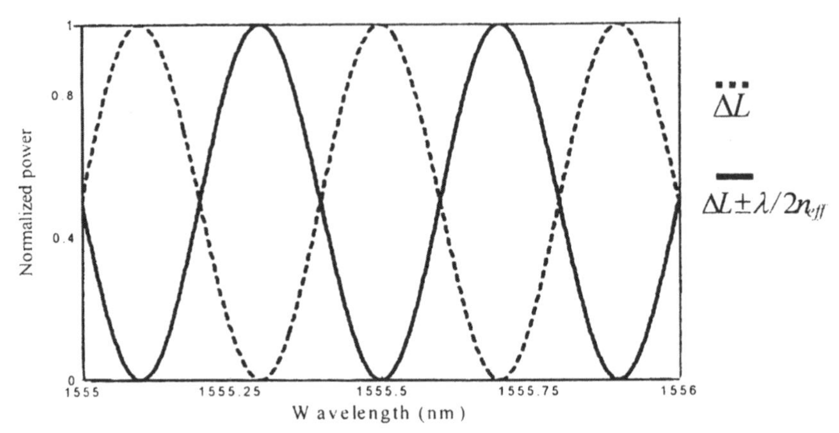

Figure 8 Typical simulated spectral response at Port 3 of the MZI based wavelength interleaver when the differential delays are ΔL and $\Delta L \pm \lambda / 2n_{eff}$.

In practical communication systems, one requires a flattop response around the central passband frequency of each channel. When evaluating interleavers, the most important parameters examined are the insertion losses, passband shape, and cross talk performance for the desired channel spacing. A uniform flattop passband besides lower insertion losses is also very desirable to minimize variations in the signal power as the source wavelength drifts. Any initial imbalance in channel power will cascade as the signal passes through multiple amplifiers and affect the signal to noise performance of the system. Often a compromise is made between achieving a flattop passband and the insertion loss, since additional filtering elements are required to achieve a uniform flattop response from a sinusoidal passband shape. To meet the requirement of a flattop response, two-stage MZI configuration has been proposed.[24] A two-stage MZI configuration consists of cascading three couplers in tandem with two delay lines in their arms; schematic of such a configuration is shown in Fig.9. A typical simulated spectral response of this configuration with optimized values[22] of

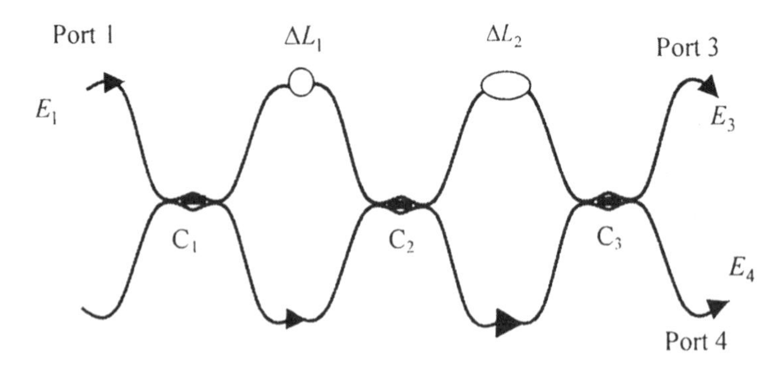

Figure 9 Schematic of an all-fiber two-stage MZI configuration used as wavelength interleaver with flattop response; ΔL_1 and ΔL_2 denotes the two delay lines, and C_1, C_2, and C_3 represent three couplers.

splitting ratios of the couplers and the delay lines is shown in Fig. 10. The flattop region has a maximum loss of 0.1 dB within the 46 dB isolation bandwidth of 30 GHz, when the adjacent channel spacing is 100 GHz as shown in the Fig.10.

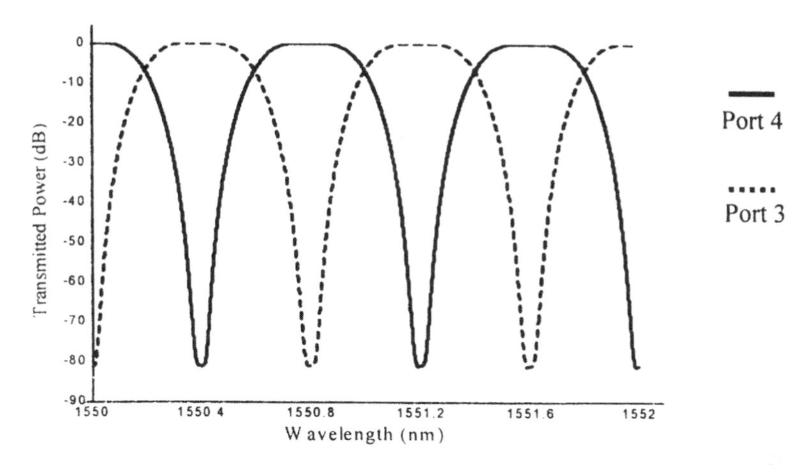

Figure 10 Typical flattop spectral response of a two-stage MZI configuration; the optimized splitting ratios[22] of couplers C_1, C_2, and C_3 are 50:50, 68:32, 4:96, respectively.

6. All-fiber loop reflector

All-fiber loop reflector, also called fiber loop mirror (FLM), is generally formed by joining/splicing the two output ports of a 3 dB fiber coupler and the schematic is shown in Fig. 11. Fiber loop reflectors find applications as a resonant cavity in a fiber laser, passive fiber Fabry-Perot devices, and also in duplex transmission along a single fiber with a light source at only one end of the link.[25-26] It has been exploited in *soliton switching* with a fiber nonlinear mirror,[27] in ultrafast all-optical demultiplexing,[28] in all optical demultiplexing of TDM data at 250 Gb/s,[29] as

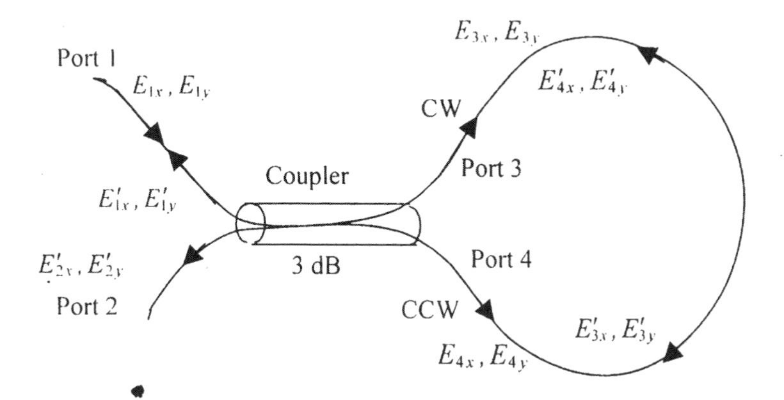

Figure 11 Schematic of a fiber loop reflector showing amplitude and phase distributions along various paths.

nonlinear amplifying loop mirror,[30] as an all-optical loop mirror switch employing an asymmetric amplifier/attenuator configuration,[31] and so on.

6.1 Theoretical background

How a fiber loop reflector functions as a mirror can be understood from the following analysis with respect to Fig.11. Let the input light be injected into Port-1 of the coupler. At the coupler, half of the input power couples into Port-4 so that it propagates along the fiber loop in a counter clockwise (ccw) direction; the other half remains in Port-3, and it propagates through the fiber loop in the clockwise (cw) direction. Thus, 50% of the input light propagates cw and the rest 50% propagates ccw around the loop (Fig.11). The relative phase difference between these two counter-propagating signals at the end of propagation through this loop is $\pi/2$ (which is the phase difference introduced by the coupler), the common phase experienced within the loop being independent of direction. However, each of these cw and ccw signals again encounters the 3-dB coupler and once again undergoes a 3-dB split on propagation through the coupler. This induces an additional phase shift of $\pi/2$ between the two signals, as a consequence of which they add in phase at Port-1, leading to retrieval of the entire power at the input port. The loop is thus said to function like a perfect mirror and hence its name 'fiber loop mirror.' On the other hand, signal components emerging from Port-2 are out of phase by π and cancel each other. In other words, a loop with an ideal 3-dB coupler behaves as a perfectly reflecting mirror. Indeed, from the coupled mode theory, it can be shown that the reflected field intensity towards Port-1 is

$$|R|^2 = \sin^2 2\kappa z \,. \tag{11}$$

Similarly, the transmitted intensity towards Port-2 is given by

$$|T|^2 = \cos^2 2\kappa z \,, \tag{12}$$

where κ is the coupling coefficient. Thus for a 3-dB coupler since $L = \pi/4\kappa$, we obtain $|R|^2 = 1$ and $|T|^2 = 0$, which is consistent with the energy conservation because the device is otherwise assumed to be loss-less.

In an FLM configuration, polarization plays an important role because some amount of inherent birefringence is induced in the fiber loop due to the presence of bends and twist etc. In order to take into account the effect of this birefringence, we assume the presence of a hypothetical wave plate[26,32] at some position in the loop. The amount of birefringence introduced by the fiber loop is equivalent to that of this wave plate[26,32] with retardation ϕ and oriented at an angle θ, and depends on the orientation of the plane of the fiber loop. This birefringence modulates the phases of the fields traveling round the loop in opposite direction. The input light is totally or partially reflected and transmitted, depending on the nature of interference between the clockwise and counterclockwise propagating fields through the loop. Thus, rotating the wave plate can alter the reflectance and the transmittance of the FLM. Mortimore[26] and Morishita[32] have reported detailed mathematical formulation to study the effect of rotating the plane of the loop on the reflectance/ transmittance of an FLM. A brief mathematical description is given below:

Using the coupled–mode theory, the output field amplitudes $E_{3x,y}$ and $E_{4x,y}$ of a directional coupler is expressed as[22]

$$\begin{pmatrix} E_{3x,y} \\ E_{4x,y} \end{pmatrix} = \begin{pmatrix} \cos(\kappa_{x,y} z) & -j\sin(\kappa_{x,y} z) \\ -j\sin(\kappa_{x,y} z) & \cos(\kappa_{x,y} z) \end{pmatrix} \cdot \begin{pmatrix} E_{1x,y} \\ E_{2x,y} \end{pmatrix} \tag{13}$$

where $E_{1x,y}$ and $E_{2x,y}$ are the input field amplitudes at the ports 1 and 2, for the x and the y polarizations; $\kappa_{x,y}$ and z are the coupling coefficients for x, y polarizations and interaction length, respectively (see Fig.11). The clockwise and the anti-clockwise fields, $E_{3x,y}$ and $E_{4x,y}$ propagate through the fiber loop and transform into $E'_{3x,y}$ and $E'_{4x,y}$, respectively, just before meeting each other at the coupler. Using the Jones matrix for the equivalent waveplate, these fields can be expressed as:

$$\begin{pmatrix} E'_{3x} \\ E'_{3y} \end{pmatrix} = \begin{pmatrix} J_{xx} & J_{xy} \\ J_{yx} & J_{yy} \end{pmatrix} \begin{pmatrix} -E_{3x} \\ E_{3y} \end{pmatrix} \tag{14}$$

$$\begin{pmatrix} -E'_{4x} \\ E'_{4y} \end{pmatrix} = \begin{pmatrix} J_{xx} & J_{yx} \\ J_{xy} & J_{yy} \end{pmatrix} \begin{pmatrix} E_{4x} \\ E_{4y} \end{pmatrix} \tag{15}$$

where

$$J_{xx} = e^{j\phi} \sin^2\theta + \cos^2\theta, J_{xy} = J_{yx} = (e^{j\phi} - 1)\sin\theta\cos\theta,$$

and $J_{yy} = e^{j\phi} \cos^2\theta + \sin^2\theta$.

The arbitrary phase term is ignored in the above analysis. The reflected and the transmitted fields, $E'_{1x,y}$ and $E'_{2x,y}$, are given by

$$\begin{pmatrix} E'_{1x,y} \\ E'_{2x,y} \end{pmatrix} = \begin{pmatrix} \cos(\kappa_{x,y} z) & -j\sin(\kappa_{x,y} z) \\ -j\sin(\kappa_{x,y} z) & \cos(\kappa_{x,y} z) \end{pmatrix} \cdot \begin{pmatrix} E'_{4x,y} \\ E'_{3x,y} \end{pmatrix} \tag{16}$$

The reflectance and the transmittance, R and T are given by[32]

$$R = E'_{1x} E'^{*}_{1x} + E'_{2y} E'^{*}_{1y} \tag{17}$$

$$T = E'_{2x} E'^{*}_{2x} + E'^{*}_{2y} E'^{*}_{2y} \tag{18}$$

where * denotes the complex conjugate. Please note that $E_{1x} E^{*}_{1y} = E'_{1y} E'^{*}_{1x} = 0$ because there is no correlation between x- and y- polarized input lights, E_{1x} and E_{1y}. Thus, using Eqs.(13-18), the expression for T and R can be determined as

$$T = |J_{xx}|^2 \cos^2(2\kappa_x z) P_{xi} + |J_{yy}|^2 \cos^2(2\kappa_y z) P_{yi} + |J_{xy}|^2 \cos^2(\kappa_x z - \kappa_y z)(P_{xi} + P_{yi}) \tag{19}$$

and $R = 1 - T$, where $|J_{xx}|^2 = |J_{yy}|^2 = 1 - \sin^2(2\theta)\sin^2(\phi/2)$, $|J_{xy}|^2 = 1 - |J_{xx}|^2$, $P_{xi} = E_{1x} E^{*}_{1x}$, and $P_{yi} = E_{1y} E^{*}_{1y}$. In the case of $\theta = 0$, the birefringence caused by fiber bends, i.e., ϕ has no effect on the performance of the FLM. Thus, $T = \cos^2(2\kappa_{x,y} z)$ and $R = \sin^2(2\kappa_{x,y} z)$. In such cases, if the coupler forming the FLM is a 3 dB, then whole of the light is reflected back to the input port making it a perfect mirror.

6.2 Wavelength Response

The spectral response of an over-coupled fiber coupler is like an aperiodic sine wave. Without any birefringence present in the loop, the response (of transmittance/reflectance) of an FLM formed with an over-coupled coupler should be periodic with transmission minima of equal amplitude. But, in the experimentally observed responses, some modulation was found[26, 32] and this can be explained by assuming that the loop has some birefringence. Thus, the wavelength response of an FLM is dependent upon the degree of birefringence in the loop and the spectral characteristics of the coupler. Typical simulated transmittance of an FLM formed with over coupled coupler, having FSR = 100 nm, is shown in Fig.12. If on traveling around the loop, orthogonal components of the fields of a particular wavelength λ acquire a relative phase difference $\Delta\varphi$, then this can be written in terms of effective refractive indices of the slow and fast axis, n_o and n_e and the path length l:

$$\Delta\varphi = \frac{2\pi\, l \left| n_o - n_e \right|}{\lambda} \tag{20}$$

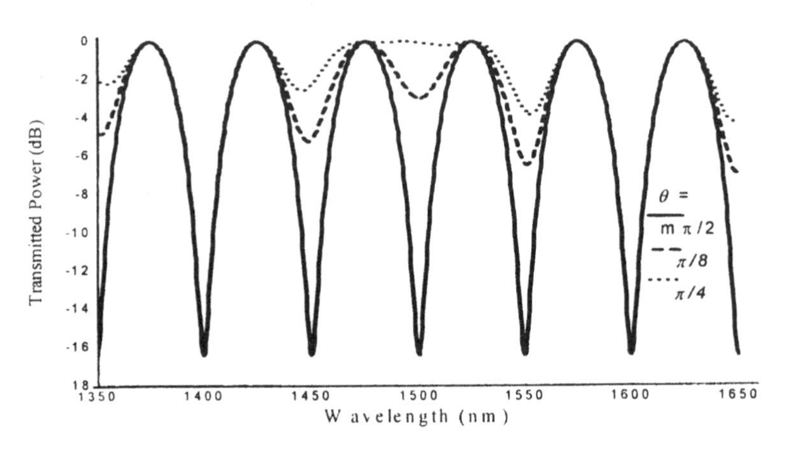

Figure 12 Typical simulated responses at the transmitted port of an FLM formed with an over-coupled coupler, having FSR = 100 nm, corresponding to different orientation angles of the fiber loop.

With $\Delta\varphi = (2m+1)\pi$ and $\Delta\varphi = 2m\pi$, (m = 0,1,2...) the loop behaves as a half- and/or as a full waveplate, respectively. The Jones matrix for full waveplate is the unit matrix, and therefore at this particular wavelength the loop birefringence has no effect and the reflectivity is maximum, (i.e., transmission is minimum). From Fig. 12, we can estimate the two wavelengths at which this condition occurs as $\lambda_1 \approx 1400$ nm and $\lambda_2 \approx 1600$ nm. We can therefore write

$$2m\pi = \frac{2\pi\, l \left| n_o - n_e \right|}{\lambda_1} \tag{21}$$

and

$$2(m+1)\pi = \frac{2\pi\, l \left| n_o - n_e \right|}{\lambda_2} \tag{22}$$

From Eqs. (21-22), we obtain

$$m = \text{INT}\frac{\lambda_2}{\lambda_1 - \lambda_2} \text{ and } \Delta\varphi = \frac{2\pi m \lambda_1}{\lambda} \tag{23}$$

where INT stands for the *integer function*. Between λ_1 and λ_2 the loop birefringence affects the maximum reflectivity attainable and this is reduced to a minimum when the loop appears as a half waveplate. The effective orientation angle θ of the waveplate axis determines the actual value of this minimum reflectivity. If $\theta = m\pi/2$, the reflectivity is a maximum and equal to that at λ_1 and λ_2. At $\theta = (2m+1)\pi/4$ the reflectivity is zero.

The transmitted power for unpolarized light, i.e. $P_{xi} = P_{yi}$, becomes minimum and maximum when the coupling coefficients and the interaction length satisfy the following conditions[32]

$$2\kappa_x z = m\pi\frac{\lambda - \lambda_{1min}}{\lambda_{2min} - \lambda_{1min}} + \frac{\pi}{2} + \delta \tag{24}$$

and

$$2\kappa_y z = m\pi\frac{\lambda - \lambda_{1\min}}{\lambda_{2\min} - \lambda_{1\min}} + \frac{\pi}{2} - \delta \tag{25}$$

where λ_{1min} and λ_{2min} are the two wavelengths corresponding to minimum transmission, and $\delta = \kappa_x z - \kappa_y z$, is a measure of the polarization dependence of the coupler. From Fig.12, we can observe flatness in the transmittance around 1500 nm for $\theta = \pi/4$. This particular feature has been used in gain flattening of the amplified spontaneous emission (ASE) spectrum of erbium-doped fiber (EDF), and is discussed in the next section.

6.3 Gain flattening of EDFA

A modern DWDM system operates in the low-loss window around 1550 nm with EDFAs forming an integral part of it. The gain spectrum of a typical EDFA[10] (see Fig.13) shows a peak around 1530 nm, which reduces the available flat gain-band. This imposes a serious limitation on the number of channels to be incorporated in an

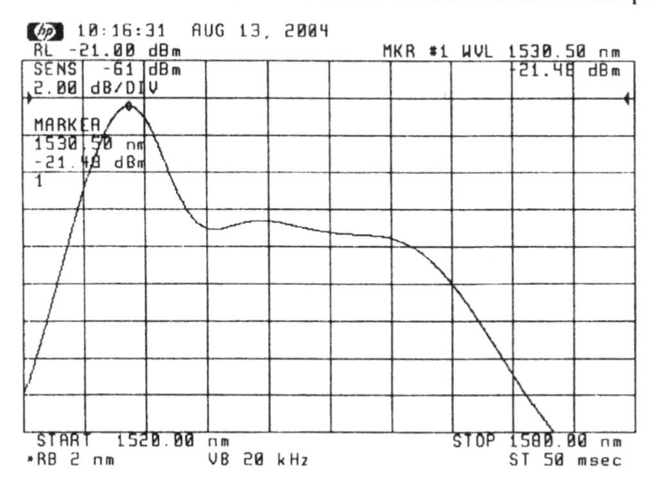

Figure 13 Typical Amplified spontaneous emission (ASE) spectral response

efficient DWDM system. Recently, use of high birefringent fiber loop mirror (HiBi FLM) has been proposed[33] for gain flattening of EDFA. A HiBi-FLM can be realized by incorporating a number of sections of HiBi fiber in the loop mirror configuration. By appropriately adjusting the orientations of the polarization controllers placed in the fiber loop of the HiBi-FLM, fan-Solc filters[34] and Lyot-type filters[35] have also been reported. But, the involvement of Hi-Bi fiber leads to higher losses and makes the technique less attractive.

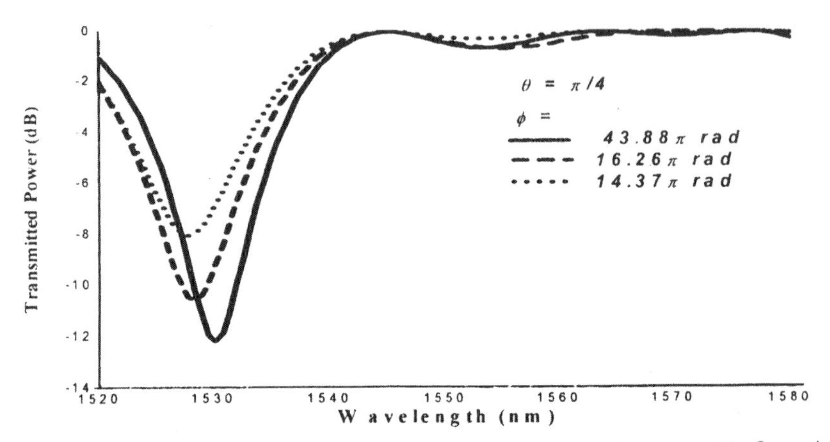

Figure 14 Wavelength response of a fiber loop mirror at the transmitted port with $\theta = \pi/4$.

If the coupler forming FLM is fabricated with a telecommunication grade fiber (such as SMF-28), the inherent birefringence present in the loop is not sufficient to produce wavelength-filtering action as in the case of HiBi-FLM. Since *over-coupled* couplers have been fabricated with wavelength filtering properties and low losses, it makes them suitable candidate for achieving gain flattening of EDFA. The FSR of the transmittance/reflectance of the loop reflector formed with an over-coupled coupler is half the FSR of the coupler. Thus, the optimized FSR of the coupler for the gain flattening of ASE spectrum of EDFA over the range of 35 nm (from 1525 – 1560 nm) is =70 nm. Typical simulation of the transmittance of an FLM, formed

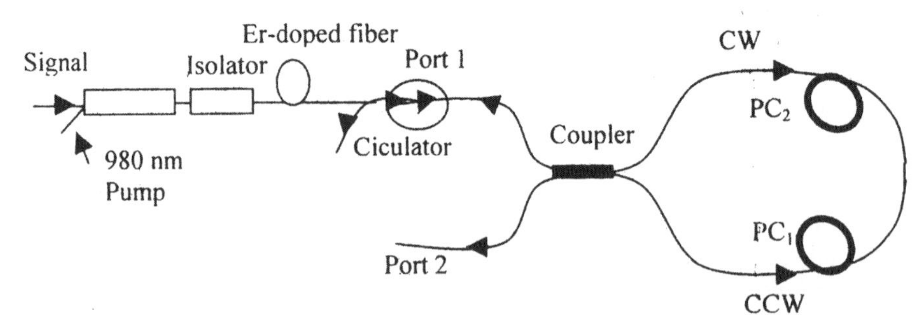

Figure 15 Schematic of the experimental set-up used in gain flattening of the ASE spectrum of EDFA.

using an over-coupled coupler with FSR 70 nm and a dip at the wavelength 1530 nm, for different retardation and orientation ($\theta = \pi/4$), is shown in Fig.14. Based on

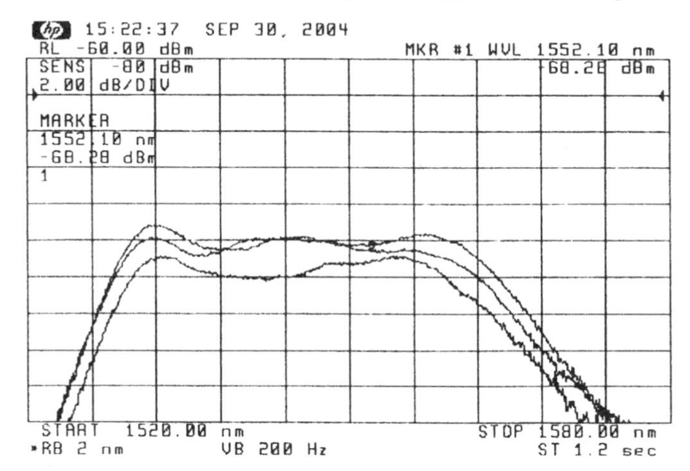

Figure 16 Flattened ASE spectra of an EDFA for different orientations of two polarization controllers in the loop mirror configuration.

this study, we realized an FLM formed with over-coupled coupler for experimentation. The schematic of the experimental set-up is shown in Fig.15. We introduced an appropriate bend-induced birefringence inside the loop layout and flattened[36] the ASE spectrum of an EDFA (see Fig.16). By adjusting the orientation of the two polarization controllers (PC_1 and PC_2), flatness within ± 0.5 dB has been achieved over a range of 30 nm in the C-band.

7. Summary

The basic device in the FBT fiber coupler technology is a *2× 2 fused fiber coupler*, which can be used to realize a variety of in-line fiber components such as power splitter/combiners, wavelength division multiplexers, interleavers, and fiber loop reflectors. These components have become extremely important in today's high-speed optical fiber communications systems and networks. We have briefly discussed the basic configuration, functional principle and characteristics of some of these devices based on the FBT coupler technology. In particular, use of all-fiber MZI as a wavelength interleaver and an FLM for gain-flattening of EDFAs has been detailed. For the sake of completeness, the essential theory behind the working of these devices has also been presented.

Acknowledgements: Partial support provided by the Ministries of Information Technology (MIT) and Human Resource Development (MHRD), Government of India, and OPTEL Telecom Ltd (Bhopal) for this activity in the initial stages is gratefully acknowledged. They would also like to place on record their appreciation of very fruitful academic interactions with a number of colleagues from the Fiber Optics Group, IIT Delhi, and also several past students who have worked on the FBT coupler fabrication rig from time to time as part of their dissertation work. In particular, contributions of Dr. P. Roy Chaudhuri, Mr. Naveen Kumar and Mr. Saju Thomas to the work on FLM based devices are acknowledged. Thanks are also due to Mr. Naveen Kumar for his help in the preparation of the manuscript.

References

1. See, for *e.g.*, B.P. Pal and G.R. Chakravarty, "All-fiber guided wave optical components: A state of-art review", *Journal of Advance Science* (Japan) **10**, 1 (1998).
2. R.A. Bergh, G. Kotler, and H.J. Shaw, "Single-mode fiber optic directional coupler", *Electron. Lett.*, **16**, 260 (1980).
3. B.S. Kawasaki, and R.G. Lamont, "Biconical-taper single-mode fiber directional coupler", *Opt. Lett.* **6**, 327 (1981).
4. D.B. Mortimore, "Monolithic 4×4 single mode fused coupler", *Electron. Lett.* **25**, 682 (1989).
5. B.P. Pal, "Electromagnetics of all-fiber components" in *Electromagnetic Fields in Unconventional Structures and Materials*, Singh, O. N. and Lakhtakia, A. (Eds.), John Wiley, New York, p. 359 (2000).
6. F.P. Payne, "Fused single-mode optical fiber couplers", *J. Inst. Electron. Telecom. Eng.* (India), **32**, 319 (1986).
7. B.P. Pal, P. Roy Chaudhuri, and M.R. Shenoy, "FBT fiber coupler process technology: a precise model for software driven fabrication of components", *Proc. National Fiber Optic Engineers Conference*, Baltimore, MD, USA, July 8-12, 734-741 (2001); see also, "Fabrication and modeling of fused bi-directional tapered fiber couplers", *Fiber and Integrated Optics*, **22**, 97 (2003).
8. D.C. Johnson, and K.O. Hill, "Control of wavelength selectivity of power transfer in fused biconical monomode directional couplers", *Appl. Opt.*, **25**, 3800 (1986).
9. P. Roy Chaudhuri, B.P. Pal, and M.R. Shenoy, "Modeling of fused 2x2 all-fiber coupler components", *Proc. National Symposium on Advances of Microwave and Lightwave Technology*, University of Delhi, March 25-28, 26 (2000).
10. See, for *e.g.*, A.K. Ghatak and K.Thyagarajan, *Introduction to Fiber Optics*, (Cambridge University Press, Cambridge), 1998, Ch.17.
11. B. Culshaw, "Interferometric optical fiber sensors", in *Fundamentals of Fiber Optics in Telecommunication and Sensor Systems*, B. P. Pal, (Ed.), (John Wiley, New York and Wiley Eastern, New Delhi, 1992), p. 584.
12. J.D.C. Jones, "Signal processing in monomode fiber optic sensor systems", *ibid*, p. 657 (1992).
13. C.M. Lawson, P.M. Kopera, T.Y. Hsu, and V.J. Tekippe, "In-line single-mode wavelength division multiplexer/demultiplexer", *Electron Lett.* **20**, 963 (1984).
14. S.S. Orlov, A. Yariv, and S.V. Essen, "Coupled-mode analysis of fiber-optic add-drop filters for dense wavelength-division multiplexing", *Opt. Lett.* **22**, 688 (1997).

15. F. Gonthier, D. Ricard, S. Lacroix, and J. Bures, "2×2 multiplexing couplers for all-fiber 1.55 μm amplifiers and lasers", *Electron. Lett.* **27**, 42 (1991).
16. J.D. Minelly, and M. Suyama, "Wavelength combining fused-taper couplers with low sensitivity to polarisation for use with 1480 nm-pumped erbium-doped fiber amplifiers", *Electron Lett.* **26**, 523 (1990).
17. M.N. McLandrich, R.J. Orazi, and H.R. Marlin, "Polarisation independent narrow channel wavelength division multiplexing fiber couplers for 1.55 μm", *J. Lightwave Technol.* **9**, 442 (1991).
18. I. J. Wilkinson, and C. J. Rowe, "Close-spaced fused fiber wavelength division multiplexers with very low polarisation sensitivity", *Electron Lett.* **26**, 382 (1990).
19. R. Ramaswamy and K.N. Sivarajan, *Optical Networks: A Practical Perspective*, Harcourt Publishers, Singapore 2000, Ch. 3.
20. S. Bourgeois, "Fused-fiber developments offer passive foundation for optical slicing", Special Report in *Lightwave* (PennWell Corporation) 2000, 17, no.3
21. J. Chon, A. Zeng, P. Peters, B. Jian, A. Luo, and K. Sullivan, "Integrated interleaver technology enables high performance in DWDM system", in *Tech. Digest NFOEC* (2001) 1410.
22. Naveen Kumar, M.R. Shenoy, and B.P. Pal, "All-fiber wavelength interleaver/slicer with flattop response", in Proc. *Int. Conf. Fiber Opt. and Photonics - Photonics-2002*, (Mumbai, Dec. 14-18, 2002), Paper FBR P 15, 81.
23. H. Yonglin, Li Jie, Ma Xiurong, K. Guiyun, Yuan Shuzhong, D. Xiaoyi, 'High extinction ratio Mach-Zehnder interferometer filter and implementation of single-channel optical switch' *Opt. Commun.*, **222**, 191 (2003).
24. B.H. Verbeek, C.H. Henry, N.A. Olosson, K.J. Orlowski, R.F. Kazarinov, and B.H. Johnson, "Integrated four-channel Mach-Zehnder multi/demultiplexer fabricated with phosphorous doped SiO_2 waveguides on Si", *J. Lightwave Technol.*, **6** 1011 (1988)
25. I. D. Miller, D. B. Mortimore, W.P. Urquhart, B.J. Ainslie, S.P. Craig, C.A. Millar, and D.B. Payne, "A Nd^{3+} -doped cw fiber laser using all-fiber reflectors", *Appl. Opt.*, **26**, 2197 (1987).
26. D.B. Mortimore, "Fiber loop reflectors", *J. Lightwave Technol.*, **6**, *no.* 7, 1217, (1988).
27. M.N. Islam, E.R. Sunderman, R.H. Stolen, W.Pleibal, and J.R. Simpson, "Soliton switching with a fiber non linear mirror", *Opt. Lett.* **15**, 811 (1989).
28. K. J. Blow, N.J. Doran, and B.P. Nelson, "Demonstration of the non-linear fiber loop mirror as an ultrafast all-optical demultiplexer", *Electron. Lett.* **26**, 262 (1990).
29. I. Glask, J.P. Sokoloff, and P.R. Prucnal, "Demonstration of all-optical of TDM data at 250 Gb/s", *Electron. Lett.* **30**, 339 (1990).
30. M.E. Fermann, F. Haberl, M. Hofer, and H. Hochreiter, "Non-linear amplifying loop mirror", *Opt. Lett.* **15**, 752 (1990).

31. A.W. Oneill, and R.P. Woff, "All-optical loop mirror switch employing an asymmetric amplifier/attenuator configuration", *Electron. Lett.* **26**, 2008 (1990).
32. K. Morishita and K. Shimamto, " Wavelength –selective fiber loop mirrors and their wavelength tenability by twisting", *J. Lightwave Technol.* **13**, *2276* (1995).
33. S. Li., K.S. Chiang, and W.A. Gambling, " Gain flattening of an erbium-doped fiber amplifier using a high-birefringence fiber loop mirror", *IEEE Photon. Technol. Lett.* **13**, 942 (2001).
34. X. Fang, and R.O. Claus, "Polarisation independent all-fiber wavelength-division multiplexer based on a sagnac interferometer", *Opt. Lett.* **20**, 2146 (1995).
35. X. Fang, H. Ji, C.T. Allen, K. Demarest, and L. Pelz, "A compound high order polarization independent filter using sagnac interferometer", *IEEE Photon. Technol. Lett.* **9**, 458 (1997).
36. B.P. Pal, G. Thursby, Naveen Kumar, and M.R. Shenoy, "Fiber loop reflector as a versatile all-fiber component", in Proc. Photonics-2004, Dec. 8-11, Cochin, India, 2004.

9

Optical Fiber Gratings and Their Applications

K Thyagarajan

1. Introduction

Optical fiber gratings which consist of a periodic modulation of refractive index along the core of an optical fiber have very interesting spectral properties and are finding many applications in fiber optic communications and sensing. There are two main types of fiber gratings, namely fiber Bragg gratings (FBG) and long period gratings (LPG). FBGs couple light from a forward propagating guided mode to a backward propagating guided mode in the fiber while LPGs couple light from a guided mode to another guided mode or a cladding mode propagating along the same direction. The former corresponds to contra directional coupling while the latter to co-directional coupling. FBGs have periods of the order of half a micrometer for operation around the 1550 nm window while LPGs have periods of a few hundred micrometers.

In this article we will describe the coupled mode theory used for the analysis of both FBG and LPG and then discuss the properties of gratings and some of their applications. For more details, readers are referred to Refs. [1-3].

2. Fiber Bragg gratings

When light encounters another dielectric of a different refractive index it suffers partial reflection. By having a thin film coating of appropriate thickness, one can reduce the reflection by destructively interfering the waves reflected from the two interfaces formed by the thin film. These films are referred to as anti reflection coatings and find applications in various optical instruments such as cameras, binoculars etc. Instead of a single thin film, if we consider a multiple layer structure consisting of a number of alternate layers of higher ($n_0 + \Delta n$) and lower ($n_0 - \Delta n$) refractive indices with $\Delta n << n_0$, then when a light wave enters this medium, it undergoes minute reflections from every interface. If all the individual reflections are in phase, then in such a case the medium will strongly reflect the incident wave. If the reflected waves are not in phase, then the net reflection would be weak. Since the phase difference between adjacent reflections is dependent on the wavelength, this implies that the overall reflection from such a medium would be very strongly wavelength dependent.

Assuming the thickness of each medium to be $\Lambda/2$, the phase difference between reflections 1 and 2 (see Fig. 1) would be (assuming $\Delta n << n_0$)

$$\Delta\phi = \pi - \frac{2\pi}{\lambda_0} n_0 \Lambda \tag{1}$$

The extra phase difference of π comes about due to the phase change on reflection suffered during reflection from a denser medium. Similarly the phase difference between waves 2 and 3 is given by

$$\Delta\phi = \pi + \frac{2\pi}{\lambda_o} n_o \Lambda \tag{2}$$

For a wavelength given by

$$\lambda_B = 2 n_0 \Lambda \tag{3}$$

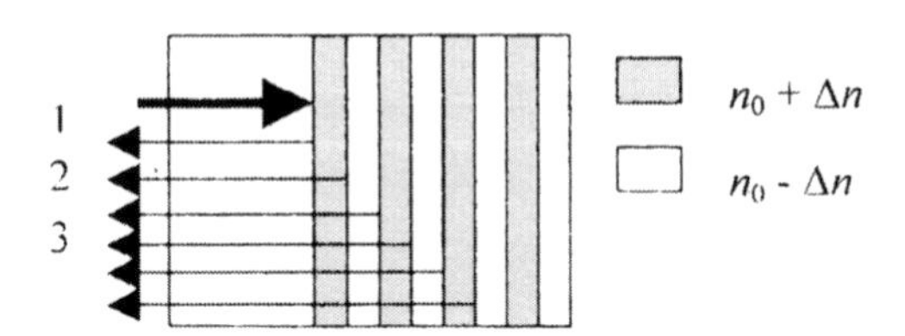

Figure 1 A periodic structure consisting of alternate layers of high and low refractive indices. When the reflections from all interfaces (1, 2, 3 etc.) add constructively then it leads to a strong reflection.

constructive interference will take place between waves 1,2,3, etc. The above equation is referred to as the Bragg condition (reminiscent of X-ray diffraction from atomic planes in crystals) and the specific wavelength λ_B satisfying Eq. (3) is referred to as the Bragg wavelength. As the wavelength deviates from the wavelength specified by Eq. (3) the waves reflected from the layers will not be in phase and thus the reflection would drop down. The wavelength given by Eq. (3) is the longest wavelength for constructive interference and other wavelengths having constructive interference would be smaller.

If instead of a bulk medium, we now consider a sinusoidal refractive index modulation with a period Λ within the *core of a single mode optical fiber*, when a guided mode propagates through such a grating then the forward propagating mode can get strongly coupled to a backward propagating mode with the same effective index when the following Bragg condition is satisfied:

$$\lambda_B = 2\Lambda n_{eff} \tag{4}$$

where n_{eff} is the effective index of the propagating mode and λ_B is the Bragg wavelength. Such a periodic index modulation within the fiber is referred to as a fiber Bragg grating (FBG). Thus when broadband light or a set of wavelengths are incident on an FBG, then only the wavelength corresponding to λ_B will get reflected; the other wavelengths just get transmitted to the output.

2.1 Coupled mode theory

One of the standard methods of analysis of FBG is using the coupled mode theory[3]. In this, the total field at any value of z is written as a superposition of the two interacting modes and the coupling process results in a z-dependent amplitude of the two coupled modes. If we represent the amplitudes of the forward and backward propagating modes (assumed to be modes of the same order) as $A(z)$ and $B(z)$ then we can derive the following coupled mode equations[3] describing their variation with z:

$$\frac{dA}{dz} = \kappa B e^{i\Gamma z}$$
$$\frac{dB}{dz} = \kappa A e^{-i\Gamma z} \quad (5)$$

Here $\Gamma = 2\beta - K$ where β represents the propagation constant of the modes, and $K = 2\pi/\Lambda$ represents the spatial frequency of the grating, which is described by the following refractive index variation:

$$n_g^2(x, y, z) = n^2(x, y) + \Delta n^2(x, y)\sin(Kz) \quad (6)$$

In Eq. (5) κ represents the coupling coefficient defined by

$$\kappa = \frac{\omega\varepsilon_0}{8}\iint \psi^* \, \Delta n^2(x, y)\, \psi \, dx\, dy \quad (7)$$

were $\psi(x,y)$ represents the normalized transverse modal field distribution. If the perturbation in the refractive index is constant and finite only within the core of the fiber, we obtain the following simple expression for the coupling coefficient[3]:

$$\kappa \approx \frac{\pi\, \Delta n\, I}{\lambda_B} \quad (8)$$

where, under the Gaussian approximation with modal width w_0 the overlap integral I is given by[3]

$$I \approx \left(1 - \exp\left\{-2a^2/w_0^2\right\}\right) \quad (9)$$

where a is the core radius. If we consider a grating of length L and assume that at the input of the grating $z = 0$, only the forward propagating mode is incident, then the boundary conditions are $A(z = 0) = 1$ and $B\,(z = L) = 0$. We then obtain the following expression for the reflectivity of the grating[3]:

$$R = \frac{\kappa^2 \sinh^2(\Omega L)}{\Omega^2 \cosh^2(\Omega L) + \dfrac{\Gamma^2}{4}\sinh^2(\Omega L)} \quad (10)$$

where

$$\Omega^2 = \kappa^2 - \frac{\Gamma^2}{4} \quad (11)$$

For a given set of parameters, the reflection coefficient can be shown to be maximum when $\Gamma = 0$, i. e., $2\beta = K$, which implies Eq. (4), i. e., the Bragg condition. In such a case the reflection coefficient given by Eq. (10) becomes

$$R = \tanh^2 \kappa L \quad (12)$$

Hence as κL increases, the grating reflectivity increases and approaches unity. (see Fig. 2).

If $\Gamma \neq 0$ then we have the non-phase matched case and the reflectivity is smaller than that given by Eq. (12). Thus peak reflectivity appears when the Bragg condition is satisfied.

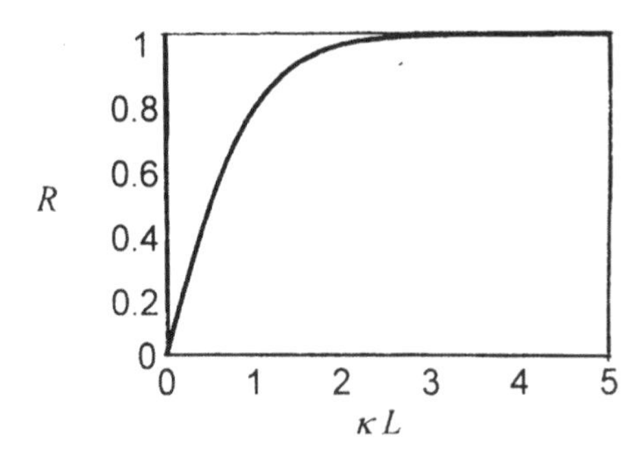

Figure 2 Variation of reflectivity with length of an FBG.

Figure 3 shows the vector diagram for reflection from FBG for the cases when the Bragg condition is satisfied and not satisfied. The wave vectors of the incident mode and the reflected mode are denoted by $+\beta$ and $-\beta$ respectively and that of the grating by the spatial frequency vector K. As seen from the figure phase matched interaction corresponds to $\Gamma = 0$.

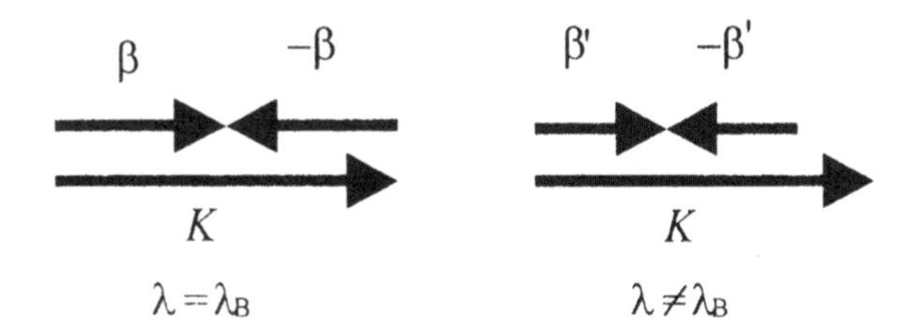

Figure 3 Vector diagrams when the incident wavelength matches the Bragg wavelength and when it deviates from it.

Figure 4 shows a measured reflection spectrum of a typical FBG with a length of 2 cm, with peak reflection at 1549.91 nm and a peak reflection coefficient of about 92 %. As can be seen the reflectivity peaks at the Bragg wavelength and on either side the reflectivity drops and becomes oscillatory. One can define the spectral bandwidth of the grating in terms of the grating parameters as (see e.g. Ref. [3])

$$\Delta\lambda \approx \frac{\lambda_B^2}{n_{eff}\,L}\left(1+\frac{\kappa^2 L^2}{\pi^2}\right)^{1/2} \tag{13}$$

From Eqs. (12) and (13) it also follows that one can design gratings having same peak reflectivity but different bandwidths by appropriate choice of the peak index modulation and grating length. A given peak reflectivity implies a given value of κL (i.e., $\Delta n\, L$). The bandwidth can be increased or reduced by decreasing or increasing the grating length while keeping the product $\Delta n\, L$ constant. It is also possible to deduce the characteristics of the grating (length, period of the grating and the index modulation) by measuring the reflectivity spectrum (peak reflection wavelength, peak reflectivity and the bandwidth).

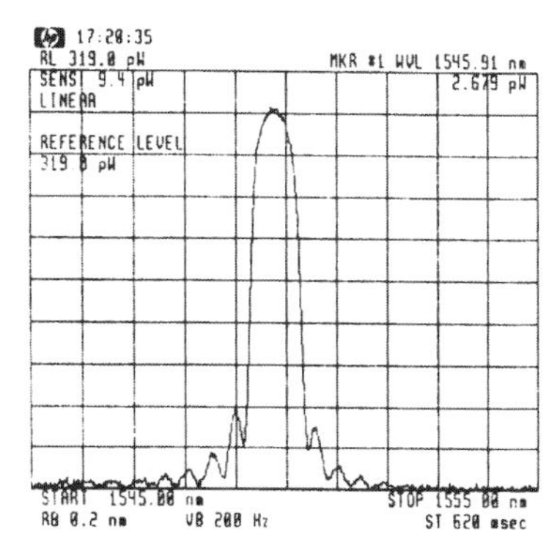

Figure 4 Measured reflection spectrum of an FBG as observed in an optical spectrum analyzer.

3. Some applications of FBGs

Fiber Bragg gratings find many applications. These include applications in add/drop multiplexers, fiber grating sensors, to provide external feedback for laser diode wavelength locking, dispersion compensation etc. Here we discuss the application of chirped FBG to dispersion compensation and to sensors.

3.1 Dispersion compensation

It is well known that when pulses of light propagate through an optical fiber link they suffer from dispersion due to the dependence of group velocity of the mode on the wavelength[3]. This dispersion needs to be compensated in the link and this can be achieved either through the use of dispersion compensating fibers (see e.g. Ref. [3]) or with the help of chirped FBGs. In chirped FBGs the period of the grating varies with the position along the fiber length and this leads to a variation of Bragg wavelength along the grating (see Fig. 5). Thus different wavelengths will get reflected at different positions along the grating leading to different time delays for different wavelengths. By using an appropriate chirped FBG one can compensate for the differential delay of different wavelengths accumulated while propagating through the fiber. As shown in Fig. 5, while propagating through the fiber wavelength λ_1 suffers larger delay than wavelength λ_2 and λ_3. In the anomalous dispersion region of a single mode fiber (wavelength greater than the zero dispersion wavelength), $\lambda_1 > \lambda_2 > \lambda_3$ The chirped grating is designed so that λ_1 reflects from the near end of the grating while λ_3 reflects from the far end so as to compensate for the differential delay between λ_1 and λ_3 and all intermediate wavelengths and thus leading to dispersion compensation.

If we consider a linearly chirped FBG with spatial frequency variation as given by

$$K(z) = K_0 + \frac{Fz}{L_g^2} \tag{14}$$

where K_0 is the grating spatial frequency at $z = 0$ the input position of the grating, L_g is the length of the grating and F is the chirp parameter then the bandwidth over which the chirped FBG will operate will be approximately given by

$$\Delta\lambda = \frac{F\,\lambda_0^2}{4\,\pi\,n_{eff}\,L_g} \tag{15}$$

and the corresponding dispersion introduced by the grating is given by

$$\frac{d\tau}{d\lambda} = \frac{8\,\pi\,n_{eff}^2}{c\lambda_0^2}\frac{L_g^2}{F} \tag{16}$$

Note that for the case shown in Fig. 5 longer wavelengths suffer less delay in returning from the grating and hence this corresponds to normal dispersion; such a normal dispersion can compensate for the anomalous dispersion of the fiber if the fiber is operated at wavelengths greater than the zero dispersion wavelength. If we consider a link fiber with dispersion coefficient D and length L_f, the accumulated dispersion in the link would be $D\,L_f$. In order to compensate for this accumulated dispersion we require

$$DL_f = -\frac{8\pi n_{eff}^2\,L_g^2}{c\lambda_0^2 F} \tag{17}$$

The spectral bandwidth over which dispersion compensation will take place is given by Eq. (15).

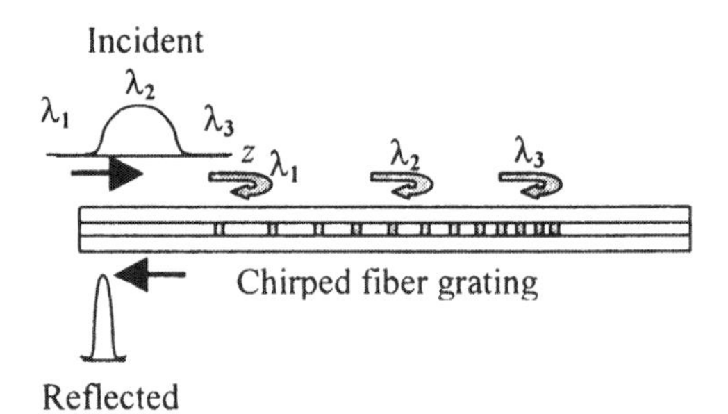

Figure 5 Application of a chirped FBG in dispersion compensation.

As an example let us consider a chirped grating of length 11 cm with the chirp parameter $F = 640$, operating at an average wavelength of 1550 nm. If the effective index of the fiber mode is 1.45, then using Eqs. (15) and (16) we obtain for the dispersion of the grating as 1380 ps/nm operating over a bandwidth of 0.77 nm. This grating can compensate for dispersion accumulated over a fiber with a dispersion coefficient of 17 ps/km-nm and of length 81 km over a bandwidth of 0.77 nm which approximately corresponds to 96 GHz of frequency bandwidth. It is interesting to note that the difference in period between the front and back end of the 11 cm long grating is only about 0.25 nm while the average period of the grating is about 0.534 μm.

In order to achieve dispersion compensation over larger bandwidths one can employ multiple chirped gratings over the same section of the fiber[4]. Chirped dispersion compensating gratings are commercially available for compensation of accumulated dispersion of up to 80 km of G.652 fiber for up to 32 wavelength channels. Unlike dispersion compensating fibers, chirped FBGs provide with the possibility of tweaking the required dispersion compensation especially for 40 Gbps systems where the margin of dispersion available is rather small. Also by using nonlinearly chirped FBGs it has been shown that delay variation from -200 ps to –1200 ps is achievable[5].

3.2 Sensing

Since the Bragg wavelength λ_B depends on the refractive index of the fiber (which determines the effective index of the fiber) as well as on the period of the grating, any external parameter which changes any of these would result in a change in the reflected wavelength. Thus by measuring the changes in the reflected wavelength, the external perturbations could be sensed. This is the basic principle of their application in sensing of mechanical strain, temperature, acceleration etc. [6]. FBG sensors are specially attractive for quasi distributed sensing applications wherein FBGs with different center wavelengths can be used to sense signals at different points along the same fiber. Since different FBGs reflect light at different wavelengths, the signals coming from different points along the fiber can easily be differentiated.

We can write for the change in the Bragg wavelength because of a change ΔT in the temperature and $\Delta\varepsilon$ in strain as

$$\Delta\lambda_B = 2\Lambda\left(\frac{dn_{eff}}{dT}\Delta T + \frac{dn_{eff}}{d\varepsilon}\Delta\varepsilon\right) + 2n_{eff}\left(\frac{d\Lambda}{dT}\Delta T + \frac{d\Lambda}{d\varepsilon}\Delta\varepsilon\right) \quad (18)$$

Figure 6 shows a measured variation of Bragg wavelength as a function of strain of the grating. The grating shown has a strain sensitivity of 1.3 pm/με at 1550 nm. The temperature sensitivity of FBGs is about 6 pm/°C. The changes in the peak wavelength are indeed very small and hence special techniques are needed for sensing such small changes. Also techniques to deconvolve the changes in temperature and strain by measuring the changes in the Bragg wavelength need to be implemented for precise sensing.

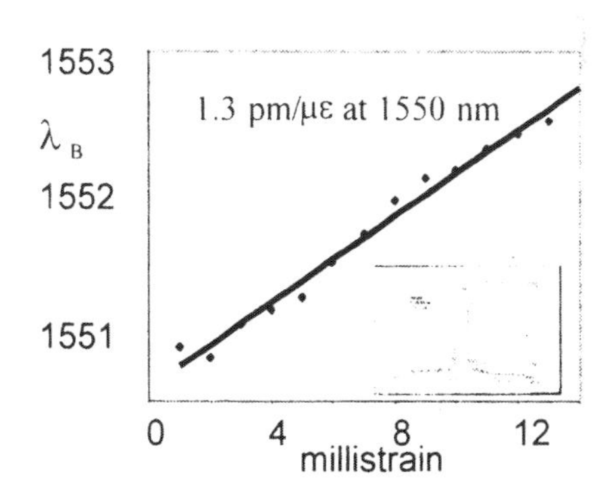

Figure 6 Measured variation of Bragg wavelength as a function of strain on the FBG. The inset shows the spectra as observed in the OSA for the unstrained and strained cases.

4. Long period gratings

Long period gratings (LPG) are periodic perturbations along the length of the fiber with periods of greater than 100 μm which induce coupling between two co-propagating guided core modes or between a guided core mode and a cladding mode propagating in the same direction. Coupling due to periodic perturbation being wavelength selective, these gratings act as wavelength dependent loss components. This makes them attractive candidates for applications in wavelength filters with specific application in gain flattening of erbium doped fiber amplifiers (EDFA), band rejection filters, WDM isolation filters or as polarization filtering components, sensors etc.[7-9].

4.1 Coupled mode theory

An LPG corresponds to a periodic perturbation of the core refractive index as given by Eq (6). Using the coupled mode theory one can obtain the following expression for the fractional power remaining in the core and cladding when at $z = 0$ only the core mode is excited [3]:

$$P_{co} = 1 - \frac{\kappa^2}{\gamma^2}\sin^2\gamma z \qquad P_{cl} = 1 - P_{co} \tag{19}$$

Here $\Gamma = \beta_1 - \beta_2 - K$, where β_1 and β_2 are the propagation constants of the two interacting modes, and K represents the spatial frequency of the grating. The coupling coefficient is now defined by

$$\kappa = \frac{\omega\varepsilon_0}{8}\iint \psi_1 \Delta n^2 \psi_2 \, dxdy \tag{20}$$

whre $\psi_1(x,y)$ and $\psi_2(x,y)$ represent the normalized field distributions of the interacting modes and β_1 and β_2 represent their propagation constants. In Eq. (19)

$$\gamma = \sqrt{\kappa^2 + \frac{\Gamma^2}{4}} \tag{21}$$

If we now consider coupling from the guided core mode to a cladding mode, then it can be seen from Eq. (19) that unlike the case of FBG the exchange of energy between the core mode and the cladding mode is periodic with distance and complete transfer of energy from the guided mode (β_{co}) to the cladding mode (β_{cl}) is possible if $\Gamma = 0$, *i.e.*, if the following condition is satisfied:

$$\beta_{co} = \beta_{cl} + K \tag{22}$$

If n_{eff}^{co} and n_{eff}^{cl} represent the effective indices of the interacting core and cladding modes respectively then from Eq. (22) we obtain for the required period of the LPG

$$\Lambda = \frac{\lambda_0}{n_{eff}^{co} - n_{eff}^{cl}} \tag{23}$$

Typical difference between the effective indices of the core and a low order cladding mode would be approximately equal to the index difference between the core and the cladding which for typical telecommunication fibers is approximately 0.003. Thus for an LPG to operate at a wavelength of 1550 nm the required grating period is approximately 520 μm which is much larger than that of an FBG.

If Eq. (22) is not satisfied then the exchange of energy is less efficient. Figure 7 shows the vector diagram corresponding to phase matched interaction between a core mode with propagation constant β_{co} and a cladding mode with propagation constant β_{cl} propagating along the same direction.

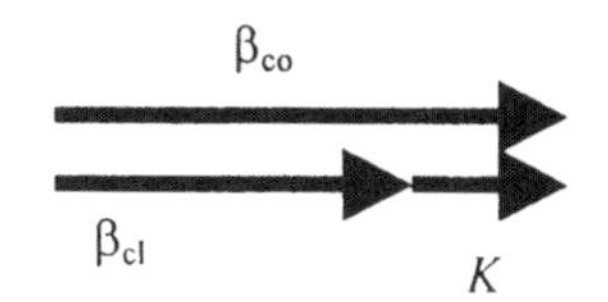

Figure 7 Vector diagram showing coupling from the core mode to a cladding mode propagating in the same direction

Figure 8 shows the comparison of the measured and simulated transmission spectra of an LPG fabricated using the fiber fusion splice machine [10]. One can see multiple dips in the transmission spectrum corresponding to the coupling of the LP_{01} guided mode to individual cladding modes. Note that if the perturbation is azimuthally symmetric, then coupling from the LP_{01} core mode will take place only to LP_{0m} cladding modes since only in these cases κ given by Eq. (20) will be non-zero.

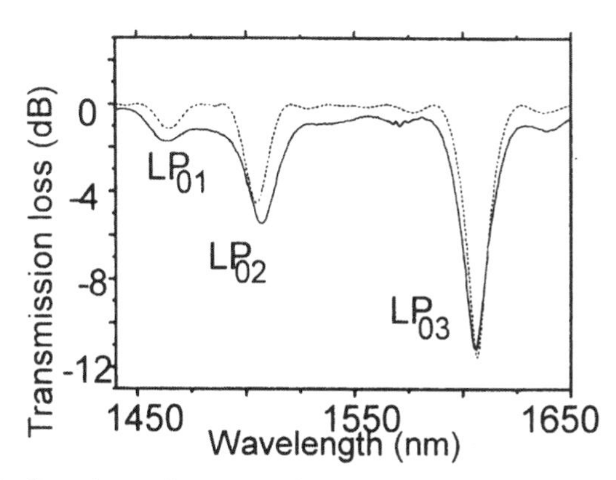

Figure 8 Experimentally measured (solid curve) and simulated (dashed curve) transmission spectra of an LPG in SMF-28 showing multiple dips due to coupling from the core LP_{01} mode to various cladding modes (After Ref. [10]).

5. Some applications of LPGs

LPGs have found many applications and here we will discuss some of them.

5.1 EDFA gain flattening

As discussed in Chapter 3, the gain spectrum of an EDFA is not flat and this causes problems in their use in a communication system. Flattening of the gain spectrum is possible by using appropriate external wavelength filters which compensate for the gain variation by having a transmission spectrum which is just the inverse of the gain spectrum. Since LPGs can be designed to have different transmission spectra, they find applications as gain flattening filters.

Figure 9 shows the transmission profile of a designed LPG for gain flattening of an EDFA and Fig.10 shows the gain spectrum of an EDFA with and without the filter. As can be seen the LPG filter flattens the gain to within ±0.75 dB.

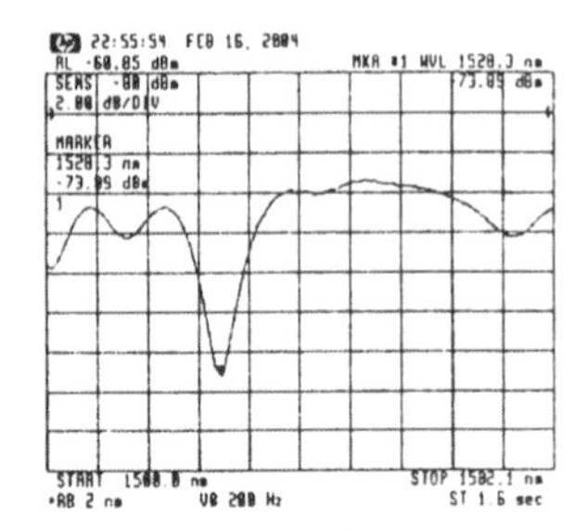

Figure 9 The transmission spectrum of an LPG used for gain flattening of an EDFA.

5.2 WDM filter

A chirped LPG has its period varying along the length of the grating. Thus if we combine two concatenated chirped LPGs the resulting device will act as a Mach-Zehnder interferometer for the range of wavelengths for which coupling is enabled by the gratings. The first grating is designed to couple 50 % of the core mode into

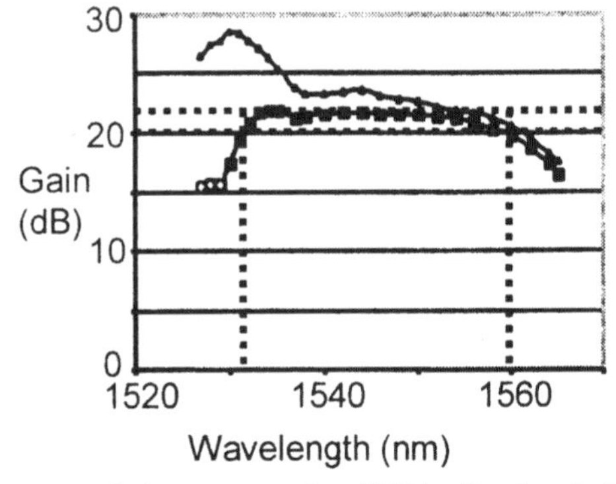

Figure 10 Gain spectrum of an EDFA without and with the fabricated LPG gain flattening filter.

the cladding while the second grating recombines them. Depending on the phase difference accumulated between the core and the cladding mode, the output power is either in the core or in the cladding. Hence a periodic transmission spectrum is

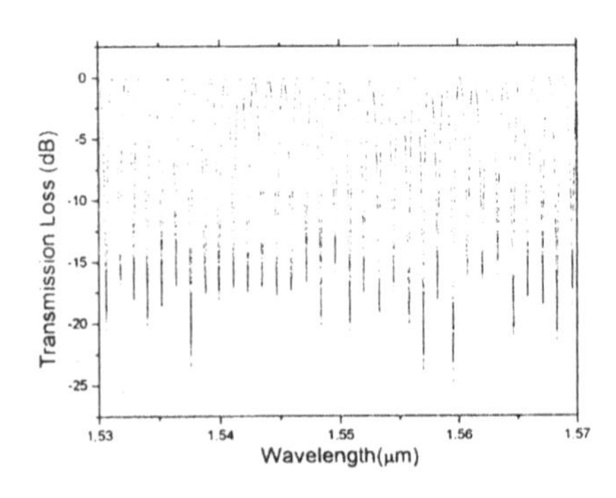

Figure 11 Transmission spectrum of two concatenated chirped LPGs (After Ref. [11]).

expected from the output of the device[11]. Figure 11 shows a typical transmission spectrum of two concatenated chirped LPGs. There are about 32 pass bands with average wavelength spacing of 1.175 nm and an isolation at the stop bands of greater than 15 dB. Such filters can find applications in various DWDM based devices[12].

5.3 Broadband LPGs

Uniform LPGs usually have a spectral bandwidth of only a few nanometers since as the wavelength changes, the modes get out of phase matching and the power exchange reduces. However, many applications such as polarization dependent loss compensators and WDM filters require broad spectral bandwidth of the transmission dips.[13] It is indeed possible to tailor the refractive index profile of the fiber in which the LPG is fabricated to achieve large spectral bandwidths. This is achieved by having the phase matching condition satisfied over a large wavelength range and at the same time having a coupling coefficient that does not vary significantly over the wavelength range. By proper design of the refractive index profile of an optical fiber it is possible to achieve LPGs with 20 dB spectral bandwidth of about 100 nm.[14]

6. Conclusions

Fiber Bragg gratings and long period fiber gratings have interesting spectral characteristics and are finding applications in many devices such as add/drop multiplexers, wavelength lockers, dispersion compensators, sensors, optical amplifier gain flatteners etc. By tailoring the grating properties and the fiber properties on which the gratings are fabricated, gratings with different spectral characteristics can be achieved.

Acknowledgements: The author would like to thank Dr. Parthasarathy Palai, Tejas Networks, India Ltd., Dr. M N Satyanarayan, BITS, Pilani, Dr. Mini Das, USA and Mr. Mandip Singh for their help in fabrication and characterization of long period gratings using the splice machine and its applications in gain flattening of EDFAs. The work on fiber gratings was partially supported by a research project sponsored by the Department of Information Technology, Govt. of India.

References

1. R. Kashyap, *Fiber Bragg Gratings*, Academic Press, San Diego, 1999.
2. A. Othonos and K Kalli, *Fiber Bragg Gratings*, Artech House, Boston, 1999.
3. A. Ghatak, K. Thyagarajan, *Introduction to Fiber Optics*, Cambridge University Press, UK. (1998).
4. R L Lachance, Y Painchaud and A Doyle, Fiber Bragg gratings and chromatic dispersion, Teraxion.
5. A E Willner, K. M. Feng, J. Cai, S. Lee, J. Peng and H. Sun, Tunable compensation of channel degrading effects using nonlinearly chirped passive fiber Bragg gratings, *IEEE J. Sel. Topics in Quant. Electron.*, **5**, 1298 –1311 (1999).

6. A. D. Kersey, M. A.. Davis, H. J. Patrick, M. LeBlac, K. P. Koo, C. G. Askins, M. A. Puman and E. J. Friebele, Fiber grating sensors, *J. Lightwave Tech.* **15**, 1442-1463 (1997).

7. A.M. Vengsarkar, P.J. Lemaire, J.B. Judkins, V. Bhatia, T. Erdogan J.E. Sipe, Long period fiber gratings as band rejection filters, *Journal of Lightwave Technology*, **14**, 58-65 (1996).

8. S. W. James and R. P. Tatam, Optical fiber long period grating sensors: characteristics and applications, *Meas. Sci. Tech.* **14**, R49-R61 (2003).

9. Ashish M. Vengsarkar, J. Renee Pedrazzani, Justin B. Judkins, Paul J. Lemaire, Neal S. Bergano, Carl R. Davidson, Long period fiber grating based gain equalizers, *Opt. Letts.*, **21**, 336-338 (1996).

10. P. Palai, M N Satyanarayan, M Das, K Thyagarajan and B P Pal, Characterization and simulation of long period gratings fabricated using electric discharge *Opt. Comm.* **193**, 181-185 (2001).

11. M. Das and K Thyagarajan, Wavelength-division multiplexing isolation filter using concatenated chirped long period gratings, *Opt. Comm.* **197**, 67-71 (2001).

12. K. Thyagarajan and J K Anand, Gain enahanced EDFA for WDM applications, Proc. International Conference On Optical Communication and Networks (ICOCN) 2002, Singapore, Paper 12D2, 73 (2002).

13. S. Ramachandran, Z. Wang and M. Yan, Bandwidth control of long period grating based mode converters in few mode fibers, *Opt. Letts.*, **27**, 698-700 (2002).

14. Charu Kakkar and K Thyagarajan, Novel fiber design for broadband long period gratings, *Optics Comm.* **220**, 309-314 (2003).

10

Optical Fiber Sensors: Opportunities, Underlying Principles and Examples

B.D. Gupta

1. Introduction

Over the past two decades the optoelectronics and fiber-optic communication industries have expanded rapidly. The developments of compact disk players, laser printers and the high performance and reliable telecommunication are the result of this expansion. The optical fiber sensor technology is a direct outgrowth of the revolutions that have taken place in optoelectronics and fiber optic communication industries. Many of the components associated with these industries are used for optical fiber sensor applications. In the beginning the main hurdle in the development of sensors was the cost of components such as optical sources and detectors. With time, components prices have fallen and the quality of these components has improved. It is expected that as time progresses the optical fiber sensors will replace conventional devices for the measurements of various physical, chemical and biological parameters such as rotation, acceleration, electric and magnetic fields, temperature, pressure, acoustics, vibration, position, strain, humidity, viscosity, pH, glucose, heavy metals, gases, viral infection and pollutants etc.

Optical fiber sensor technology is very promising and is growing fast. At present there are about 100 companies dealing in optical fiber sensor products and more than 150 products are in the developed stage. This shows the utility of the technology and acts as a thrust for further development in the field. The current market for optical fiber sensor systems has been reported to be in the region of $550 million and is expected to reach more than $50 billion by 2008. The tremendous interest and activity in optical fiber sensors may be attributed to one or more of the following reasons:

1. In optical fiber sensors, the primary signal is optical, therefore there is no risk of electrical sparking or fire. Thus, they can be used in medical sciences without presenting any risk to patients. Further, they are safe for operation in hazardous environments such as oil refineries, grain bins, mines, and chemical processing plants.
2. Since the fibers are made of dielectrics there is no need of electrical isolation of patients in medicine, or, elimination of conductive paths in high voltage environments. Further, the signal cannot be interfered with by the static electricity of the body. Optical fiber sensors can be used in electrically noisy

environments and strong magnetic fields without any electromagnetic interference.

3. The optical fibers are manufactured from non-rusting materials such as plastics or glasses, therefore, these have excellent stability when in permanent contact with electrolyte solutions, ionizing radiation etc. Further, fibers can withstand temperatures as high as 350°C. Special fibers can extend sensor operation beyond 350°C to as high as 1200°C.
4. The light guiding core of the fiber typically ranges from 5-600 μm in diameter. Further, a 200 μm plastic clad silica fiber can be bent round a 1 cm radius mandrel. This implies that a sensor with a very small probe can be manufactured. This is advantageous, at least, in the case of minute samples and for invasive sensing in clinical chemistry and medicine. Further, due to lightweight and small size, these can be used in aircrafts.
5. With the availability of low loss optical fibers, the optical signal can be transmitted up to long distances (≈ 10-1000 m). Thus, remote sensing is possible with the optical fiber. This is important when the samples are hard to reach, dangerous, too hot or too cold, in harsh environments, or radioactive.
6. Optical fiber sensors are highly sensitive and have large bandwidth. When multiplexed into arrays of sensors, the large bandwidths of optical fibers themselves offer distinct advantages in their ability to transport the resultant data.
7. Distributed sensing is one of the advantages where there is no competition. It enables the measurement of a physical or a chemical parameter as a function of position along the length of an optical fiber, and hence provides a unique capability to measure spatial variations of these quantities using a compact, inert and non-electrical sensing cable. Such systems are finding applications in industrial and environmental sensing.
8. With the availability of solid-state configurations (small size sources and detectors) it is possible to design a compact optical fiber sensor system.

2. Generic Optical Fiber Sensor

The block diagram of a generic optical fiber sensor is shown in Fig. 1. The basic instrumentation required for the sensor is light source, detection system, referencing

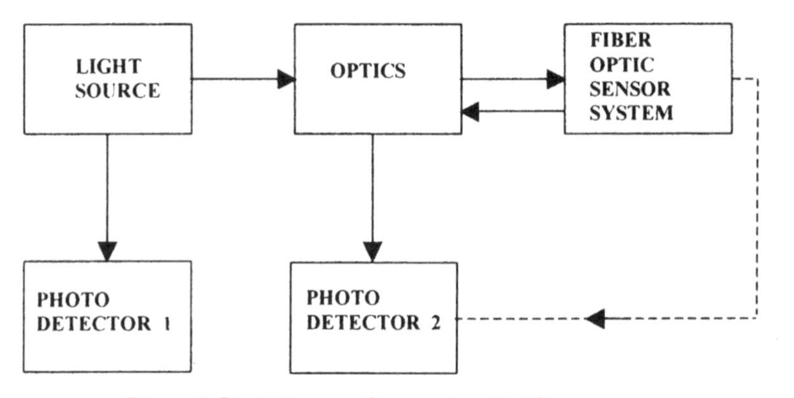

Figure 1 Block diagram of a generic optical fiber sensor.

scheme and sensor geometry. In the diagram, photodetector 1 is used to check the light source output power fluctuation and hence, is for referencing. The light from the source is coupled into the fiber using suitable optics. The sensing action occurs either within the fiber or external to the fiber. To communicate the change in output power taking place due to the measurand, either the same or different optics can be used. The change in output power is detected by the photodetector 2. The instrumentation can be simple or complex. It depends on the particular application. The choice of light source, the characteristics of the optical fiber and the detection system used are critical in determining the device capabilities such as the sensitivity and dynamic range. For example, the material of the fiber determines the usable range of wavelengths.

3. Modulation Schemes

Optical fiber sensors may be divided in accordance with the modulation scheme used. The divisions are as follows :

1. Intensity modulated sensors,
2. Phase modulated sensors,
3. Polarization modulated sensors,
4. Frequency modulated sensors, and
5. Wavelength modulated sensors.

In intensity modulated sensors, change in transmitted light is measured as a function of perturbing environment. These sensors require intense light to function. Generally, multimode fibers are used for intensity modulated sensors. Further, these sensors are simple and cheap. Phase modulated sensors are based on interference phenomenon. In this class of sensors, the phase of light in the sensing fiber is compared with the phase of light in the reference fiber. Multimode fibers are rarely used for phase modulated sensors because the multiplicity of paths within the reference and measurement arms makes it difficult to decipher the relative phase relationships. These sensors are more generally based on single mode fibers and have higher accuracy and larger dynamic range. But they are more expensive than intensity modulated sensors. In the polarization modulated sensors, plane polarized light is launched in the fiber and the change in the state of polarization is measured as a function of the perturbing parameter of interest. The frequency modulated sensors use Doppler effect, while the wavelength modulated sensors are based on fluorescence phenomenon.

4. Fields of Applications

Optical fiber sensors are finding their way into an ever-increasing number of applications as engineers and designers take advantage of their unique capabilities. In petroleum engineering, for example, these sensors are used for remote monitoring of the structure and function of an oil well to prevent hazards or malfunction. As the search for oil intensifies, wells are being drilled deeper, and monitoring equipment faces great demands in terms of pressure and temperature. Optical fiber sensors are also being used in squeezing the last ounce of production from old oil wells by optimizing the amount of water or steam injection used to efficiently remove reserves.

Optical fiber sensors also find application in structural monitoring. By measuring stress and strain in bridges, buildings, dams and tunnels – where failure can mean disaster – these sensors provide crucial analysis. The sensors are small enough to be embedded into the structure of buildings or vehicles, so they can passively monitor on a continuous basis, providing information that can be tracked over time to indicate important structural changes. These sensors can provide early warning of weaknesses in structure so that minor repairs can be made before a major disaster strikes.

Optical fiber sensors can be used to detect a large number of toxic substances and other chemical constituents such as formaldehyde, ammonia, nitrogen oxides, chloroform, hydrogen sulfide, sulfur dioxide, hydrocarbons, etc. in air and sea water. Since the signal is optical rather than electrical, there will not be any electrical interference that is normally present in a typical shipboard environment.

The greatest field of application is sensing clinically and biochemically important analytes. Accurate and rapid measurement of the level of these analytes in blood are critical to good medical practice and patient care. To measure, blood samples are withdrawn from the patient and sent to the clinical laboratory to determine their contents. The delays and potential errors that can be introduced because the laboratory is far from the patient and the physician who interprets the test results, may cause therapeutic decisions to be made without adequate information. If a large number of measurements are to be made per day then withdrawing sample and taking to the laboratory each time is not practical. Further, elderly, critically ill, or infants simply cannot afford to lose any blood. Optical fiber sensors with miniaturized probes can overcome these problems. The sensor can be fitted into a catheter that can be inserted into a blood vessel and hence pH, pO_2, pCO_2, hemoglobin, glucose, urea etc. in blood can be measured. Further, various forms of cancer and infection diseases can be detected using optical fiber sensors.

The main problem with conventional electrodes or electrode-based biosensors is that of sterilization. The measurement of physical and chemical parameters in bioreactors requires sterilized sensors. For optical fiber sensors, sterilization is not a problem. The fibers can be sterilized by steam at 115-130°C without compromising their performance. Thus optical fibers can be used in sensing O_2, pH, pCO_2, glucose, glutamate etc. in fermentation plants.

Because of public concern about the quality of drinking water, continuous monitoring of ground water has become a necessity. By introducing fibers down to the ground water level one can monitor the pH of the water and the amount of chloride, uranium, organic pollutants and tracer substances present in water before digging a well. This will save drilling cost because fiber hold (1-2 cm diameter) allows the use of small bore-holes. Further, using optical fibers the quality of the water can be monitored *in situ* and in real time. Thus, there will not be any need to take samples and get them analyzed in the laboratory.

Production efficiency and product quality depend on process control. Optical fiber sensors can do on-line measurements within the factory itself. During the process, temperature, pressure, flow, liquid level, analytical parameters etc. can be measured with optical fiber sensors.

There are a large number of other applications of optical fiber sensors. Optical fibers can be used in gyroscopes for rotation measurement, flight controls, engine monitoring, nuclear radiation testing, security systems etc. The temperature of various equipments (such as transformers) in power plants, and the current and voltage in power stations can also be sensed using optical fibers. The position, vibration and strain can be measured using optical fibers.

In this article, we briefly highlight some of the optical fiber sensors utilizing different modulation schemes. We discuss sensors for the measurements/detection of temperature, humidity, pH, refractive index, gases, rotation, pressure, strain and various chemical parameters. Some of these sensors have been extensively studied in our laboratory. We also discuss the underlying spectroscopic techniques or principles of these sensors.

5. Temperature Sensors

Several measurement principles have been used for optical fiber temperature sensors. These include the temperature dependence of the spectral intensity distribution of a black body[1], absorption (using thermochromic materials)[2,3], optical reflectance of liquid crystals[4], optical absorption edge in semiconductors[5], period of fiber Bragg gratings[6], absorption spectrum of rare-earth-doped fiber[7], fluorescence decay time,[8] etc. The temperature sensor developed using a thermochromic material as a transducer[3] uses cobalt chloride solution in 15% water and 85% isopropyl alcohol with a net concentration of 0.1M. The solution absorbs light strongly at 660 nm. As the temperature increases the absorbance increases. A schematic diagram of the optoelectronic system constituting the sensor is shown in Fig. 2. Light from a He-Ne laser is launched into two identical fibers of 200 μm core diameter. One fiber is used for the reference signal while, the other, for the measuring signal. The two fibers are equal in length except that the reference fiber is connected to a detector directly,

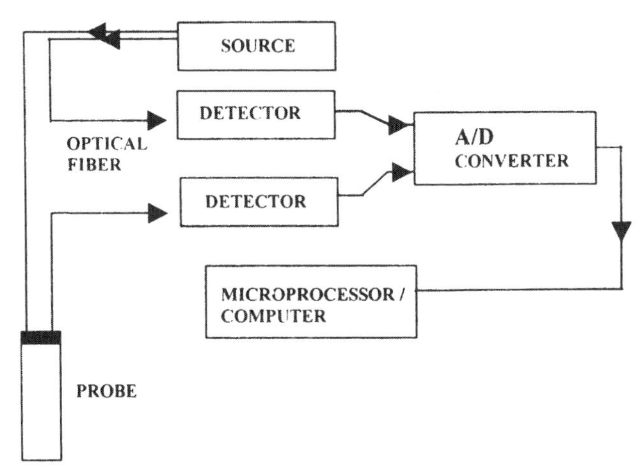

Figure 2 Schematic diagram of the optoelectronic system constituting an optical fiber temperature sensor (adapted from Refs.[3,4])

while, the other fiber is inserted into the cover of the probe containing thermochromic solution. The light exiting from this fiber passes through the solution and is reflected from the mirrored bottom of the probe. The reflected light after passing through the solution is recollected by a second fiber that is connected to another detector. The signals from the two detectors are fed into an A/D converter followed by a computer where the ratio of the two intensities is calculated. The ratio gives the temperature of the probe. The reference fiber eliminates the source fluctuations. The sensor operates in the temperature range 25 to 60°C.

The second temperature sensor, we describe here, is based on the temperature dependent optical reflectance of liquid crystals[4]. The experimental arrangement of the sensor is similar to that shown in Fig. 2. The only difference is in the design of the probe. It consists of two concentric glass tubes with one end sealed and the other end opened (Fig. 3A). The diameters of two tubes are such that the inner diameter of the outer tube is equal to the outer diameter of the inner tube. The liquid crystal is placed at the bottom between the two tubes. The inner tube contains seven fibers as

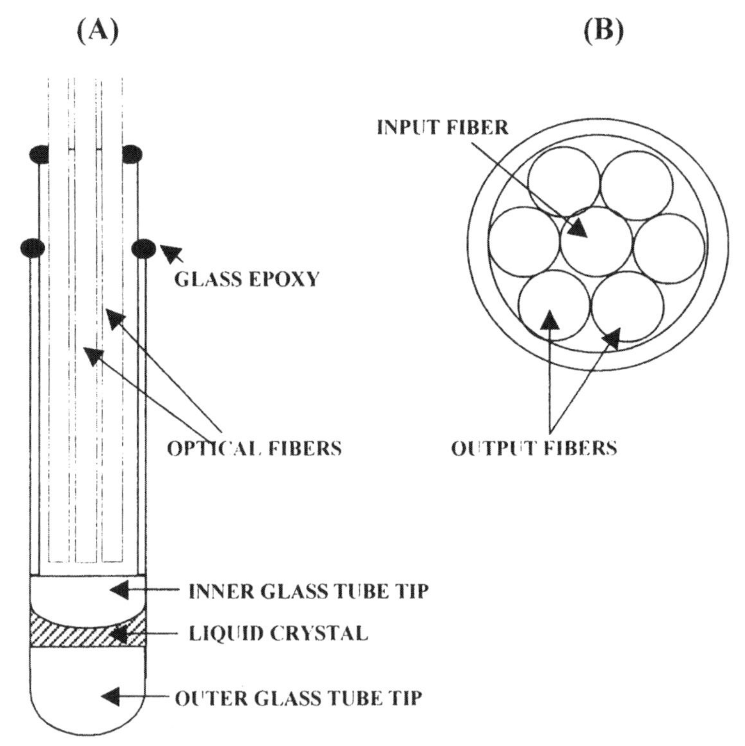

Figure 3 Liquid crystal based optical fiber probe (A) side view, (B) cross-sectional View (adapted from Ref.[4]).

shown in the cross section of the probe (Fig. 3B). The light from a He-Ne laser is launched into the middle fiber. The reflected light from the liquid crystal is collected by the remaining six fibers which are connected to the detector. The rest of the procedure is the same. The mixture of three cholesteric liquid crystals, namely cholesteryl chloride, cholesteryl nonanoate and cholesteryl carbonate is used in a certain ratio to avoid hysteresis during heating and cooling of the probe. The sensor operates in the range 32-42°C with an accuracy of 0.02°C.

The temperature sensors described above are mainly useful only for biomedical application. For very high temperature measurements fluorescent materials are used. One can measure temperatures as high as 1500 °C by using the fluorescence decay time in sapphire[9], Yb doped yttrium aluminum garnet[8] and single crystal ruby fiber[10].

6. Evanescent Wave Absorption Sensor

Evanescent wave absorption spectroscopy is a powerful technique for studying the absorption spectra of liquid and solid samples. It is also used to monitor, in real time and *in situ*, concentrations of reactants in chemical transformations of liquid and pastes. The technique utilizes evanescent wave penetration at the boundary between two dielectric media in conditions of attenuated total reflections (ATR). One of the media, a thin slab-shaped non-absorbing crystal layer, is called the waveguide, while, the other medium is the absorbing sample being studied, and has a lower refractive index. The penetration depth of the evanescent field in this medium is given by

$$d_p = \frac{\lambda}{2\pi n_1 (\sin^2\theta - n_{21}^2)^{1/2}} \tag{1}$$

where λ is the wavelength of light in free space, n_1 is the refractive index of the first medium, θ is the incident angle of the ray with respect to the normal to the interface, $n_{21} = n_2/n_1$, and n_2 is the refractive index of the second (absorbing) medium. In place of a slab-shaped waveguide, if an optical fiber is used then one can have a large number of reflections at the interface. The number of ray reflections per unit length of the fiber of core radius ρ is given by

$$N = \frac{\cot\theta}{2\rho} \tag{2}$$

Thus, smaller the radius the larger is the number of reflections. Increase in number of reflections increases the absorption and hence, the sensitivity. Much of the research with optical fibers utilizing evanescent wave has used plastic-clad silica fibers. The plastic cladding is removed from the middle portion of the fiber and an absorbing fluid is placed there. The presence of the fluid is detected by absorption of evanescent wave of light launched into the fiber. The sensitivity of such sensor depends on the launching condition of the ray and the radius of the fiber core. If the angle θ approaches the critical angle $\theta_c (= \sin^{-1} n_{21})$ of the unclad region (or, the sensing region), penetration depth increases and hence, the absorption increases, which results in the increase in the sensitivity of the sensor. Similarly, if the core radius decreases, the number of reflections increases, and hence, the absorption, and thereby the sensitivity, increases. In the case of a uniform core fiber, the angle of the ray launched at the input end of the sensing region remains constant throughout the sensing region (Fig. 4A). Therefore, the minimum value of θ that one can have in the sensing region is equal to the critical angle of the fiber $(= \sin^{-1} n_{cl}/n_1)$; n_{cl} is the refractive index of the fiber cladding. In practice $n_{cl} > n_2$, and hence, θ cannot approach θ_c in the case of a uniform core fiber. However, θ can approach θ_c if the probe is tapered[11-16] as shown in Fig. 4B. The first advantage of the taper is that the radius of the core decreases along the direction of light propagation, thereby, increasing the number of reflections. Its second advantage is that the angle of the ray with the normal to the interface decreases as the ray propagates, and thus, θ approaches θ_c. The sensitivity increases as the tapering increases. The angle of the ray can also approach the critical angle of the sensing region if the sensing region is

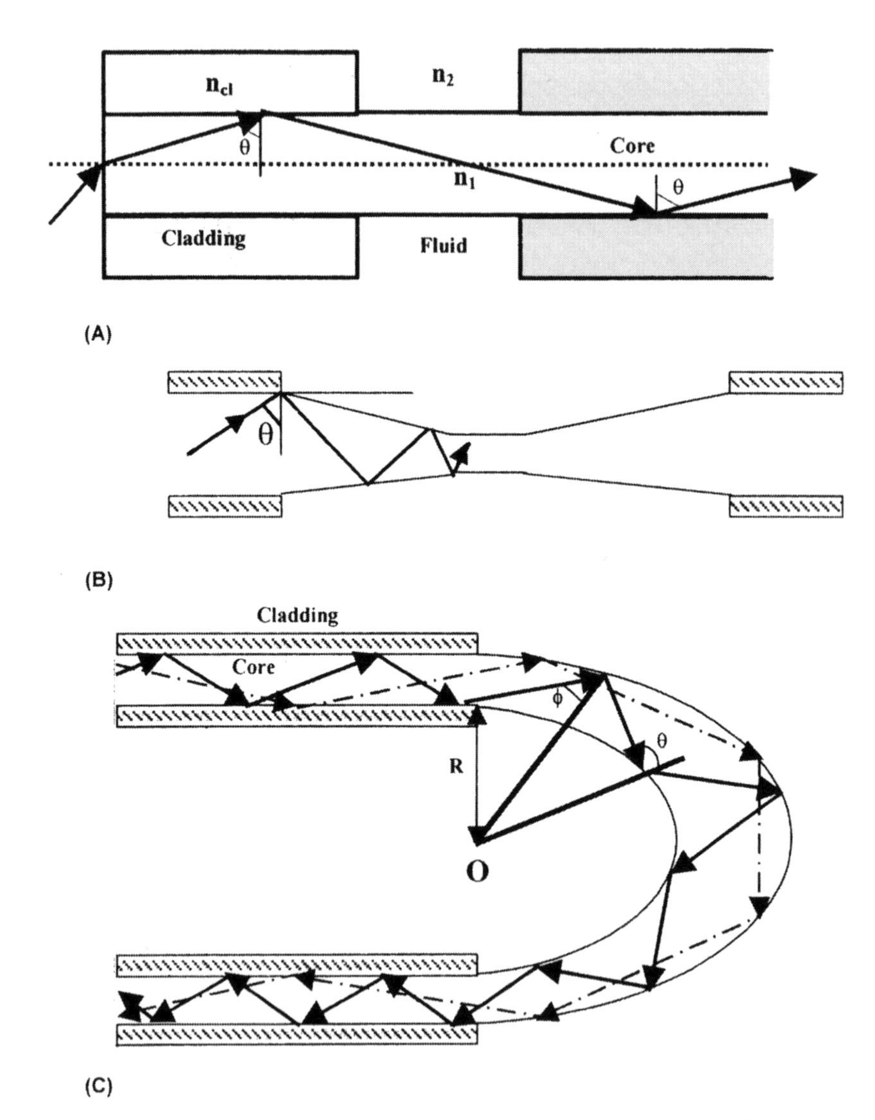

Figure 4 Fiber optic evanescent field absorption sensor probes. (A) Uniform and straight core probe, (B) tapered probe and (C) U-shaped probe.

made U-shaped[17-19] as shown in Fig. 4C. Theoretical and experimental studies on the sensitivity of these sensors have been reported in the literature. It has been shown that the U-shaped probe can be made more sensitive than the tapered probe by reducing the bending radius of the probe. The tapered probe is more fragile and difficult to handle. The additional advantage of the U-shaped probe is that it can be used as a point sensor.

7. Gas Sensor

Evanescent wave absorption spectroscopy can be used to detect gases. The fiber, with its unclad core, is placed in the area where the gas is to be detected. This allows the gas to interact with the evanescent waves. In the case of gases, the fractional power carried by the evanescent wave in the sensing region is very small because the refractive index of a gas is much smaller than that of the core. Hence, the interaction of the gas with the evanescent field is weak, which implies that the sensitivity of the sensor is poor. To enhance the sensitivity of this sensor, porous materials such as fluorocarbon polymers and silicones have been used as fiber cladding. These materials have refractive indices slightly less than that of the fiber core and thereby increase the fractional power carried by the evanescent wave in the cladding region. Furthermore, these materials are porous and the gas can diffuse through and interacts with the evanescent field present. This increases the sensitivity of the sensor. To further improve the sensitivity and the response time of the gas sensor utilizing this approach, the use of the tapered sensing region (both core and porous cladding) in place of the uniform sensing region has been proposed[20,21]. This further increases the evanescent field in the porous cladding, while the tapered cladding decreases the response time of the sensor.

8. pH Sensor

There are a variety of fiber optic sensors that use evanescent wave absorption spectroscopy indirectly for the detection of various chemical parameters. In these sensors a suitable chemical reagent is immobilized on the surface of the fiber core. When the analyte comes in contact with the reagent, it affects the absorption properties of the reagent. In other words, the reagent acts as a chemical transducer for the analyte that is not directly measurable by optical techniques. Certain conditions for the use of reagents in sensors must be met. First, the reagent must be highly sensitive, second, a substantial change in the absorption spectrum of the reagent should occur for small concentration change of the analyte and, third, it should be easily detected. In addition, the reagent should be specific for a given determination. pH is one of the chemical parameters that cannot be detected using direct absorption spectroscopy.

The measurement and control of pH is required in practically all kinds of sciences including chemical, biomedical and environmental. In recent years, numerous efforts have been directed toward the development of fiber optic pH sensors[22-28]. This is because fiber optic pH sensors offer many advantages over other types of sensors, such as, immunity to electromagnetic interference, no need for electrical insulation, the absence of reference electrodes and liquid junction, minute size and flexibility. Optical fiber pH sensors are based on pH-induced reversible changes in optical or spectroscopic properties such as, absorbance, reflectance, fluorescence and energy transfer. The major component in an optical fiber pH sensor is a pH sensitive layer/film that is prepared by immobilizing a pH sensitive dye onto the tip or sides of the core of the fiber. To immobilize, many research groups have used either covalent chemical linking or simple physical encapsulation techniques. For covalent linking of the dye molecules, surface modification is required[22-24]. The covalent linking method is a long, complex and tedious method, and may lead to loss of dye sensitivity, or, result in poor absorption and fluorescence properties[25]. In the non-covalent immobilization techniques, the dye is immobilized on the polymer support and entrapped behind semi-permeable membranes[26-28]. The disadvantage of the non-covalent immobilization methods is that these suffer from leachability of the dye that makes the long-term use of the sensor impractical.

Recently, sol-gel technique has been used for the immobilization of pH-sensitive dyes[29-37]. In this technique, a thin film of glass, with dye entrapped in it, is prepared from the hydrolysis and condensation polymerization of a metal alkoxide solution followed by the densification process. The sol-gel technique has several advantages over other techniques of film deposition. Apart from simplicity, the film produced is tough, inert and more resistant than polymer films in aggressive environments. Most of the fiber optic pH sensors developed and reported in the literature have a dynamic range of 3-4 pH unit. To increase the dynamic range of the sensor, a mixture of three suitably selected pH-sensitive dyes has been used[34]. The sensor was based on evanescent wave absorption spectroscopy. The probe was prepared by immobilizing pH-sensitive dyes on the surface of the unclad core of the fiber using sol-gel technique. The mixture of pH sensitive dyes enhanced the dynamic range of the sensor to 8-9 pH unit. To increase the pH range further, a number of pH sensitive dyes and their combinations have been tried. One of the dyes tested includes ethyl violet. Recently, fabrication and characterization of a fiber optic pH sensor based on evanescent wave absorption spectroscopy and immobilization of ethyl violet dye on the surface of the unclad core using sol-gel technique has been reported[37]. The sensor is found to operate in the pH range 2 to 13 covering a range of 11 pH units, the maximum range ever reported in the literature. The influence of ionic strength and temperature of the fluid on the response characteristics of the fiber optic pH sensor has been reported[37]. The sensitivity of the fiber optic pH sensor was increased by using an U-shaped probe[35].

Many fiber optic components such as couplers, polarizers, modulators, filters and amplifiers. have been developed using side-polished single mode optical fibers[38-41](see also Chapter 7). In the side polished fiber, the cladding is partially removed from one side of the fiber. If this side is brought in contact with some medium, the mode field of the fiber is modified. This principle is used to sense physical and chemical parameters of a fluid. The fabrication and characterization of a fiber optic pH sensor based on the side polished single mode optical fiber has been reported[36]. The sensor is prepared by fixing the fiber in a groove made on one of the surfaces of a fused silica block. To make the surface sensitive to pH, it is polished up to the core of the fiber, and a film containing pH sensitive dyes is coated on it using the sol-gel technique. The experimental set up of the sensor is shown in Fig. 5. Light from a He-

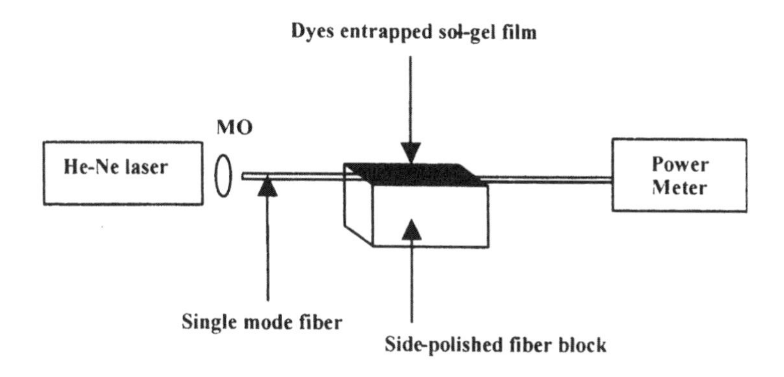

Figure 5 Experimental set up of the fiber optic pH sensor based on side-polished single mode optical fiber half block (adapted from Ref. [36]).

Ne laser operating at wavelength 632.8 nm is launched into one of the ends of the side polished fiber with the help of a microscope objective. The power coming out from the other end of the side polished single mode fiber is measured by using a silicon detector along with a power meter. The fluid is kept on the surface of the film. When light is launched in the fiber, the evanescent field present in the film interacts with the dyes in the presence of H^+ ions. The strength of interaction depends on the pH of the fluid and accordingly, the transmitted power at the other end of the fiber changes. The calibration curve of the sensor is shown in Fig. 6. The influence of the radius of curvature of the groove on the sensitivity of the sensor is also reported. It has been shown that as the radius of curvature of the groove decreases, the sensitivity increases[36]. The advantage of the side-polished single mode optical fiber sensor is that only one drop of the sample is sufficient to measure the pH of the fluid. This is important when the procurement of the sample in large quantities is not possible, for example, blood in the case of an infant.

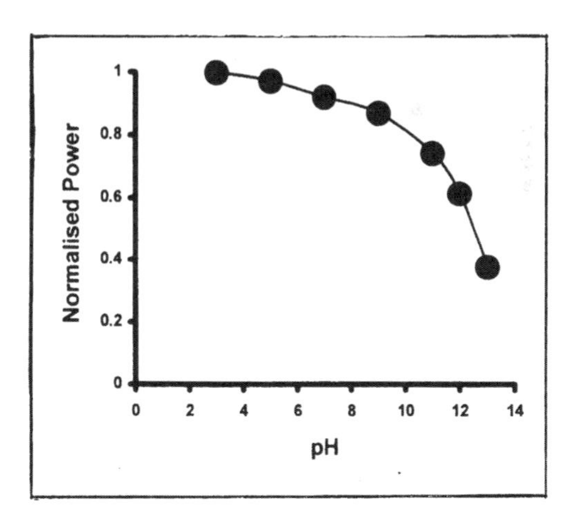

Figure 6 Calibration curve of the side polished single mode optical fiber pH sensor for half block with radius of curvature of the groove 22 cm (•) (adapted from Ref. [36]).

9. Humidity Sensor

In recent years the need for humidity sensors has greatly increased because of their applications in control of air conditioning, quality control of food products in a wide range of industries, paper and textile industries, optimal functioning of modern solid state electronic equipments, civil engineering, etc. Due to these applications, the development of simple, inexpensive, highly sensitive and quick to respond humidity sensors is required. There are a number of fiber optic humidity sensors employing different concepts that have been reported in the literature. These sensors use either expensive methods of probe fabrication or fragile probes. In the case of some sensors, the probe is less sensitive or the response is slow. When a polymethylmethacrylate (PMMA) film containing phenol red dye is exposed to moisture environments, its absorption coefficient around 530 nm changes remarkably. Based on this, a highly sensitive fiber optic humidity sensor has been developed[42]. A PMMA film containing phenol red dye is deposited over the unclad core of a highly multimode plastic clad silica (PCS) fiber. The refractive index of the PMMA film is greater than that of the fiber core. When the light launched in the fiber reaches the coated portion of the fiber a fraction of light guided in the core transmits into the film after each reflection at the core-film interface as it propagates along the fiber axis. If the wavelength of the light launched in the fiber is close to the peak absorption wavelength of the phenol red dye present in the film, the absorption of the transmitted light in the film occurs. The absorption depends on the moisture present in the air around the phenol red doped PMMA film. After traversing the film, the light couples back to the fiber core and is detected by a power meter at the other end of the fiber. For a given length of the coated probe, the amount of absorption depends on the angle of incidence of the ray at the core-film interface. Smaller the angle of incidence, larger will be the transmission. The angle of incidence at the core-film interface can be decreased by using a U-shaped probe. Gupta and Ratnanjali[42] used a U-shaped probe for the fabrication of a humidity sensor. The experimental set up is shown in Fig. 7. Light from a He-Ne laser operating at 544 nm is focused using a microscope objective on the input face of the fiber. The probe is fixed inside a cylindrical plastic container (humidity chamber). A hygrometer is placed inside the chamber to calibrate the sensor. The output power is measured as a function of humidity inside the chamber. The sensor operates from 20 to 80% RH and its response time is reported to be about 5 sec. The sensitivity of the sensor increases as the bending radius of the probe decreases[42].

Recently a novel concept utilizing fiber Bragg gratings has been reported for a relative humidity sensor[43] (see Section 12 for fiber Bragg grating). The bare silica fiber, which is not sensitive to humidity, is coated with polyimide. Polyimide polymers are hygroscopic and swell in aqueous media as water molecules diffuse into them. In the sensor, the fiber Bragg grating is coated with polyimide. The swelling of the polyimide coating strains the fiber which modifies the Bragg condition and thus serves as the basis of the humidity sensor. The sensor is reported to have linear, reversible, and accurate response behaviour in the range 10-90% RH and at 13-60°C. The RH and temperature sensitivities are also studied as a function of coating thickness.[43]

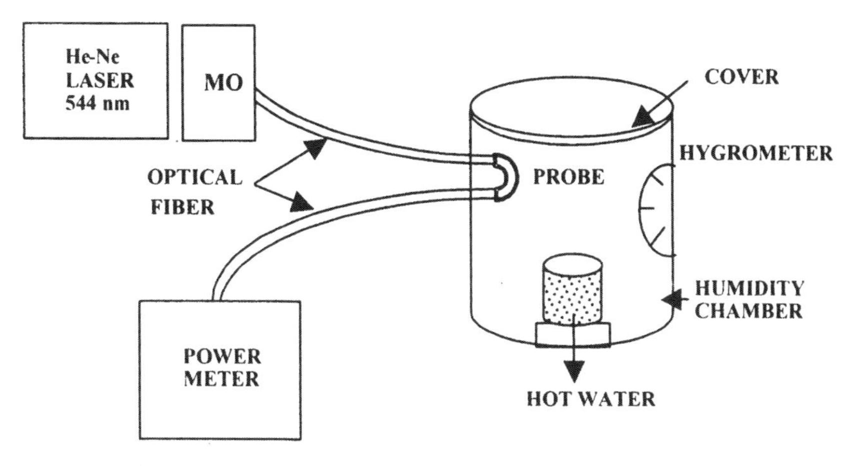

Figure 7 Fiber optic humidity sensor based on U-shaped probe (adapted from Ref. [42]).

10. Fiber optic Gyroscope

The fiber gyroscope was first reported in 1976 and first appeared in the civil aircraft Boeing 777 well over a decade later. The fiber optic gyroscope (FOG) is now a regular production item. These are now being used for aerospace applications, off-shore exploration systems, automobile navigation systems, satellite antennas and telescopes. The FOG is based on the Sagnac effect that senses inertial rotation[44]. The device relies on the production of a phase difference between two counter-propagating light beams in a fiber loop as a result of system rotation. The device facilitates precise measurement of angular rate. The basic FOG contains a multi-turn coil of fiber (Fig. 8). Light from a broadband source such as a light-emitting diode (LED) or a superluminescent diode is coupled into a directional coupler (DC1). The

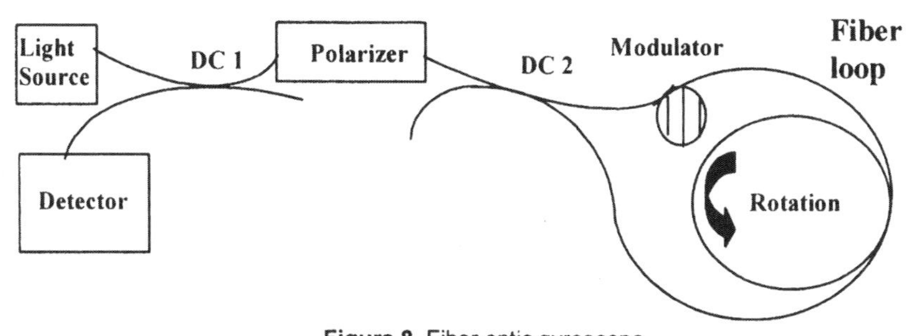

Figure 8 Fiber optic gyroscope.

light beam then passes through a polarizer to ensure the reciprocity of the counter-propagating light beams through the fiber coil. The second directional coupler (DC2) splits the two light beams into the fiber optic coil where they pass through a modulator that is used to generate a time varying signal for ease of processing. When the loop rotates, light traveling in the same direction as the rotation, sees the coupler move away from it during the transit time. Consequently, the light stays a little longer in the loop than the light traveling against the rotation. This time difference is measured as a fringe drift at the output from the coupler. The sensitivity of the FOG increases with the increase in the length of the fiber. The FOG typically has a detection threshold of 0.1 to 0.01° per hour with a fiber loop area of the order of 10 cm^2. One of the main advantages of the FOG is that there are no moving parts in the design. Further, the cost is less than the equivalent mechanical gyroscopes.

11. Surface Plasmon Resonance Sensor

Among a variety of optical evanescent wave techniques, spectroscopies based on the resonant excitation of surface plasmons have matured into a widely accepted surface analytical method used in different fields. Liedberg et al.[45] were the first to demonstrate the exploitation of surface plasmon resonance (SPR) for chemical sensing. Since then the SPR sensing principle has received much attention.

Surface plasmons are the surface localized electromagnetic waves produced by the collective resonating oscillation of free electrons on the plasma surface (for example, a metal). The surface plasmon is a transverse wave propagating along the plasma surface with the oscillating electric field vector normal to the surface. Because the surface plasmon is a transverse magnetic mode, it can be excited by p-polarized light. When the wave vector and the frequency of the incident light coincide with those of the surface plasmon, the light resonantly excites the surface plasmon. The wave vector K_{sp} of the surface plasmon on the boundary between a metal and a sample at frequency ω is given approximately by[46]

$$K_{sp} = \frac{\omega}{c}\sqrt{\frac{\varepsilon_m \varepsilon_s}{\varepsilon_m + \varepsilon_s}} \tag{3}$$

where ε_s and ε_m are the real parts of the dielectric constants of the sample and the metal, respectively and c is the velocity of light in vacuum. The wave vector of the light at frequency ω propagating through the sample medium is given by

$$K_s = \frac{\omega}{c}\sqrt{\varepsilon_s} \tag{4}$$

Since $\varepsilon_m < 0$ and $\varepsilon_s > 0$, for a given frequency, the wave vector of surface plasmon is greater than the wave vector of the light in sample medium. Therefore, to excite SPR the momentum and, hence, the wave vector of the exciting light in sample medium should be increased. Kretschmann[47] proposed a way to excite SPR by an evanescent wave from a high refractive index prism at the total reflection condition (Fig. 9). A high refractive index prism is coated with a thin metal film (around 50 nm thick) touching the sample. When a light beam is incident through the prism on the surface of thin metal film at the total reflection angle θ, the evanescent wave interacts with the sample, and at θ_{sp} angle, it couples with the K_{sp} of surface plasmon. Figure 10 shows the dispersion curves of the surface plasmon and the lightwaves. The wave vector K_{ev} of the evanescent wave is equal to the lateral component of the incident lightwave vector (K_g) in the prism. Thus,

$$K_{ev} = K_g \sin\theta = \frac{\omega}{c}\sqrt{\varepsilon_g}\,\sin\theta \quad (5)$$

where ε_g is the dielectric constant of the prism. The excitation of surface plasmon occurs when the wave vector of the evanescent wave of frequency ω_o matches that

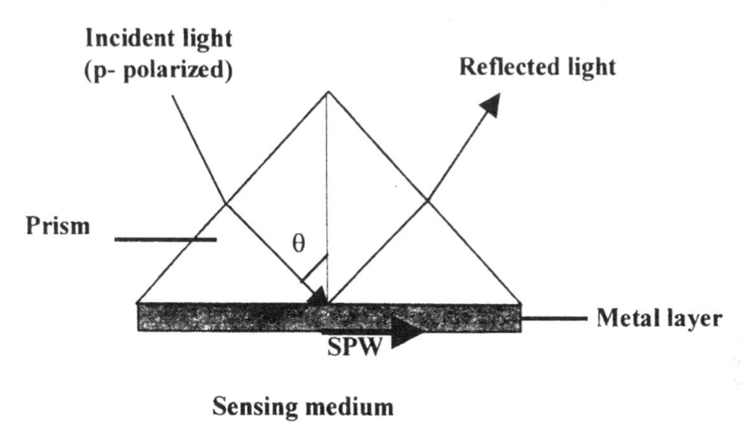

Figure 9 Kretschmann configuration of a prism based SPR sensor.

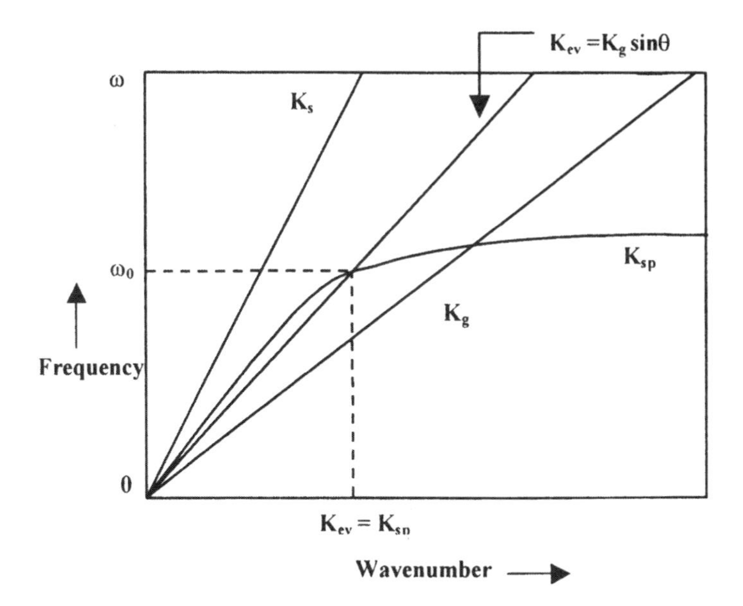

Figure 10 Dispersion curves of surface plasmon (K_{sp}), sensing medium (K_s), evanescent wave (K_{ev}) and prism (K_g).

of the surface plasmon of frequency ω_o. This occurs at a particular angle of incidence, θ_{sp}. This results in the transfer of energy to surface plasmons, which reduces the energy of the reflected light. If reflectance R is measured as a function of incident angle, θ, a sharp dip is observed at the resonance angle, θ_{sp}. Knowing this angle, the dielectric constant of the sample ε_s can be determined using Eqs. (3) and (5). The resonance angle is very sensitive to variations in the refractive index (or, dielectric constant) of the sample medium. Moreover, the drop in the resonance is very sharp when the imaginary part of the dielectric constant of the metal is small. The narrower the dip, the better the detection of the variation in refractive index.

Both silver and gold have traditionally been used to excite SPR in the visible and near-infrared regions. Silver produces sharper resonance than gold because of the imaginary part of its dielectric constant being smaller. The permittivity coefficient of the silver and gold layers are $18.5+0.75i$ and $10.5+2.5i$, respectively. Nevertheless, silver has not been used extensively owing to the reactivity of its surface, particularly its susceptibility to oxidation on exposure to laboratory air. The prism based SPR sensing instrument has a number of drawbacks. One of the drawbacks is its bulky size and the presence of various optical and mechanical (moving) parts, which make it difficult to optimize and commercialize on a large scale. Further, it is inapplicable for remote sensing applications. Miniaturization of SPR components is a key factor that can be achieved by the use of optical fibers. One of the biggest advantages of using optical fibers is the small diameter of the core which allows them to be used in very small areas. The other advantages of fiber optic SPR sensing configurations are simplified optical design and the capability for remote sensing. Basically, three main approaches have been used in SPR sensors. These are the measurement of SPR-induced intensity change[48], angular interrogation[49] and spectral interrogation[50,51] of SPR.

In the fiber optic SPR sensor the coupling prism is eliminated by depositing the metal film directly on the core of the fiber as shown in Fig. 11. The SPR sensing method of fixed angle of incidence and modulated wavelength is selected

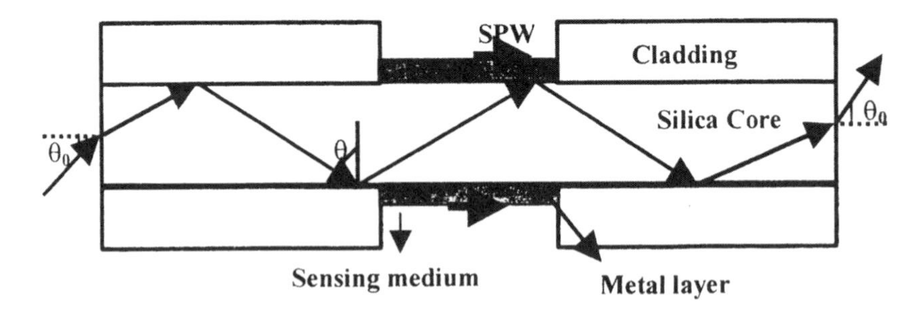

Figure 11 Fiber optic SPR sensor probe.

(wavelength interrogation). This is because the wavelength intensity distribution may be preserved in an optical fiber, whereas the angular intensity distribution of light will be indistinguishable due to mode mixing as a result of the inherent bending of the multimode optical fiber in practical sensing applications. Further, the sensor is fabricated on a multimode fiber there is not one fixed angle of incidence, rather a range of incident angles are allowed to propagate in the fiber.

The SPR sensors based on multimode fibers[52,53] exhibit a rather limited resolution mainly due to the modal noise present in multimode fibers causing the strength of the interaction between the fiber-guided light wave and the SPW to fluctuate. To overcome this inherent limitation of SPR sensing devices based on multimode fibers, SPR sensors based on a single mode optical fibers were proposed[54,55]. They include SPR sensors based on tapered[55] and side-polished[54] single mode optical fibers. Out of these two intensity modulated SPR sensors, side polished single mode optical fiber based SPR sensor offers superior sensitivity[54], although it suffers from adverse sensitivity to fiber deformations because any deformation in fiber changes the state of polarization of the fiber mode, and consequently, also the strength of its interaction with SPW. To overcome this, a side polished fiber optic SPR sensor based on spectral interrogation using depolarized light was proposed[56]. The applicability of the sensor to refractometry and affinity biosensing was demonstrated. Recently, an optical fiber SPR sensor based on polarization maintaining fibers and wavelength modulation was reported[57]. The design provides superior immunity to deformation of optical fibers of the sensor and hence stability. The sensor was able to resolve refractive index changes as low as 4×10^{-6} under moderate fiber deformations.

As mentioned above, the metallic layer used in SPR measurements consists of either silver or gold. Gold demonstrates a higher shift of resonance parameter to change in refractive index of the sensing layer, and is chemically stable. Silver, on the other hand, displays a narrower resonance curve, causing a higher signal-to-noise ratio (SNR) of SPR chemical sensor, but has a poor chemical stability. The oxidation of silver occurs as soon as it is exposed to air and especially to water, which makes it difficult to get a reliable sensor for practical applications. Treatment of the silver surface by a thin and dense cover is therefore required. Recently, Zynio et al.[58] suggested a new structure of resonant metal film based on bimetallic layers (gold as outer one) on the prism base with angular interrogation sensing. They showed that the new structure displayed high shift of resonance angle as gold films, but also showed narrower resonance curve as silver film, thus providing a higher SNR in addition to protecting silver against oxidation. Recently, the above structure with spectral interrogation in place of angular interrogation has been studied. The investigation is extended to optical fibers for the selected, as well as, all guided rays launching configuration[59]. The sensitivity and signal-to-noise ratio (SNR) of the optical fiber based bimetallic SPR sensor are evaluated numerically. The effects of fiber parameters like numerical aperture (NA), core diameter and sensing parameters like sensing region length and metallic film thickness have been studied for different ratios of silver and gold layer thickness. The analysis has been extended to remote sensing applications[59].

Recently, attention has also been focused on introducing optical absorption technique into SPR. An optical absorption-based SPR sensor was proposed and its theoretical aspects were discussed in terms of mathematical descriptions and numerical simulations of the SPR curve[60]. The response theory of the absorption-based SPR sensor was based on the expansion of Kretschmann's SPR theory into the case in which optical absorption in the sensing layer was expressed by the Lorentz model. The theory has now been applied to fibers with single metallic layer[61] and bimetallic layer[62].

12. Fiber Bragg Grating Based Sensors

Fiber Bragg gratings[63-65] (FBGs) are fabricated by writing an index grating directly into a doped optical fiber. Two intense ultraviolet beams are angled to form an interference pattern with the desired periodicity and are incident upon the side of the bare fiber after the external coatings have been stripped. The intense bright and dark fringes cause local changes in the refractive index through migration of the dopants in the fiber. After the grating is written, the fiber is recoated with polyamide. When writing many FBGs on a single optical fiber, careful consideration of the specifications of each FBG must be made. Each FBG is allotted its own wavelength segment so that their signals do not overlap.

When light is coupled to a fiber with a Bragg grating written on it, part of it is reflected and the rest is transmitted (see Chapter 9). The wavelength of the light reflected is called Bragg wavelength (λ_B). It is related to the refractive index of the fiber (n) and the spacing of the grating planes (Λ) by the simple formula

$$\lambda_B = 2n\Lambda \qquad (6)$$

Temperature and strain directly affect the period of the grating as well as the effective refractive index. Thus, any change in temperature or strain directly affects the Bragg wavelength. To measure the wavelength shift, FBG sensor systems must consist of an optical source that continuously interrogates the reflection spectrum, and a detection system that records the shift in the peak reflectivity as a function of wavelength (see Fig. 12). The sensitivity of the FBG sensor depends on the wavelength accuracy. The simplest FBG sensor uses a broadband light source with a tunable filter and a detector. Because detectors are wavelength insensitive, a tunable wavelength filter is required to scan the wavelength range of the FBG sensor to determine the sensor's Bragg wavelength. The main advantage of these systems is that they are low cost, but because the output power from a white light source is low, only a limited number of in-line gratings can be measured and with a limited dynamic range. Moreover, an external wavelength filter limits the accuracy and scan frequency of these systems.

Replacing the broadband white light source with a laser provides higher output powers and thus an increase in both the number of FBG sensors that can be monitored and the available dynamic range. In addition, a laser emits only a narrow spectrum of optical radiation, thus eliminating the need for tunable filter. Narrow-line width, low noise, swept wavelength external-cavity diode lasers (ECDL) provide even higher output powers and increase both the number of FBG sensors and the dynamic range. FBG sensors that use low-noise swept-wavelength lasers are being investigated for monitoring stress in airplanes, efficient oil-well recovery, and flood control. In flood control, FBGs are placed in rivers upstream of populated areas. Early-warning systems constantly monitor the changes in pressure. Large changes enable these systems to provide enough warning to evacuate people from the affected areas safely.

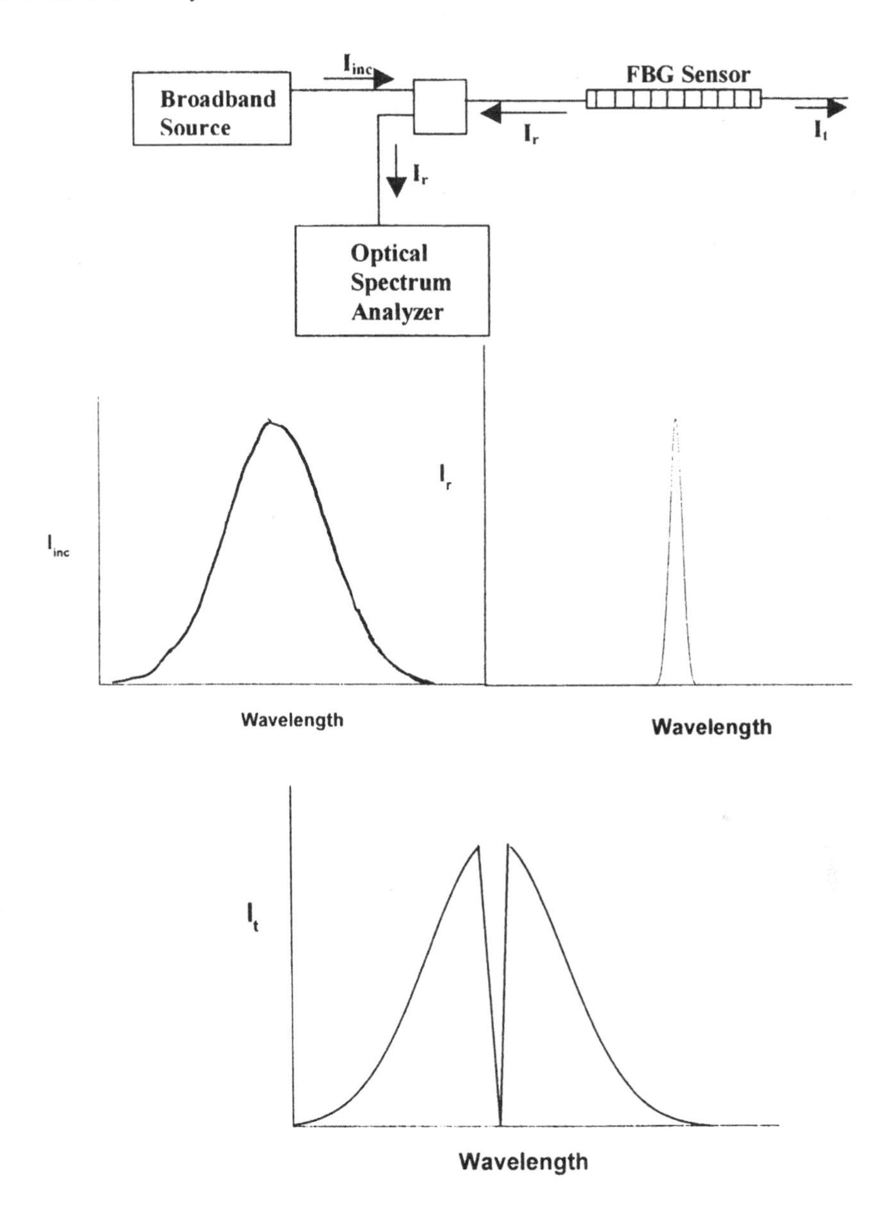

Figure 12 Basic fiber Bragg grating sensor with transmitted and reflected spectra.

Fiber Bragg gratings are also being used during the production of oil and gas in a variety of sensing applications such as temperature and pressure inside the well. The FBGs are spaced over a long stretch of fiber thereby continuous readings are taken from many locations simultaneously, even over distances of many kilometers. A fiber optic sensor can be sent miles down an oil well inside a spaghetti-thin wire capillary. The associated electronic read-out instrumentation remains in the equipment room at the wellhead. This is the one of the main advantages over traditional sensor technologies in which the electronics is located close to the transducer. In an oil well, temperatures could be in excess of 250°C and pressure to 25,000 psi - hardly the place for electronics, but perfect for fiber optic sensors. FBG based distributed temperature sensor has been accepted into the oil and gas industry to date. At temperature above 200°C, free hydrogen exists in the well bore. The FBG-based sensors operating around 1550 nm wavelength are less affected by the free hydrogen. Thus, they can last longer at these high temperatures. The FBGs are not immune to the hydrogen effects, but higher signal-to-noise ratios and wavelength utilization away from the hydrogen peak means a sensor can be designed to remain effective for more than 10 years.

13. Photonic-Crystal Fiber Sensor

In recent years microstructured or "holey" fibers have generated a wave of excitement because they promise properties that cannot be achieved in conventional optical fibers (see Chapter 6). The light-guiding properties of conventional fibers are limited by the properties of the materials from which they are made. Internal microstructures add another degree of freedom in controlling the light-guiding properties of fibers, opening up new possibilities in designing fiber-based devices. Recently a gas sensor has been reported using holey fiber with an air cavity at its core to sense gas molecules[66]. The core of the fiber acts as the cell. This gives a very long interaction length between the guided mode of the light and the gas while requiring only a small total volume of the gas.

In the sensor, light from a tunable laser is focused into a vacuum chamber. About 1 m long and 10-15 μm core diameter holey fiber is placed in the chamber with one end near the light-input window. The other end of the fiber is spliced to a single mode fiber connected to an optical spectrum analyzer. When a gas at low pressure is introduced into the chamber, it enters the core of the fiber. The light traveling through the fiber is almost entirely confined in the core. In addition to the long interaction length, microstructured fibers with air at the core offer higher sensitivity than index-guiding photonic crystal fibers because more than 99% of the main mode is in the air core.

The guiding capabilities of the holey fiber make it highly insensitive to bending, allowing the fiber to be coiled into a relatively compact package. Further, low power light sources such as LEDs can be used to reduce the cost of the sensor. The tunable laser provides higher resolution, which would be useful, for example, in distinguishing the multiple lines in a spectrum.

14. Issues

Optical fiber sensors have their limitations and these need to be appreciated if their advantages are to be exploited. Areas of concern include

1. Cost: The optical fiber sensors are costlier than the conventional electrical-based sensor systems. Therefore, optical fiber sensors should be used in situations where conventional sensor systems cannot be deployed.
2. Robustness: With the exception of polymeric fibers, optical fibers and sensors tend to be brittle, and adequate protection must be provided prior to their deployment. For example, for monitoring concrete structures, the sensor system has to be chemically protected from the alkaline matrix and physically protected from potential damage from the aggregates.
3. Transduction: This is the efficiency with which the measurand of interest is transferred to the sensing region. Thus the measurand should be transferred efficiently from the substrate to the sensor. Various factors can influence this transfer.
4. Data interpretation: The interpretation of data from the sensors may not be straightforward because the output can be influenced by other parameters. Thus the effect of these parameters on the output should be controlled.
5. Stability and reliability: Optical fiber sensor is an emerging technology, there is little information available on the long-term performance, drift and reliability.

***Acknowledgement*:** The author is thankful to Mr. Anuj Kumar Sharma for helping me in the preparation of the manuscript.

References

1. R.R. Dils, "High temperature optical fiber thermometer", *J. Appl. Phys.* **54**, 1198-1200 (1983).
2. M. Bassi, M. Brenci, G. Conforti, R. Falciai, A.G. Mignani and A.M. Scheggi, "Thermochromic transducer optical fiber thermometer", *Appl. Opt.* **25**, 1079-1081 (1986).
3. B.D. Gupta, A. Sharma and S.K.S. Nair, "Fiber optic temperature sensor for biomedical applications", *J. IETE*, **38**, 368-370 (1992).
4. B.D. Gupta, A. Sharma and R. Goyal, "Optical fibre temperature sensor based on liquid crystals", *Int. J. Optoelectron.* **8**, 13-19 (1993).
5. Y. Zhao and Y. Liao, "Compensation technology for a novel reflex optical fiber temperature sensor used under offshore oil well", *Opt. Commun.* **215**, 11-16 (2003).
6. A.D. Kersey and T.A. Berkoff, "Fiber-optic Bragg-grating differential – temperature sensor", *IEEE Photon. Technol. Lett.* **4**, 1183-1185 (1992).
7. K.W. Quoi, R.A. Lieberman, L.G. Cohen, D.S. Shenk and J.R. Simpson, "Rare earth doped optical fibers for temperature sensing", *J. Lightwave Technol.* **10**, 847-852 (1992).
8. J.L. Kennedy a nd N. Djeu, "Operation of Yb:YAG fiber-optic temperature sensor up to 1600 °C", *Sensors and Actuators A* **100**, 187-191 (2002).
9. Y. Shen, L. Tong, Y. Wang and L. Ye, "Sapphire-fiber thermometer ranging from 20 to 1800 °C", *Appl. Opt.* **38**, 1139-1143 (1999).

10. H.C. Seat, J.H. Sharp, Z.Y. Zhang and K.T.V. Grattan, "Single-crystal ruby fiber temperature sensor", *Sensors and Actuators A* **101**, 24-29 (2002).
11. B.D. Gupta, A. Sharma and C.D. Singh, "Evanescent wave absorption sensors based on uniform and tapered fibers : a comparative study of their sensitivities", *Int. J. Optoelectron.* **8**, 409-418 (1993).
12. B.D. Gupta and C.D. Singh, "Evanescent-absorption coefficient for diffuse source illumination : uniform and tapered –fiber sensors", *Appl. Opt.* **33**, 2737-2742 (1994).
13. B.D. Gupta, C.D. Singh and A. Sharma, "Fiber optic evanescent field absorption sensor : effect of launching condition and the geometry of the sensing region", *Opt. Engn.* **33**, 1864-1868 (1994).
14. B.D. Gupta and C.D. Singh, "Fiber-optic evanescent field absorption sensor : a theoretical evaluation", *Fiber Integrat. Opt.* **13**, 433-443 (1994).
15. B.D. Gupta, A.K. Tomar and A. Sharma, "A novel probe for an evanescent wave fiber-optic absorption sensor", *Opt. Quant. Electron.* **27**, 747-753 (1995).
16. S.K. Khijwania and B.D. Gupta, "Fiber optic evanescent field absorption sensor based on a tapered probe : effect of fiber parameters on the response curve" *Proc. SPIE* **3666**, 578-584 (1999).
17. B.D. Gupta, H. Dodeja and A.K. Tomar, "Fibre-optic evanescent field absorption sensor based on U-shaped probe", *Opt. Quant. Electron.* **28**, 1629-1639 (1996).
18. S.K. Khijwania and B.D. Gupta, "Fiber optic evanescent field absorption sensor : effect of fiber parameters and geometry of the probe", *Opt. Quant. Electron.* **31**, 625-636 (1999).
19. S.K. Khijwania and B.D. Gupta, "Maximum achievable sensitivity of the fiber optic evanescent field absorption sensor based on the U-shaped probe", *Opt. Commun.* **175**, 135-137 (2000).
20. C.D. Singh and B.D. Gupta, "Gas sensor response time analysis for tapered porous-clad fibers", *Fiber Integrat. Opt.* **14**, 171-177 (1995).
21. C.D. Singh and B.D. Gupta, "Detection of gases with porous –clad tapered fibers", *Appl. Opt.* **34**, 1019-1023 (1995).
22. O.S. Wolfbeis, N.V. Rodriguez and T. Werner, "LED-compatible fluorosensor for measurement of near-neutral pH values", *Mikrochim Acta* **108**, 133-141 (1992).
23. R. Wolthuis, D. McCrae, E. Saaski, J. Hartl and G. Mitchell, "Development of a medical fiber-optic pH sensor based on optical absorption", *IEEE Trans. Biomed. Eng.* **39**, 531-537 (1992).
24. J.W. Parker, O. Laksin, C. Yu, M. Lau, S. Klima, R. Fisher, I. Scott, B.W. Atwater, "Fiber optic sensors for pH and carbon dioxide using a self referencing dye", *Anal. Chem.* **65**, 2329-2334 (1993).
25. K. Buchholz, N. Buschmann and K. Cammann, "A fibre-optical sensor for the determination of sodium with a reversible response", *Sensors and Actuators B* **9**, 41-47 (1992).

26. J.I. Peterson, S.R. Goldstein and R.V. Fitzgerald, "Fiber optic pH probe for physiological use", *Anal. Chem.* **52**, 864-869 (1980).
27. R.B. Thomas and J.R. Lakowicz, "Fiber optic pH sensor based on phase fluorescence lifetimes", *Anal. Chem.* **65**, 853-856 (1993).
28. T.L. Blair, T. Cynkowski and L.G. Bachas, "Fluorocarbon-based immobilization of a fluoroionophore for preparation of fiber optic sensors", *Anal. Chem.* **65**, 945-947 (1993).
29. B.D. MacGraith, V. Ruddy, C. Potter, B. O'Kelly and J.F. McGilp, "Optical waveguide sensor using evanescent wave excitation of fluorescent dye in sol-gel glass", *Electron. Lett.* **27**, 1247-1248 (1991).
30. J.Y. Ding, M.R. Shahriari and G.H. Sigel, "Fibre optic pH sensors prepared by sol-gel immobilisation techniques", *Electron. Lett.* **27**, 1560-1562 (1991).
31. S.A. Grant and R.S. Glass, "A sol-gel based fiber optic sensor for local blood pH measurements", *Sensors and Actuators B* **45**, 35-42 (1997).
32. B.D. Gupta and D.K. Sharma, "Evanescent wave absorption based fiber optic pH sensor prepared by dye doped sol-gel immobilization technique", *Opt. Commun.* **140**, 32-35 (1997).
33. O. Ben-David, E. Shafir, I. Gilath, Y. Prior and D. Avnir, "Simple absorption optical fiber pH sensor based on doped sol-gel cladding material", *Chem. Mater.* **9**, 2255-2257 (1997).
34. B.D. Gupta and S. Sharma, "A long-range fiber optic pH sensor prepared by dye doped sol-gel immobilization technique", *Opt. Commun.* **154**, 282-284 (1998).
35. B.D. Gupta and N.K. Sharma, "Fabrication and characterization of U-shaped fiber-optic pH probes", *Sensors and Actuators B* **82**, 89-93 (2002).
36. N.K. Sharma and B.D. Gupta, "Fabrication and characterization of pH sensor based on side polished single mode optical fiber", *Opt. Commun.* **216**, 299-303 (2003).
37. N.K. Sharma and B.D. Gupta, Fabrication and characterization of a fiber optic pH sensor for the pH range 2 to 13", *Fiber Integrat. Opt.* **23**, 327-335 (2004).
38. R.A. Bergh, G. Kotler and H.J. Shaw, "Single-mode fiber optic directional coupler", *Electron. Lett.* **16**, 260-261(1980).
39. D. Gruchmann, K. Petermann, L. Staudigel and E. Weidel, "Fibre-optic polarizers with high extinction ratio", Proc. 9th Eur. Conf. Optical Commun., Geneva, Switzerland, North Holland, Amsterdam, 1983, p. 305.
40. W.V. Sorin, K.P. Jackson and H.J. Shaw, "Evanescent amplification in a single-mode optical fiber", *Electron. Lett.* **19**, 820-821(1983).
41. C. Millar, M. Brierley and S. Mallinson, "Exposed-core single-mode fiber channel-dropping filter using a high index overlay waveguide", *Opt. Lett.* **12**, 284-286 (1987).
42. B.D. Gupta and Ratnanjali, "A novel probe for a fiber optic humidity sensor", *Sensors and Actuators B* **80**, 132-135 (2001).

43. P. Kronenberg, P.K. Rastogi, P. Giaccari and H.G.Limberger, "Relative humidity sensor with optical fiber Bragg gratings", *Opt. Lett.* **27**, 1385-1387 (2002).
44. Fiber-optic gyros : 15th Anniversary Conference, edited by S. Ezekial and E. Udd, Proc. SPIE, **1585** (1991).
45. B. Liedberg, C. Nylander and I. Sundstrom, "Surface plasmon resonance for gas detection and biosensing", *Sensors and Actuators B* **4**, 299-304 (1983).
46. H. Raether, *Physics of thin films* (Academic : New York, 1984).
47. E. Kretchmann, "Die bestimmung optischer konstanten von metallen durch anregung von oberflachenplasmaschwingungen", *Z. Phys.* **241**, 313-324 (1971).
48. M.B. Vidal, R. Lopez, S. Aleggret, J. Alonso-Chamarro, I. Garces and J. Mateo, "Determination of probable alcohol yield in musts by means of an SPR optical sensor", *Sensors and Actuators B* **11**, 455-459 (1993).
49. K. Matsubara, S. Kawata and S. Minami, "Optical chemical sensor based on surface plasmon measurement", *Appl. Opt.* **27**, 1160-1163 (1988).
50. L.M. Zhang, and D. Uttamchandani, "Optical chemical sensing employing surface plasmon resonance", *Electron. Lett.* **23**, 1469-1470 (1988).
51. J. Homola, G. Schwotzer, H. Lehmann, R. Willsch, W. Ecke and H. Bartelt, "Fiber optic sensor for adsorption studies using surface plasmon resonance", *Proc. SPIE* **2508**, 324-333 (1995).
52. R.C. Jorgenson and S.S. Yee, "A fiber-optic chemical sensor based on surface plasmon resonance", *Sensors and Actuators B* **12**, 213-220 (1993).
53. A. Trouillet, C. Ronot-Trioli, C. Veillas and H. Gagnaire, "Chemical sensing by surface plasmon resonance in a multimode optical fibre", *Pure Appl. Opt.* **5**, 227-237 (1996).
54. J. Homola and R. Slavik, "Fibre-optic sensor based on surface plasmon resonance", *Electron. Lett.* **32**, 480-482 (1996).
55. A.J.C. Tubb, F.P. Payne, R.B. Millington and C.R. Lowe, "Single-mode optical fiber surface plasma wave chemical sensor", *Sensors and Actuators B* **41**, 71-79 (1997).
56. R. Slavik, J. Homola, J. Ctyroky and E. Brynda, "Novel spectral fiber optic sensor based on surface plasmon resonance", *Sensors and Actuators B* **74**, 106-111 (2001).
57. M. Piliarik, J. Homola, Z. Manikova and J. Ctyroky, "Surface plasmon resonance sensor based on a single-mode polarization-maintaining optical fiber", *Sensors and Actuators B*, **90**, 236-242 (2003).
58. S.A. Zynio, A.V. Samoylov, E.R. Surovtseva, V.M. Mirsky and Y.M. Shirshov, "Bimetallic layers increase sensitivity of affinity sensors based on surface plasmon resonance", *Sensors* **2**, 62-70 (2002).
59. A.K. Sharma and B.D. Gupta, "On the sensitivity and signal to noise ratio of a step-index fiber optic surface plasmon resonance sensor with bimetallic layers", *Opt. Commun.* (In press, 2004)

60. K. Kurihara and K. Suzuki, "Theoretical understanding of an absorption –based surface plasmon resonance sensor based on Kretchmann's theory", *Anal. Chem* **74**, 696-701 (2002).
61. A.K. Sharma and B.D. Gupta, "Absorption-based fiber optic surface plasmon resonance sensor : a theoretical evaluation", *Sensors and Actuators B* **100**, 423-431 (2004).
62. B.D. Gupta and A.K. Sharma, "Sensitivity evaluation of a multi-layered surface plasmon resonance based fiber optic sensor: a theoretical study", *Sensors and Actuators B* (In press, 2004).
63. G. Meltz, W.W. Morey and W.H. Glenn, "Formation of Bragg gratings in optical fibers by a transverse holographic method", *Opt. Lett.* **14**, 823-825 (1989).
64. F.P. Payne, "Photorefractive gratings in single-mode optical fibers", *Electron. Lett.* **25**, 498-499 (1989).
65. A. Othonos, "Fiber Bragg gratings", *Rev. Sci. Instrum.* **68**, 4309-4341 (1997).
66. T. Ritari, J. Tuominen. H. Ludvigsen, J.C. Petersen, T. Sorensen, T.P. Hansen and H.R. Simonsen, "Gas sensing using air-guiding photonic bandgap fibers", *Opt. Express* **21**, 4080-4087 (2004).

11

Modified Airy Function Method For Solving Optical Waveguide Problems

I.C.Goyal

1. Introduction

In this chapter we revisit a basic equation of mathematical physics,

$$\Psi''(x) + \Gamma^2(x)\Psi(x) = 0 \tag{1}$$

where prime denotes differentiation with respect to x and $\Gamma^2(x)$ is an arbitrary but known function of x except for a possible eigenvalue constant. This equation is encountered in many areas of physics and engineering. In quantum mechanics, for example, the one dimensional Schrödinger equation is of the same form as Eq.(1) with

$$\Gamma^2(x) = \frac{2m}{\hbar^2}[E - V(x)]$$

where m is the mass of the particle, E the total energy, $V(x)$ the potential energy function and $\hbar \equiv h/2\pi$, h being Planck's constant. In the above equation, E appears as an eigenvalue in a bound state problem and as a given parameter in a scattering problem. Similarly, in optical waveguide theory, for a medium characterized by a refractive index distribution $n^2(x)$, the y –component of the electric field can be written in the form

$$E_y(x, y, z, t) = \Psi(x)\exp[i(\omega t - \beta z)]$$

where $\Psi(x)$ satisfies Eq. (1) with $\Gamma^2(x) = k_0^2 n^2(x) - \beta^2$. Thus,

$$\Psi''(x) + \left(k_0^2 n^2(x) - \beta^2\right)\Psi(x) = 0 \tag{2}$$

where $k_0 = \omega/c$ and β is the propagation constant. Equation (2) describes the TE modes of a slab waveguide.

For some specific profiles, that is, for some specific forms of $\Gamma^2(x)$, one can obtain analytical solution of Eq.(1). However, for profiles, which do not lead to exact solutions, we usually resort to one of the three approximate methods: The perturbation method (see Chapter 12), the variational method (see Chapters 13 and 14) or the WKB Method. The perturbation method is based on a closely related problem, which yields an exact solution, and we usually resort to first-order perturbation; even then it is extremely difficult to calculate the perturbed eigenfunctions, as it would involve summation of an infinite series. On the other hand the variational method can give a good estimate for the lowest-order mode by choosing an appropriate trial function and carrying out an optimization; the method becomes quite cumbersome when one has to apply it to higher-order modes. The WKB method has been widely used but the wave function diverges near the turning point.

We will describe here another approximate method which is much more powerful than any other method and uses the Airy functions. The method was first suggested by Langer[1]. This is applicable to both the *initial value* as well as the *eigenvalue* problems. The advantages of the method will be illustrated by means of examples.

2. The Modified Airy Function Method

The use of the WKB method is well established in problems of quantum mechanics and optical waveguides. Following the WKB methodology[2], we assume a solution of Eq.(1) of the form

$$\Psi(x) = F(x)Ai[\xi(x)] \quad \text{or} \quad G(x)Bi[\xi(x)] \tag{3}$$

where $Ai(x)$ and $Bi(x)$ are solutions of the Airy equation :

$$f''(x) - xf(x) = 0 .$$

Substitution of Eq. (3) in Eq. (1) gives

$$\begin{gathered} F''(x)Ai(\xi) + 2F'(x)Ai'(\xi)\xi'(x) + F(x)Ai'(\xi)\xi'' + \\ Ai(\xi)F(x)[\xi(\xi'(x))^2 + \Gamma^2(x)] = 0 \end{gathered} \tag{4}$$

where prime denotes differentiation with respect to the argument. Equation (4) is rigorously correct. We choose $\xi(x)$ so that

$$\xi[\xi'(x)]^2 + \Gamma^2(x) = 0 , \tag{5}$$

the solution of which gives [2]

$$\xi(x) = \frac{3}{2}\left\{\int_x^{x_0} \sqrt{-\Gamma^2}\, dx\right\}^{2/3} , \tag{6}$$

with x_0 being the turning point where $\Gamma^2(x)$ has a zero of order one. If we neglect the term proportional to F'' in Eq.(4) (that is the only approximation we will make), it becomes

$$2F'(x)Ai'(\xi)\xi'(x) + F(x)Ai'(\xi)\xi''(x) = 0 ,$$

the solution of which is

$$F(x) = \frac{\text{cosnst.}}{\sqrt{\xi'(x)}} . \tag{7}$$

The approximate solution that we will discuss, then, is

$$\Psi(x) = C_1 \frac{Ai[\xi(x)]}{\sqrt{\xi'(x)}} + C_2 \frac{Bi[\xi(x)]}{\sqrt{\xi'(x)}} , \tag{8}$$

where C_1 and C_2 are constants to be determined from the initial or boundary conditions. Before illustrating the method by a few examples from optical waveguides, let us rewrite Eq. (2) in terms of the dimensionless variables in the following form

$$\frac{d^2\Psi}{dX^2} + V^2[N^2(x) - b]\psi(x) = 0, \tag{9}$$

where $X = x/d$ (d is a suitably chosen length), $V = k_0 d\sqrt{n_1^2 - n_2^2}$ is the normalized waveguide parameter (n_1 is the maximum refractive index of the core and n_2 the index of the cladding), and

$$b = \frac{\beta^2 / k_0^2 - n_2^2}{n_1^2 - n_2^2} \quad \text{and} \quad N^2(X) = \frac{n^2(X) - n_2^2}{n_1^2 - n_2^2}$$

are the normalized propagation constant and the normalized index distribution, respectively.

As an example we consider a planar optical waveguide with refractive index variation given by (see inset of Fig. 1)

$$n^2(x) = n_2^2 + (n_1^2 - n_2^2)\exp(-x/d) \qquad \text{for } x > 0$$

$$n^2(x) = n_c^2 \qquad \text{for } x < 0$$

where d is the diffusion depth of the waveguide. In this case, Eq.(9) becomes

$$\frac{d^2\Psi}{dX^2} + V^2[\exp(-X) - b]\Psi(X) = 0 \quad \text{for } X > 0$$

$$\frac{d^2\Psi}{dX^2} - V^2[(b + B)]\Psi(X) = 0 \qquad \text{for } X < 0 \tag{10}$$

where $B = (n_2^2 - n_c^2)/(n_1^2 - n_2^2)$. For $X < 0$, the exact solution of Eq.(10) which does not diverge as $X \to -\infty$, is given by

$$\Psi(X) = \exp(XV\sqrt{b + B}) \tag{11}$$

For $X > 0$, the MAF (see Eq.(8)) solution can be written as [3-5]

$$\Psi(X) = \left(\frac{\xi_0'}{\xi'}\right)^{1/2} \frac{Ai(\xi)}{Ai(\xi_0)} \qquad \text{for } X > 0 \tag{12}$$

subscript 0 indicates the value of the function at $X = 0$. It may be noticed that in writing the solution given by Eq. (12) we have ignored the term proportional to $Bi(\xi)$ as it will diverge for large X. The proportionality constants in Eq. (11) and (12) have been so chosen as to satisfy the continuity of $\Psi(X)$ at $X = 0$ with $\Psi(0) = 1$. The eigenvalue equation can be written by using the continuity of $\Psi'(X)$ at $X = 0$:

$$\frac{V(B + b)^{1/2}}{\xi_0'} = \frac{Ai'(\xi_0)}{Ai(\xi_0)} - \frac{1}{2}\frac{\xi_0''}{(\xi_0')^2}.$$

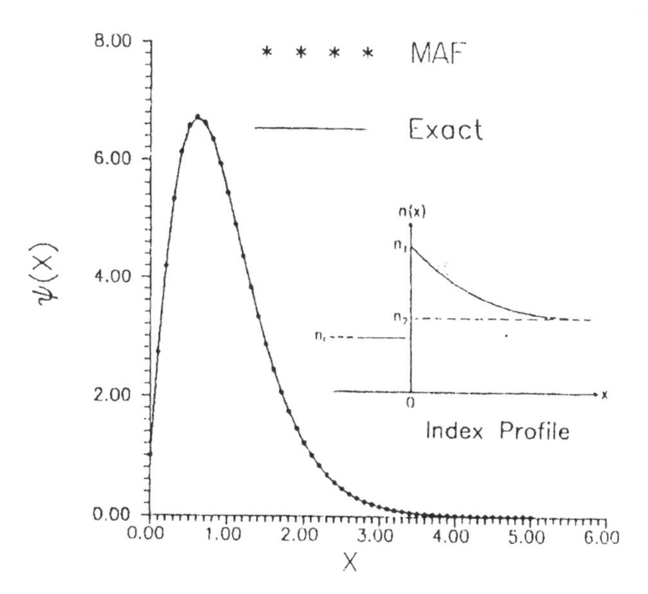

Figure 1 Variation of $\Psi(X)$ with X.

We have used $n_2 = 2.177$, $n_c = 1$ and $n_1^2 - n_2^2 = 0.187$ in our calculations and compared the MAF results with the exact values[6] and those calculated by WKB method [2]. For $V = 1.5$, $b_{\text{EXACT}} = 0.03501$, $b_{\text{MAF}} = 0.03609$ and $b_{\text{WKB}} = 0.03783$. The error in MAF values is only 3% (while the error in WKB is 8%). Figure 1 shows Ψ_{MAF} versus X for V = 4 as calculated by Eq. (12) and Ψ_{EXACT} [6]. The figure shows no discernable difference between the two curves even at the turning point.

3. Symmetric Profile

The method becomes particularly simple if the refractive index profile is symmetric, i.e., if

$$\Gamma^2(-x) = \Gamma^2(x) \tag{13}$$

For a bound mode $\Psi(x)$ must tend to 0 as $x \to \pm\infty$, so we again reject the solution proportional to Bi function. Hence,

$$\Psi(x) = \frac{\text{const.}}{\sqrt{\xi'(x)}} Ai[\xi(x)] \tag{14}$$

Equations(2) and (13) suggest that the solutions are either symmetric or antisymmetric in x, so $\Psi'(0) = 0$ for symmetric solutions and $\Psi(0) = 0$ for antisymmetric solutions. This leads to the eigenvalue equations

$$\frac{Ai'(\xi(0))}{Ai(\xi(0))} = \frac{1}{2}\frac{\xi''(0)}{[\xi'(0)]^2} \qquad \text{(symmetric solution)}$$

$$Ai[\xi(0)] = 0 \qquad \text{(antisymmetric solution)} \tag{15}$$

Let $-Z_{an}$ ($n = 1,2,3,\ldots$) denote the nth zero of the Ai function i.e.,

$$\mathrm{Ai}(-Z_{an}) = 0 \tag{16}$$

Using Eqs.(6), (15) and (16) and assuming $\Gamma^2(x)$ to be positive from 0 to x_0, we obtain for antisymmetric solutions

$$\xi(0) = -\left[\frac{3}{2}\int_0^{x_0}\Gamma(x)\,dx\right]^{2/3} = -Z_{an},$$

or,

$$\int_0^{x_0}\Gamma(x)\,dx = (\zeta_{an} + 1/2)\pi/2 \tag{17}$$

where

$$\zeta_{an} = \left(\frac{4}{3\pi}Z_{an}^{3/2} - \frac{1}{2}\right).$$

Equation (17) is similar to the WKB quantization condition [2,7]

$$\int_0^{x_0}\Gamma(x)\,dx = (n + 1/2)\pi/2$$

except that (for the WKB case) ζ_{an} is replaced by odd integers. Similarly, it can be shown that for symmetric solutions

$$\int_0^{x_0}\Gamma(x)\,dx = \left(\zeta_{sn} + \frac{1}{2}\right)\pi/2, \tag{18}$$

where

$$\zeta_{sn} = \left(\frac{4}{3\pi}Z_{sn}^{3/2} - \frac{1}{2}\right)$$

and $-Z_{sn}$ is the nth zero of the function $xAi'(x) + Ai(x)/4$. Z_{an}, ζ_{an}, Z_{sn} and ζ_{sn} are universal constants (see Ref. 2, Table 6.2).

As an example, let us consider a symmetric refractive index profile (see Fig 2(a)) given by

$$n^2(x) = n_2^2 + (n_1^2 - n_2^2)\exp(-|x|/d) \tag{19}$$

In terms of the dimensionless variables the equation to be solved (see Eq.(9)) is

$$\Psi''(X) + \Gamma^2(X)\Psi(X) = 0 \text{ with } \Gamma^2(X) = V^2[\exp(-|X|) - b]$$

the solution of which is given by Eqs.(14) and (16), and the eigenvalue for the first antisymmetric mode can be determined from the following relation[2] (see Eq.(17))

$$\int_0^{x_0}\Gamma(x)\,dx = \left(\zeta_{a1} + \frac{1}{2}\right)\pi/2 \quad \text{with} \quad \zeta_{a1}\ 1.01734.$$

We have used $n_2 = 2.177$, $n_1^2 - n_2^2 = 0.187$, $\lambda_0 = 632.8$ nm and d = 0.4236 μm in our calculations. It gives $b_{MAF} = 0.056578$. Corresponding value of $b_{EXACT} = 0.055643$ and $b_{WKB} = 0.05929$. The error by the MAF method is 1.7% compared to an error of

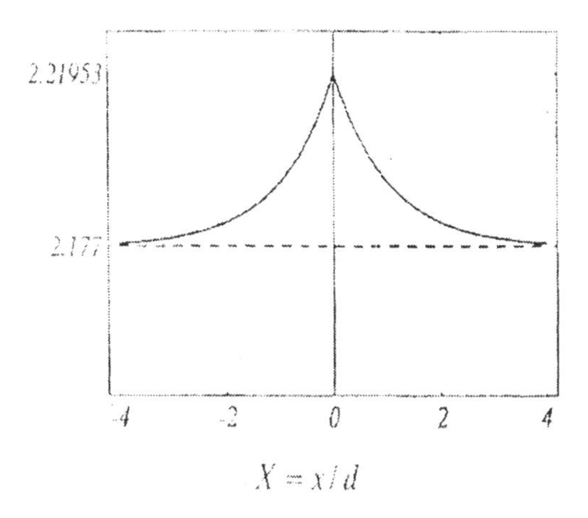

Figure 2 (a) Refractive index Profile (see Eq.(19)).

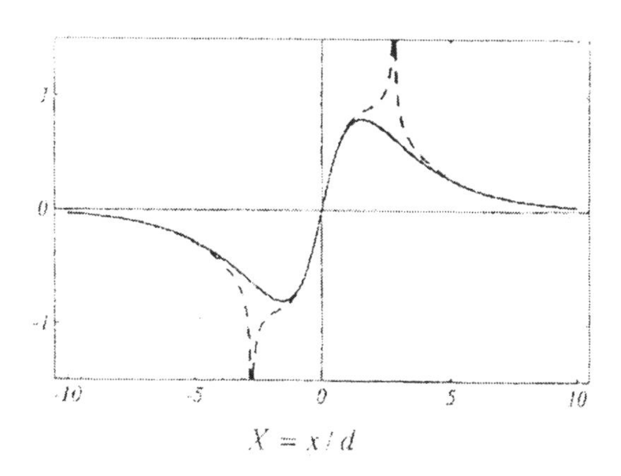

Figure 2(b) $\Psi(X)$ as a function of X. Solid, dotted and dashed curves correspond to Exact, MAF and WKB respectively.

7% with WKB. Figure 2(b) shows the variation of $\Psi(X)$ with X. Again it can be observed that while the WKB solution diverges at the turning points, the MAF solution matches very well with the exact.

4. Perturbation Correction

The fact that the MAF eigenfunction matches very well with the exact wavefunction, can be used to improve the eigenvalues with the help of the first order perturbation theory. It can be shown by simple substitution that the MAF solution given by

$$\Psi(X) = \frac{\text{const.}}{\sqrt{\xi'(X)}} Ai[\xi(X)] \tag{20}$$

is an exact solution of the following differential equation

$$\frac{d^2\Psi}{dX^2}+\Gamma^2(X)\Psi(X)+\left[\frac{1}{2}\frac{\xi'''}{\xi'}-\frac{3}{4}\left(\frac{\xi''}{\xi'}\right)^2\right]\Psi(X)=0 \qquad (21)$$

where $\Gamma^2(X)=V^2\left[N^2(X)-b\right]$ in problems of optical waveguides. Comparing Eqs. (21) and (9) and considering the last term in Eq (21) as a perturbation, we get a first order correction Δb to normalized propagation constant [5,7]

$$\Delta b \cong \frac{-\int_{-x}^{+x}\left[\frac{1}{2}\frac{\xi'''}{\xi'}-\frac{3}{4}\left(\frac{\xi''}{\xi'}\right)^2\right]\Psi^2(X)dX}{V^2\int_{-x}^{+x}\Psi^2(X)\,dX} \qquad (22)$$

If we apply the above perturbation correction to both the examples given above, we get $b_{\mathrm{MAF+Pert.}}= 0.03502$ (comparison with exact shows an error of only ~ 0.03%) in the case of the asymmetric profile considered earlier, and $b_{\mathrm{MAF+Pert.}}= 0.05565$ (comparison with exact shows an error of only ~ 0.024%), in the case of the symmetric profile. Thus, the application of the first order perturbation theory greatly improves the eigenvalues though the eigenfunctions remain the same.

5. Improved MAF

In the MAF method described so far, we have neglected the term proportional to $F''(x)$ (see Eq. (4)). If we retain this term, then, in place of Eq. (5) we will get

$$\xi(x)\xi'^2+\Gamma^2(x)=-\frac{F''(x)}{F(x)}.$$

Using Eq. (7) we get

$$\xi\xi'^2+\Gamma^2(x)=-\left[\frac{3}{4}\left(\frac{\xi''}{\xi'}\right)^2-\frac{1}{2}\frac{\xi'''(x)}{\xi'(x)}\right] \qquad (23)$$

For an arbitrary $\Gamma^2(x)$, it is not easy to find the solution for $\xi(x)$. Using $\xi(x)$ found by the MAF method (see Eq.(6)) one can determine [8] the RHS of Eq.(23), which we denote by $f(x)$. Thus,

$$\xi\xi'^2+\gamma^2(x)=0 \qquad (24)$$

where $\gamma^2(x)=\Gamma^2(x)-f(x)$ is a known function of x except for an eigenvalue constant in $\Gamma^2(x)$. Equation (24) can now be solved the same way as Eq. (5) to obtain an improved value of ξ (let us denote it by ξ^i) given by

$$\xi^i(x)=\left\{\frac{3}{2}\int_x^{x_t}\sqrt{-\gamma^2(x)}\,dx\right\}^{2/3} \qquad (25)$$

where x_t is the turning point of $\gamma^2(x)$, *i.e.*, $\gamma^2(x_t)=0$. The solution is now given (see Eq. (8)) by

$$\Psi(X) = \frac{1}{\sqrt{d\xi^i / dx}} \left[C_1 Ai(\xi^i(x)) + C_2 Bi(\xi^i(x)) \right] \tag{26}$$

C_1, C_2 and b are obtained by using the boundary conditions. As an example of this method, let us consider a truncated parabolic refractive index profile given by

$$n^2(x) = n_1^2 - (n_1^2 - n_2^2)\left(\frac{x}{a}\right)^2 \qquad \text{for } 0 < x < a$$

$$= n_c^2 \qquad \text{for } x < 0$$

$$= n_2^2 \qquad \text{for } x > a$$

The normalized frequency V is defined as

$$V = k_0 a \sqrt{(n_1^2 - n_2^2)}$$

The values of the various parameters used in the calculations [8] are n_2 = 2.177, n_c = 1, $n_1^2 - n_2^2$ =0.187 and V = 4. The calculated valued of b are given in Table-I. The error in the wavefunction obtained is so small that it is would not be noticeable if the wavefunctions are plotted. Therefore, the error in $\Psi(X)$ is plotted as a function of X in Fig. 3. The wavefunctions are normalized such that $\Psi(X=0)=1$. It can be noticed that by the improved method (IMAF), the maximum error in the wavefunction is reduced to about one fifth of the corresponding error by the MAF method.

Table-I Values of b obtained using different method for the profile defined in Eq. 27.

Method	b
WKB	0.30790
MAF	0.31691
IMAF	0.32703
Exact	0.32617

With the above improvement, it becomes an extremely accurate method to obtain the eigenvalues as well as the eigenfunctions for planar waveguides with an arbitrary index profile. Thus, the method should be very useful in the better designing of many optical waveguide based devices and components.

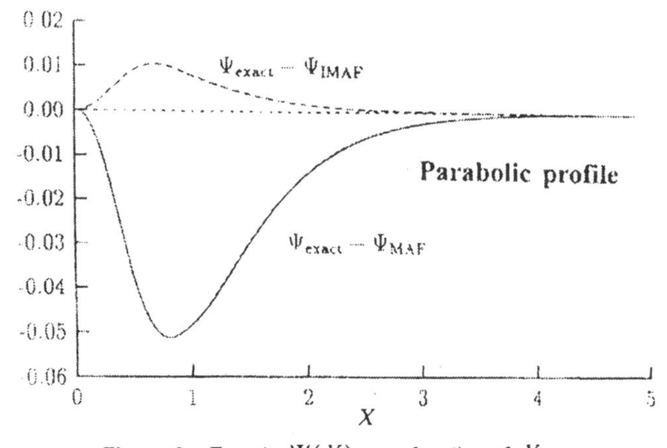

Figure 3 Error in $\Psi(X)$ as a function of X.

Acknowledgement: The author would like to acknowledge the financial support from All India Council of Technical Education during this work.

References

1. R.E. Langer, "On the asymptotic solutions of ordinary differential equations, with an applications to the Bessel functions of large order", *Trans. Am. Math. Soc.* **33** 23 (1931).
2. A.K. Ghatak, R.L. Gallawa and I.C.Goyal, "Modified Airy Function and WKB solutions to the wave equation," National Insitute of Standards & Technology Monograph 176, U.S.Govt. Printing Office, Washington (1991).
3. I.C. Goyal, R.L. Gallawa and A.K. Ghatak, "Approximate solution to the wave equation- revisited," *J. Electromagnetic Waves and Appl. Opt.* **5**, 623 (1991).
4. I.C. Goyal, R.L.Gallawa and A.K. Ghatak, "Approximate solution to the scalar wave equation for planar optical waveguides," *Appl. Opt.* **30**, 2985 (1991).
5. I.C. Goyal, R.L. Gallawa and A.K. Ghatak, "Method of analyzing planar optical waveguides," *Opt. Lett.* **16**, 30 (1991).
6. E.M. Conwell, "Modes in optical waveguides formed by diffusion," *Appl. Phys. Lett.* **23**, 328 (1973).
7. A.K. Ghatak and S. Loknathan, *Quantum mechanics: Theory and applications,* Macmillan Press, New Delhi (1984).
8. I.C. Goyal, Rajeev Jindal and A.K. Ghatak, "Planar optical waveguides with arbitrary index profile: An accurate method of analysis," *J. Lightwave Technol.* **15**, 2179 (1997).

12

Perturbation Method for Rectangular Core Waveguides and Devices

Arun Kumar

1. Introduction

The phenomenal progress in the area of Fiber and integrated optics has established the need for analyzing a wide range of optical waveguide structures. However, exact analytical analyses exist only for a few geometries such as planar or circular-core waveguides with step index profile[1]. There are other important waveguides for which either exact analytical analyses do not exist or are too cumbersome to use. The corresponding examples are rectangular-core and elliptic-core waveguides. The former are important, as they are the basic building blocks of integrated optic devices. In order to be able to design efficient integrated-optical devices such as directional couplers, filters, switches etc., it is important to understand the modal characteristics of various rectangular core wave-guides. The analysis of such waveguides has become more important because of progress in silica-based waveguide technology, which provide almost completely rectangular cross sections[2].

The elliptical-core waveguides are also finding important applications as polarization maintaining waveguides[3], dual-mode fiber sensors and devices[4,5] and dispersion compensators[6] for single-mode fiber networks. Although the propagation characteristics of elliptical-core waveguides can be obtained exactly (in the weakly guiding approximation) using a series of Mathieu functions[7], it is quite cumbersome to obtain propagation constants up to the desired accuracy because the eigenvalue equation involves infinite determinants. As shown by Kumar and co-workers[8,9], elliptical-core waveguides can also be considered as a perturbation over the appropriate rectangular geometries. Thus there is a need to have accurate methods to analyze rectangular-core waveguides and devices.

Exact analytical analyses for rectangular-core waveguides do not exist because of the corner regions where the boundary conditions cannot be satisfied exactly. As a result various approximate methods are used for obtaining the propagation characteristics of these waveguides. These methods include both analytical as well as numerical methods. Since numerical methods involve time-consuming computer calculations, approximate analytical methods may be a good choice many times. In the following, various approximate methods used to analyze rectangular-core waveguides, with emphasis on the perturbation method as proposed by Kumar et al.[10], are discussed.

2. Methods to analyze Rectangular-core waveguides

2.1 The Method of Marcatili

The earliest method in the series of approximate analytical methods is due to Marcatili[11]. In order to describe the method, let us consider a rectangular-core waveguide as shown in Fig.1, which can represent a channel ($n_2 = n_4$), a rib ($n_3 = n_4$), or a symmetric ($n_2 = n_3 = n_4$) waveguide.

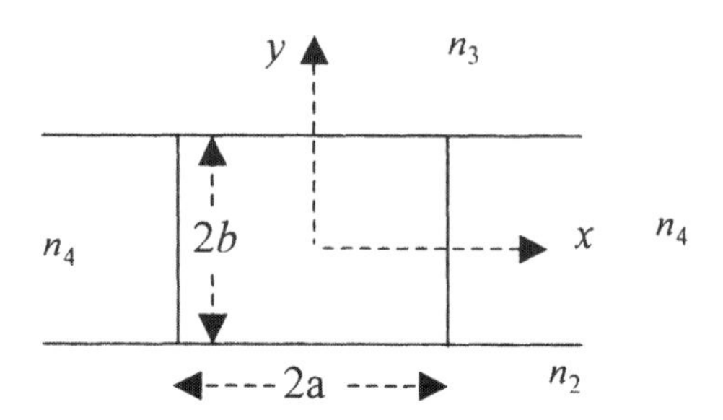

Figure 1 Transverse cross-section of a rectangular core wave-guide.

According to this method the modal field patterns are approximated by $\psi(x, y)$ which are separable in x and y co-ordinates, i.e. $\psi(x, y)$ is of the form

$$\psi(x, y) = X(x)Y(y), \tag{1}$$

where $X(x)$ and $Y(y)$ are the modal field patterns corresponding to two slab-waveguides which are obtained by extending the height (2b) and the width (2a) respectively of the rectangular-core to infinity. To be more specific $X(x)$ and $Y(y)$ represents the modal field patterns corresponding to slab waveguides shown in Figs. 2a and 2b respectively, and are of the form

$$X(x) = \begin{cases} A\cos(k_x x - \phi) & \text{for } |x| < a \\ B\exp[-\gamma_4(|x| - a)] & \text{for } |x| > a \end{cases} \tag{2}$$

and

$$Y(y) = \begin{cases} D\exp(-\gamma_3 y) & \text{for } y > b \\ E_1\cos(k_y y) + E_2\sin(k_y y) & \text{for } -b < y < b \\ F\exp(\gamma_2 y) & \text{for } y < -b \end{cases} \tag{3}$$

where

$$k_x = (k_0^2 n_1^2 - \beta_x^2)^{1/2}, \qquad \gamma_4 = (\beta_x^2 - k_0^2 n_4^2)^{1/2} \tag{4}$$

$$k_y = (k_0^2 n_1^2 - \beta_y^2)^{1/2}, \qquad \gamma_{2,3} = (\beta_y^2 - k_0^2 n_{2,3}^2)^{1/2} \tag{5}$$

and $\phi = 0$ or $\pi/2$ for modes symmetric or anti-symmetric respectively in x. The relative values of the constants A, B, D, E_1, E_2, F and the eigenvalue equations determining the values of β_x and β_y can be obtained by satisfying the appropriate boundary conditions.

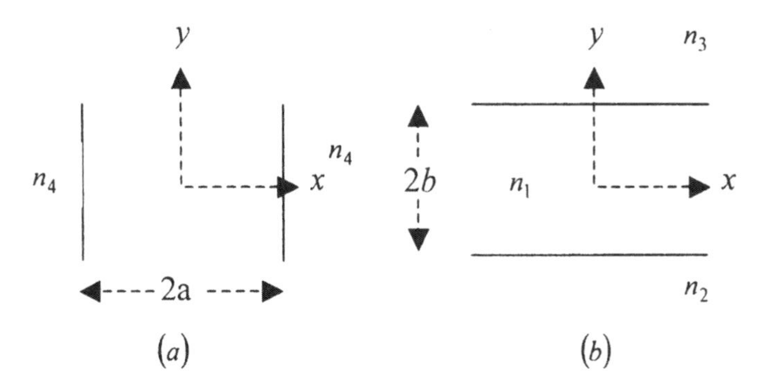

Figure 2 Planar wave-guides obtained as (a) $2b \to \infty$ (b) $2a \to \infty$.

The propagation constants of modes of the given rectangular waveguide are then given by

$$\beta^2 = \beta_x^2 + \beta_y^2 - k_0^2 n_1^2 \tag{6}$$

The modes of the waveguide shown in Fig.1 can be classified as E_{mn}^x (predominantly x-polarized) and E_{mn}^y (predominantly y-polarized) modes. In the first family, the electric field is almost polarized along the x- direction while in the second family in the y-direction. The subscripts m and n represent the number of extremas, the corresponding mode has along x and y -directions respectively. In the weakly guiding approximation i.e.,

$$\frac{n_i}{n_2} - 1 << 1 \quad \text{with} \quad i = 2,3,4,$$

it can be assumed that for E_{mn}^x (or E_{mn}^y) modes the electric field is almost completely polarized in the x (or y) direction and the corresponding magnetic field is along y (or x) direction. Under such an approximation if the approximate field components are matched along the sides of the core region ($|x| = a$, and $|y| = b$), one gets the following eigenvalue equations for determining β_x and β_y:

(a) For E_{mn}^x modes:

The eigenvalue equation determining β_x is given by,

$$\tan(\kappa_x a) = \begin{cases} \dfrac{n_1^2}{n_4^2} \cdot \dfrac{\gamma_4}{\kappa_x} & \text{for modes symmetric in } x \\[2ex] -\dfrac{n_4^2}{n_1^2} \cdot \dfrac{\kappa_x}{\gamma_4} & \text{for modes antisymmetric in } x \end{cases} \tag{7}$$

and the eigenvalue equation determining β_y is given by,

$$\tan(2\kappa_y b) = \frac{\frac{\gamma_2}{\kappa_y} + \frac{\gamma_3}{\kappa_y}}{1 - \frac{\gamma_2}{\kappa_y} \cdot \frac{\gamma_3}{\kappa_y}} \tag{8}$$

It may be noticed that Eqs.(7) and (8) are nothing but the eigenvalue equations for TM and TE modes of planar waveguides shown in Figs.2a and 2b respectively.

(b) For E^y_{mn} modes:

The eigenvalue equation determining β_x is given by,

$$\tan(\kappa_x a) = \begin{cases} \frac{\gamma_4}{\kappa_x} & \text{for modes symmetric in } x \\ -\frac{\kappa_x}{\gamma_4} & \text{for modes antisymmetric in } x \end{cases} \tag{9}$$

and the eigenvalue equation determining β_y is given by,

$$\tan(2\kappa_y b) = \frac{\frac{n_1^2}{n_2^2}\frac{\gamma_2}{\kappa_y} + \frac{n_1^2}{n_3^2}\frac{\gamma_3}{\kappa_y}}{1 - \frac{n_1^2}{n_2^2}\frac{\gamma_2}{\kappa_y} \cdot \frac{n_1^2}{n_3^2}\frac{\gamma_3}{\kappa_y}} \tag{10}$$

In this case Eqs.(9) and (10) are the same as the eigenvalue equations for TE and TM modes of planar waveguides shown in Figs.2a and 2b respectively.

Although Marcatili[11] in his original paper has mentioned that his analysis neglects the power in the corner regions; however, as shown in Ref. 10, the wavefuntions corresponding to the Marcatili's method [see Eqs.(2) and (3)] are the exact solutions, in the scalar-wave approximation, corresponding to a pseudo-rectangular-core waveguide described by the following dielectric constant distribution,

$$n_0^2(x, y) = n'^2(x) + n''^2(y) - n_1^2 \tag{11}$$

where

$$n'(x) = \begin{cases} n_1 & for\ |x| < a \\ n_4 & for\ |x| > a \end{cases} \tag{12}$$

and

$$n''(y) = \begin{cases} n_3 & for\ y > b \\ n_1 & for\ -b < y < b \\ n_4 & for\ y > -b \end{cases} \tag{13}$$

The above dielectric constant distribution is shown in Fig.3. Since $n_0^2(x, y)$ is separable in x and y, the corresponding scalar wave equation

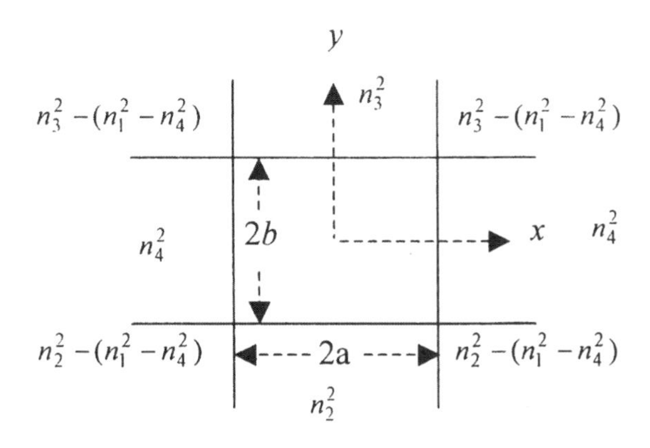

Figure 3 Dielectric constant distribution given by Eq. 11.

$$\frac{\partial^2\psi}{\partial x^2}+\frac{\partial^2\psi}{\partial y^2}+\left[k_0^2 n_0^2(x,y)-\beta^2\right]\psi(x,y)=0 \tag{14}$$

can be solved using the method of separation of variables. Accordingly $\psi(x,y)$ can be taken of the form $X(x).Y(y)$, where $X(x)$ and $Y(y)$ are the solutions of the following one dimensional equations:

$$\frac{d^2X}{dx^2}+\left[k_0^2 n'^2(x)-\beta_x^2\right]X(x)=0 \tag{15}$$

and

$$\frac{d^2Y}{dy^2}+\left[k_0^2 n''^2(y)-\beta_y^2\right]Y(y)=0 \tag{16}$$

Thus $X(x)$ and $Y(y)$ are the field patterns corresponding to the slab wave guides given by Eqs.(12) and (13) respectively and hence are the same as taken in Marcatili's method. Here β_x and β_y are two separation constants obtained by satisfying the continuity equations for $X(x)$, dX/dx and $Y(y)$, dY/dy at boundaries $x=\pm a$ and $y=\pm b$ respectively. The resulting equations would be the same as Eqs.(9) and (8) respectively in the scalar approximation. Further substituting Eqs.(15) and (16) in Eq.(14), one gets the same expression for β^2 as given by Eq.(6). Thus, Marcatili's method for the rectangular-core wave guide of Fig.1 corresponds exactly to the wave guide of Fig.3, in the scalar approximation.

The method works well in the far-from the cut off region but gives poor results in the near cutoff region.

2.2 The Effective–Index Method

In order to improve upon Marcatili's method, Knox and Toulios[12] proposed a method, which later on generalized by many authors[13-16]. This method is usually referred to as the effective-index method (EIM). The method in its original form, assumes the following,

(i) The aspect ratio $a/b \geq 2$,

(ii) The relative dielectric constant difference between the core and cladding regions is small.

In order to explain the method we again consider a rectangular-core waveguide shown in Fig.1. The method for obtaining the propagation constants of E^x_{mn} modes is summarized in the following steps:

Step1: We first determine the propagation constant (β_{ny}) of the nth TE mode for the slab wave-guide (see Fig.2b), which is obtained by letting the longer dimension of the rectangular guide approach infinity. We thus solve the following equation first

$$\frac{d^2Y}{dx^2}+\left[k_0^2 n''^2(y)-\beta_{ny}^2\right]Y(y)=0 \tag{17}$$

where $n''(y)$ is given by Eq.(13) (see Fig.2b). The solutions of the above equation are given by Eqn.3 and the β_{ny} for TE modes are obtained by solving the eigen value equation, Eq.(8).

Step 2: Having obtained β_{ny} from step 1 , the next step is to construct an equivalent slab waveguide of thickness $2a$, whose core index is taken as β_{ny}/k_0 and the cladding indices as n_4 (see Fig.4).The propagation constant of mth TM mode of this waveguide is taken to represent the original waveguide. We thus, now solve the following equation

$$\frac{d^2X}{dx^2}+\left[k_0^2 n^2(x)-\beta_{mx}^2\right]X(x)=0 \tag{18}$$

where,

$$n(x)=\begin{cases}\beta_{ny}/k_0 & \text{for } |x|<a\\ n_4 & \text{for } |x|>a\end{cases} \tag{19}$$

The solution of Eq.(18) would be given by Eq.(2) and β_{mx}/k_0 is obtained by solving Eq.(7) with n_1 replaced by β_{ny}/k_0. The resulting value of β_{mx} would represent the propagation constant of the given rectangular-core waveguide shown in Fig.1.

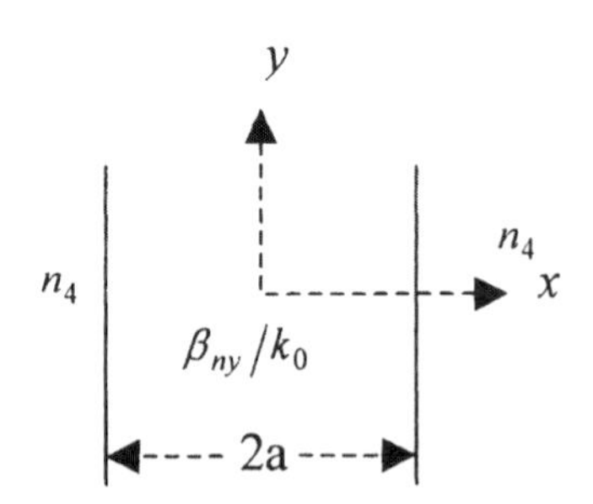

Figure 4 Equivalent slab wave-guide used in EIM.

In a similar way the propagation constants of E^y_{mn} modes can be obtained. In this case β_{ny} is taken corresponding to the nth TM mode of the planer waveguide give by Eq.(13) and hence obtained by solving Eq.(10). The planar waveguide of step 2 above is then constructed with this new value of β_{ny}/k_0. The propagation constant (β_{mx}) of the mth TE mode of this new planar wave-guide is then obtained by solving Eq.(9) with n_1 replaced by β_{ny}/k_0 and represents the propagation constant of the E^y_{mn} modes of the rectangular-core waveguide of Fig.1.

It is shown in Ref.16, that the above process is equivalent to analyzing a profile shown in Fig.5. This profile, which is exact for the effective-index method, differs from the original profile in cladding and corner regions ($|x| > a$). It is clear from this figure that the dielectric constant distribution shown in Fig.5 has a higher value in the cladding regions ($|x| > a$, $|y| < b$) than the given waveguide (Fig.1) by an amount $(n_1^2 - \beta_{ny}^2/k_0^2)$ and a lower value in the corner regions ($|x| > a$, $|y| > b$) by an amount $(\beta_{ny}^2/k_0^2 - n_4^2)$. Since the fractional power in the cladding regions is more than the corner regions, the net effect of the difference between the dielectric constant distributions of the given waveguide (Fig.1) and those of Fig.5 is that the propagation constants for the different guided modes are overestimated by the EIM method. The overestimation increases as one move towards the cutoff of a mode due to increase in the factional modal power in the cladding regions.

2.3 Perturbation Method

Having recognized the separable profile to which the Marcatili's method is exact, Kumar et al.[10] have developed a perturbation approach which gives very accurate

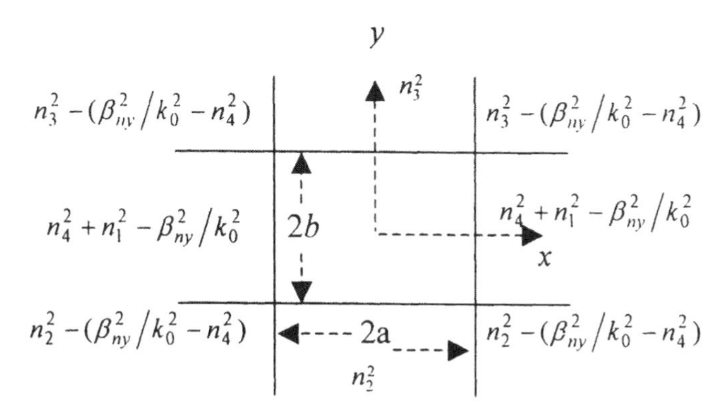

Figure 5 Separable profile to which EIM corresponds.

values for the propagation constants of rectangular core waveguides. The authors consider the given rectangular core waveguide as a perturbed form of the separable profile to which Marcatili's method is exact. The difference between the two profiles is then taken into account through a first-order perturbation approach. The method is quite simple as one obtains analytical expressions for the first-order perturbation correction to the propagation constants and is described below for a symmetric rectangular core waveguide.

2.3.1 Symmetric rectangular-core waveguide

Let us consider a symmetric rectangular-core waveguide as shown in Fig.6(a). We first discuss the scalar wave modes of this waveguide. As mentioned above, the given waveguide is considered as a perturbation over the separable dielectric constant profile $n_0^2(x, y)$, shown in Fig.6(b).

The scalar-modes for $n_0^2(x,y)$ (see Sec.2.1) are of the form $\psi(x, y) = X(x).Y(y)$ where $X(x)$ and $Y(y)$ are the solutions of one dimensional Eq.(15) and (16) respectively, with $n'(x)$ and $n''(y)$ given by

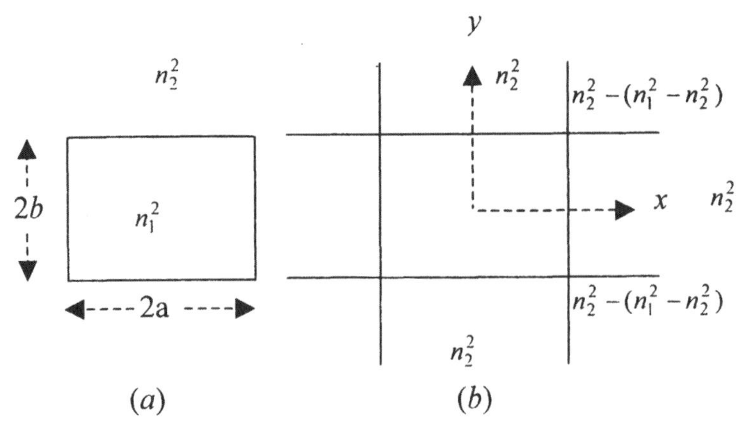

Figure 6 (a) A symmetric rectangular-core wave-guide and (b) the corresponding separable profile.

$$\begin{aligned} n'(x) &= n_1 \quad for\ |x| < a \\ &= n_2 \quad for\ |x| > a \end{aligned} \tag{20}$$

and

$$\begin{aligned} n''(y) &= n_1 \quad for\ |y| < b \\ &= n_2 \quad for\ |y| > b \end{aligned} \tag{21}$$

The unperturbed propagation constants β_0 are then given by

$$\beta_0^2 = \beta_x^2 + \beta_y^2 - k_0^2 n_1^2 \tag{22}$$

In the scalar approximation, β_x and β_y are obtained by making $X(x)$, dX/dx continuous at $x=\pm a$ and $Y(y)$, dY/dy at $y=\pm b$ respectively. The resulting eigenvalue equations are given by Eqs.(7) and (9) and can be put in the following form,

For β_x,

$$\begin{aligned} \tan\mu &= \left(\frac{V_1^2}{\mu^2}-1\right)^{1/2} \quad \text{mode symmetric in } x \\ &= -\left(\frac{V_1^2}{\mu^2}-1\right)^{-1/2} \quad \text{mode antisymmetric in } x \end{aligned} \tag{23}$$

For β_y,

$$\begin{aligned} \tan\nu &= \left(\frac{V_2^2}{\nu^2}-1\right)^{1/2} \quad \text{mode symmetric in } y \\ &= -\left(\frac{V_2^2}{\nu^2}-1\right)^{-1/2} \quad \text{mode antisymmetric in } y \end{aligned} \tag{24}$$

where,

$$\mu^2 = \left(k_0^2 n_1^2 - \beta_x^2\right) a^2 \; ; \quad v^2 = \left(k_0^2 n_1^2 - \beta_y^2\right) b^2 \tag{25}$$

and

$$V_1^2 = k_0^2 a^2 \left(n_1^2 - n_2^2\right) \; ; \quad V_2^2 = k_0^2 b^2 \left(n_1^2 - n_2^2\right) \tag{26}$$

Using the first order perturbation approach, the normalized propagation constants of the given waveguide will be given by:

$$P^2 = P_0^2 + P'^2 = \frac{\left(\beta^2 - k_0^2 n_2^2\right)}{k_0^2\left(n_1^2 - n_2^2\right)} \tag{27}$$

Where P_0^2, the unperturbed normalized propagation constant, is given by,

$$P_0^2 = \frac{\left(\beta_0^2 - k_0^2 n_2^2\right)}{k_0^2\left(n_1^2 - n_2^2\right)} = 1 - \frac{\mu^2}{V_1^2} - \frac{v^2}{V_2^2} \tag{28}$$

and P'^2, the first order perturbation correction to P_0^2, due to the difference between $n^2(x,y)$ and $n_0^2(x,y)$ in the corner regions, is given by

$$P'^2 = \frac{1}{n_1^2 - n_2^2} \frac{4\int_a^\infty \int_b^\infty \left(n_1^2 - n_2^2\right) |\Psi|^2 \, dxdy}{\int_{-\infty}^{\infty}\int_{-\infty}^{\infty} |\Psi|^2 \, dxdy} \tag{29}$$

$$= \left[1 + \left(\frac{V_1^2}{\mu^2} - 1\right)^{1/2} \left(\frac{2\mu + p\sin(2\mu)}{1 + p\cos(2\mu)}\right)\right]^{-1} \left[1 + \left(\frac{V_2^2}{v^2} - 1\right)^{1/2} \left(\frac{2v + p\sin(2v)}{1 + p\cos(2v)}\right)\right]^{-1}$$

where $p = 1(-1)$ for modes symmetric (antisymmetric) in x and $q = 1(-1)$ for modes symmetric (antisymmetric) in y.

Figure 7 shows the variation of P^2 as a function of $B = (2/\pi)V_2$ by various methods for rectangular waveguides with $a/b = 2$. This figure confirms that the

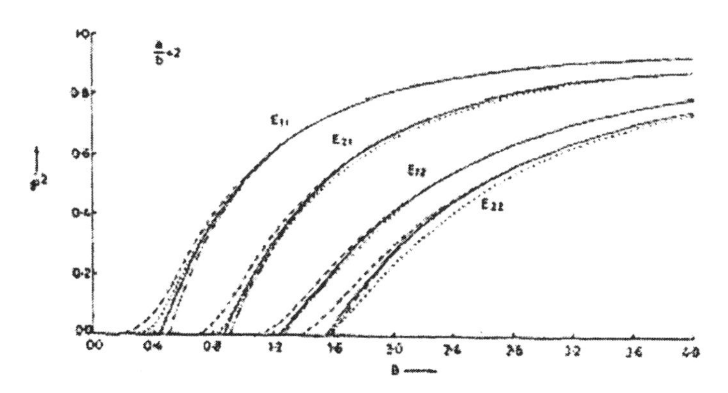

Figure 7 Variation of P^2 as a function of B for a symmetric rectangular-core waveguide with $a/b = 2$. Eq.(28) and Ref.11(-·-·-), Perturbation i.e. Eq. (29) (——), Goell's results (··········), EIM (- - - -). Reproduced from A Kumar et al. Optics Lett., **8**, 63(1983).

conventional EIM and the Marcatili's method respectively overestimate and underestimate the propagation constants. The results obtained by Marcatili's method are the same as obtained by Eq.(28). The results obtained by perturbation method (namely Eq.29) agree well with the numerical calculations of Goell[17]. It may be mentioned here that for higher order modes it becomes cumbersome to get accurate results by using the circular harmonic analysis of Goell[17]. For example, for the E_{22} mode the results reported by Goell[17] are certainly in error since even the unperturbed value of P^2 is larger than that of Goell, and the perturbation would further increase the value of P^2. In the following we will discuss, how this method can be applied to practical rectangular-core waveguides and devices.

2.3.2 A Channel Waveguide

The perturbation method described above can easily be extended to asymmetric rectangular-core waveguides. For example, let us consider an embedded channel waveguide shown in Fig.8(a).

This waveguide can be considered as a perturbation over the waveguide shown in Fig. 8(b) whose dielectric constant profile $n_0^2(x,y)$ can be written in the separable form similar to Eq.(11), where $n'(x)$ and $n''(y)$ are given by Eqs.(12) and (13) respectively with $n_4 = n_2$. This profile differs with the actual profile only in the

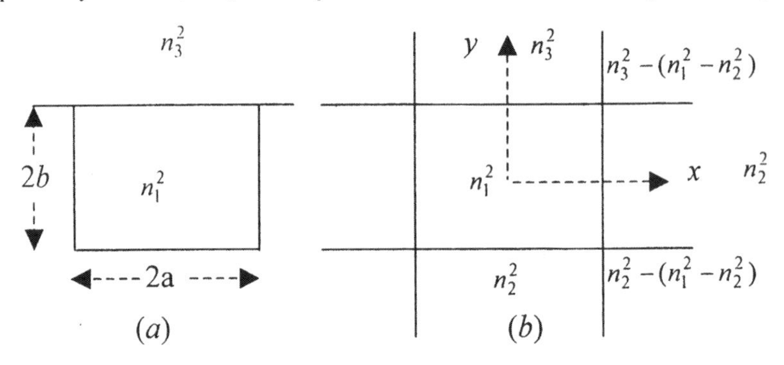

Figure 8 (a) Cross-section of a rectangular-core channel waveguide (b) the corresponding separable profile.

corner regions by $n_1^2 - n_2^2$ which is a small quantity for weakly guiding waveguides and can be taken into account using first order perturbation approach. The scalar modes can be obtained in a similar way as discussed in Sec. (2.3.1). The method can be modified to obtain the propagation constants of predominantly x-polarised (E_{mn}^x) and predominantly y-polarised (E_{mn}^y) modes in the following way:

The E_{mn}^x modes would approximately correspond to a TM mode in the x-direction and to a TE mode in the y-direction, and one can assume that ψ corresponds to E_x, which is the major electric field component. In order to get the eigen-value equations, we make E_x and $\partial E_x/\partial y$ continuous at $y = \pm b$ and $n^2 E_x$ and $\partial E_x/\partial x$ at $x = \pm a$. The continuity conditions at $y = \pm b$ will lead to TE eigen value Eq.(8) corresponding to the asymmetric planar waveguide given by Eq.(13) while the continuity conditions at $x = \pm a$ will lead to the TM eigen-value Eq. (7)

corresponding to the symmetric planar waveguide given by Eq.(12) with n_4 replaced by n_2. In a similar way one can obtain the propagation constants of the E^y_{mn} modes for the unperturbed structure. The propagation constants for the given waveguide can then be obtained by using the first order perturbation approach in a similar way as discussed in Sec. (2.3.1). The perturbation method has also been used to obtain the propagation characteristics of anisotropic rectangular waveguides described by a diagonal dielectric constant tensor[18]. Comparison with the results obtained by finite-element technique shows an excellent agreement.

2.3.3 Directional Couplers

Directional couplers made with rectangular core waveguides form the basis of many integrated-optical devices, such as switches, modulators and wavelength filters. Such devices have almost completely rectangular-cores when fabricated using high-silica (HiS) technology[2]. Thus it is quite important to have analysis of directional couplers consisting of rectangular-core waveguides. Kumar and co-workers[19] has extended the perturbation method discussed in earlier sections for singular rectangular-core waveguides to directional couplers having two rectangular waveguides as discussed below.

Let us consider a directional coupler consisting of two rectangular channel waveguides, shown in Fig.9a. Various approximate methods have been proposed to analyse such a device amongst which the method of Marcatili is again the earliest one. As was pointed out by Yeh and Taylor[20], Marcatili's field patterns correspond to the following refractive-index profile:

$$n^2(x,y) = n'^2(x) + n''^2(y) - n_1^2\,, \tag{30}$$

where

$$\begin{aligned} n'^2(x) &= n_2^2, && 0 < |x| < d, \\ &= n_1^2, && d < |x| < d+2a, \\ &= n_2^2, && |x| > d+2a, \end{aligned} \tag{31}$$

and

$$\begin{aligned} n''^2(y) &= n_s^2, && y < -b, \\ &= n_1^2, && -b < y < b, \\ &= n_2^2, && y |> b. \end{aligned} \tag{32}$$

The above dielectric constant profile [see Fig.9b] matches the given profile [Fig.9a] in all regions except the corner regions, where the difference between the two profiles is $n_1^2 - n_2^2$ (say, Δ), which is small.

In Marcatili's analysis this difference is neglected, and the scalar wave equation

$$\left[\frac{\partial^2}{\partial x^2} + \frac{\partial^2}{\partial y^2} + \left(k_0^2 n^2(x,y) - \beta^2\right)\right]\psi = 0 \tag{33}$$

is solved exactly by using the method of separation of variables. By putting $\psi(x,y) = X(x)Y(y)$, it can be shown that $X(x)$ and $Y(y)$ satisfy the following simple equations:

$$\left[\frac{d^2}{dx^2} + k_0^2 n'^2(x) - \beta_x^2\right] X(x) = 0, \tag{34}$$

and

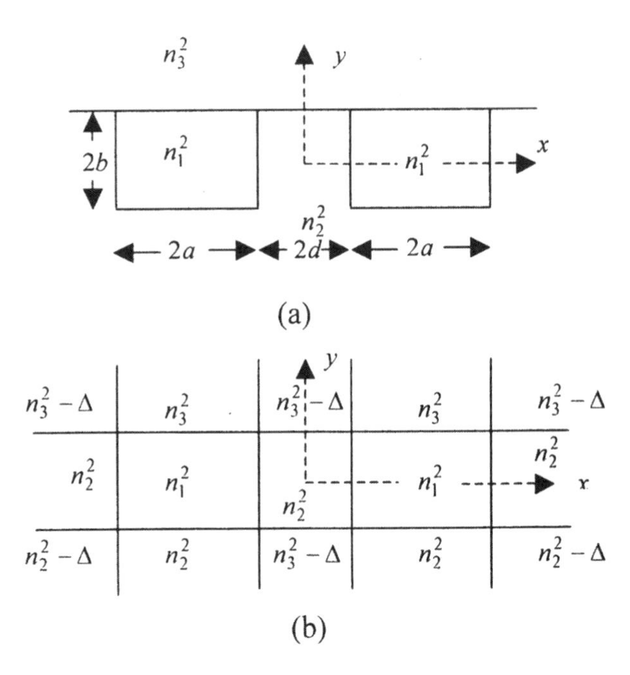

Figure 9a Cross-section of a rectangular-core waveguide coupler. **b** the corresponding separable profile.

$$\left[\frac{d^2}{dy^2} + k_0^2 n''^2(y) - \beta_y^2\right] Y(y) = 0, \tag{35}$$

where β_x and β_y represent the separation constants and

$$\beta^2 = \beta_x^2 + \beta_y^2 - k_0^2 n_1^2. \tag{36}$$

The approximate vector modes are obtained by satisfying the following boundary conditions. For example, the E_{mn}^y mode would approximately correspond to a *TE* mode in the x direction and to a *TM* mode in the y direction. Thus by assuming that $E_y = \psi(x,y)$, we make E_y and $\partial E_y/\partial x$ continuous at $x = d$ and $x = d + 2a$ and $n^2 E_y$ and $\partial E_y/\partial y$ continuous at $y = b$ and $y = -b$, which gives the following eigenvalue equation for β_x:

For the mode symmetric in x

$$\frac{\nu_1}{\nu_2}\tanh(\nu_1 d)=\frac{\left[\tan(2\nu_2 a)-\frac{\nu_1}{\nu_2}\right]}{\left[1+\frac{\nu_1}{\nu_2}\tan(2\nu_2 a)\right]}. \tag{37}$$

For the mode antisymmetric in x

$$\frac{\nu_1}{\nu_2}\coth(\nu_1 d)=\frac{\left[\tan(2\nu_2 a)-\frac{\nu_1}{\nu_2}\right]}{\left[1+\frac{\nu_1}{\nu_2}\tan(2\nu_2 a)\right]}, \tag{38}$$

and for β_y

$$\tan(2\mu_2 b)=\frac{n_1^2\mu_2\left(n_3^2\mu_3+n_2^2\mu_1\right)}{\left(n_2^2 n_3^2\mu_2^2-n_1^4\mu_1\mu_3\right)}, \tag{39}$$

where

$$\nu_1^2=\left(\beta_x^2-k_0^2 n_2^2\right),\ \nu_2^2=\left(k_0^2 n_1^2-\beta_x^2\right), \tag{40}$$

and

$$\mu_1^2=\left(\beta_y^2-k_0^2 n_2^2\right),\ \mu_2^2=\left(k_0^2 n_1^2-\beta_y^2\right),$$
$$\mu_3^2=\left(\beta_y^2-k_0^2 n_3^2\right). \tag{41}$$

After obtaining β_x and β_y, one can obtain the propagation constants β_s (for the symmetric mode) and β_a (for the antisymmetric mode) by using Eq.(36). The coupling coefficient would then be given by

$$\kappa=\left(\beta_s-\beta_a\right)/2. \tag{42}$$

A similar procedure can be used to obtain κ for the coupling in E_{mn}^x modes. Marcatili has also given an approximate formula for the calculation of κ which was later improved by Kuznetsov[21] to the following form for the coupling in E_{mn}^y modes:

$$\kappa=\frac{h^2\exp(-2pd)}{\beta\, a\left(h^2+p^2\right)\left(1+\frac{1}{pa}\right)}, \tag{43}$$

where

$$h^2=\left(k_0^2 n_1^2-\beta_x^2\right),\ p^2=\left(\beta_x^2-k_0^2 n_2^2\right). \tag{44}$$

Numerical calculations show that the effect of b on κ and hence on the coupling length $L_c(=\pi/2\kappa)$ is so small that the variation of L_c with respect to b is negligible, and as a result the slab-coupling approximation [i.e., when β and h in Eq.(43) correspond to a coupler consisting of two parallel slab waveguides] is equally good.

The results obtained by Eqs.(37-42) (which also correspond to Marcatilti's analysis) can be improved by considering the difference in the dielectric constant of the given waveguide and the profile given by Eq.(30) as a small correction for computing improved values of β by first-order perturbation theory[19]. It can easily be shown that by applying the first-order perturbation correction to β^2 the propagation constant would be given by

$$\beta_p^2 = \beta^2 + k_0^2\left(n_1^2 - n_2^2\right)\Gamma, \tag{45}$$

where

$$\Gamma = \frac{\int_b^{\infty}\int_0^{d} |\psi(x,y)|^2\, dxdy + \int_b^{\infty}\int_{(d+2a)}^{\infty} |\psi(x,y)|^2\, dxdy}{\int_0^{\infty}\int_0^{\infty} |\psi(x,y)|^2\, dxdy} \tag{47}$$

represents the fractional power flowing in the corner regions. An analytical (although lengthy) expression for Γ can be obtained, however, for the sake of brevity, we do not give that expression here.

In order to compare the results obtained by various approaches, the coupling length L_c is calculated for the E_{11}^y mode in a rectangular-core-waveguide directional

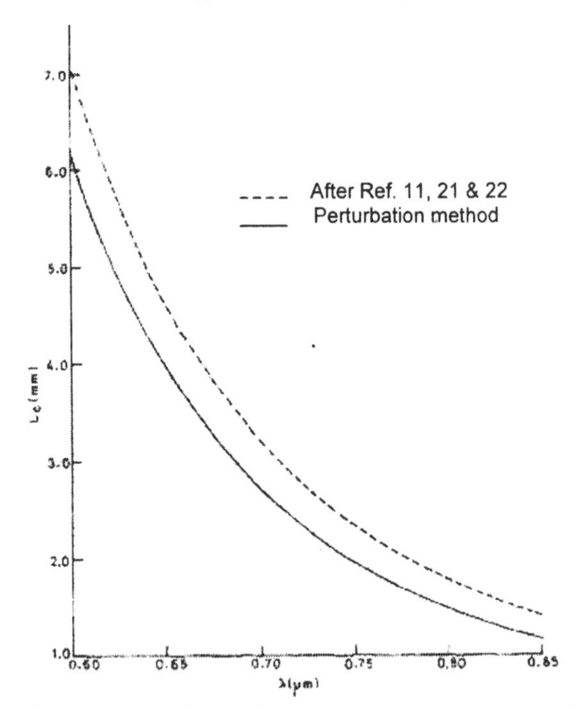

Figure 10 Variation of coupling length L_c as a function of wavelength λ for a rectangular-core directional coupler with $2a = 2b = 2d = 2.0\,\mu$m (adapted from Ref.[19]; © Optical Society of America).

coupler with parameters $n_1 = 2.211$, $n_2 = 1.0$, $n_s = 2.2$, and $2a = 2d = 2.0\mu m$. Figure10 shows the variation in coupling length as a function of wavelength λ for a coupler having $2b = 2\mu m$. The dashed curve corresponds to Marcatili's analysis[11], and the solid curve corresponds to the calculations after the first order correction is applied to β^2. The coupling length is also calculated by using Kuznetsov's approximate formula for such a directional coupler (Eq.43) and by using a similar formula for the corresponding slab waveguide problem. The difference between the coupling length obtained with these methods and that shown by the dashed curve is so small that it cannot be shown in the figure.

The figure shows that Marcatili's analysis overestimates the coupling lengths. As discussed later, this is in agreement with the results reported by Kawachi[2] on directional couplers fabricated using HiS technology, which gives almost rectangular cores.

It should be noted that for a separable profile, the boundaries in one direction (say y) would not affect the field variation in the perpendicular direction i.e. x. Thus Marcatili's method would not be able to predict any change in the coupling length of such a directional coupler if the height of the channels ($2b$) is changed. This implies that the Marcatili's method for a directional coupler is no better than a slab waveguide approximation where the given coupler is replaced by two parallel slab waveguides (of widths $2a$ each) separated by $s(=2d)$. This is clear from Fig.11 that shows the variation of coupling length L_c as a function of b for $\lambda = 0.6\ \mu m$. The dashed curve corresponds to Marcatili's analysis, and the solid curve is obtained after applying the first order correction to the propagation constants. The perturbation approach shows that the coupling length decreases as the

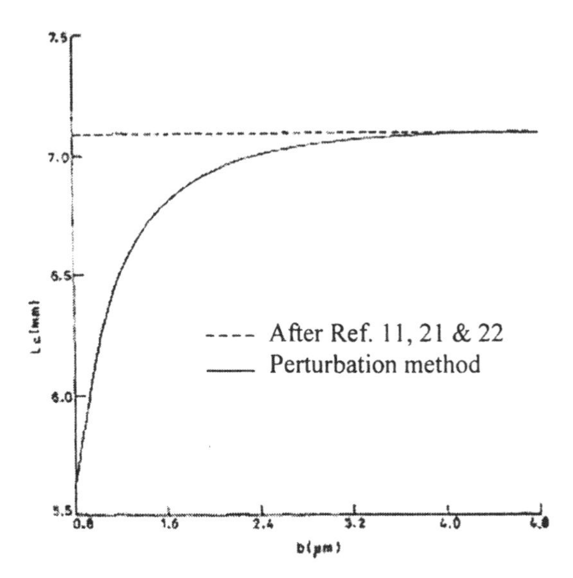

Figure 11 Variation of coupling length L_c as a function of depth b for a rectangular-core directional coupler with $2a = 2b = 2d = 2.0\mu$ m (adapted from Ref.[19]; © Optical Society of America).

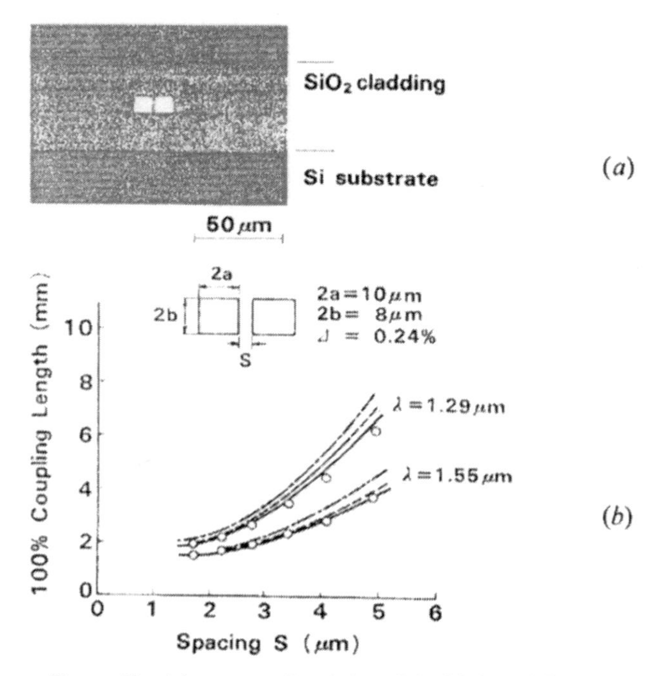

Figure 12 (a) cross-sectional view of the fabricated directional coupler (b) Variation of coupling length with waveguide spacing S (= $2d$). (——) PMM[(23)], (- - - -) Perturbation[(19)], (—·—) Marcatili[(11)], o( ) experiment. Adapted from M Kawachi, *Optical and Quantum Electronics*, **22**, 391-416 (1990).

channel depth $2b$ decreases. This is only expected because as core dimensions decrease, field should spread more outside the core resulting in a stronger coupling between the two channels. It is clear from this figure that if the difference Δ in the corner regions is neglected, the coupling length becomes almost independent of b. Thus the slab-waveguide approximation is equally good, and one need not use Marcatili's analysis for obtaining the coupling length.

The applicability of the perturbation approach for rectangular core waveguide directional couplers is also established from the results reported by Kawachi[(2)] on directional couplers fabricated using HiS technology, which gives almost rectangular cores. Figure 12 shows the variation of coupling length of a directional coupler consisting of two rectangular cores (shown in the inset) as a function of spacing $s\,(=2d)$ between them, at wavelengths 1.29 μm and 1.55 μm as reported in Ref.(2). Theoretical predictions based on (i) the Marcatili's method[(11)] (ii) the perturbation method[19] and (iii) the point matching method[23] also shown in this figure along with the experimental results. A comparison of different results shows that the Marcatili's method overestimates the value of the coupling length and also that the simple perturbation approach gives fairly accurate results.

The perturbation method can also be used to model fiber couplers. For example Yokohama et al[24] have used this method to model their fabricated fiber optic polarization beam splitters based on fused-taper couplers. They approximated the circular shapes of the cores and the stress applying parts of the PANDA fibers (used in the experiment) by approximate squares and used the perturbation method to analyse such a structure. The theoretical results on the phase difference between the symmetric and the antisymmetric modes of the coupler as a function of wavelength are found to agree very well with the experiment. Encouraged with the applicability of the rectangular-core waveguide models to fiber-optic fused couplers, Kumar and co-workers[25] have extended the perturbation approach for the analysis of 4×4 waveguide couplers. Chiang[25] has extended the same approach, in slightly different form, to waveguide arrays.

3. Summary

In summary, various approximate analytical methods with emphasis on the perturbation method to analyze rectangular-core waveguides and devices have been discussed.

Acknowledgements: The author acknowledges the help extended by Ms. Siny Antony G. in preparing this chapter.

References

1. A.K. Ghatak and K. Thyagarajan, *Optical Electronics*, Cambridge University Press (1989).
2. M. Kawachi, "Silica waveguides on silicon and their application to integrated-optic components", *Optical and Qunatum Electron.*, **22**, 391-416 (1990).
3. J. Noda, K. Okamoto and Y. Saskai, "Polarisation-maintaining fibers and their applications", *J. Lightwave Tech.*, **4**, 1071-1089 (1986).
4. B.Y. Kim, J.N. Blake, S.Y. Huang and H.J. Shaw, "Use of highly elliptical core fibers for two mode devices", *Opt. Lett.*, **12**, 729-731 (1987).
5. K.A. Murphy, M.S. Miller, A.M. Vengsarkar and R.O. Claus, "Elliptical-core two-mode optical fiber sensor implementation methods", *J. Lightwave Tech.*, **8**, 1688-1696 (1990).
6. C.D. Poole, J.M. Wiesenfeld and D.J. Digiovanni, "Elliptical-core dula-mode fiber dispersion compensator", *IEEE Photon. Technol. Lett.*, **5**, 194-197 (1993).
7. C. Yeh, "Elliptical dielectric waveguides", J. Appl. Phy., **33**, 3235-3245 (1962).
8. A. Kumar, R.K. Varshney and K. Thyagarajan, "Birefringence calculations in Elliptical-Core Fibers", *Electron. Lett.*, **20**, 112-113 (1984).
9. A. Kumar, R.K. Varshney, "Perturbation characteristics of highly elliptical-core optical waveguides: a perturbation approach", *Optical and Qunatum Electron.*, **16**, 349-354 (1984).
10. A. Kumar, K. Thyagarajan and A.K. Ghatak, "Analysis of rectangular core dielectric waveguide; an accurate perturnbation approach", *Opt. Lett.*, **8**, 63-65 (1983).
11. E.A.J. Marcatili, "Dielectric rectangular waveguides and directional coupler for integrated optics", *Bell Sys. Tech. J.*, **48**, 2071-2102 (1969).

12. R.M. Knox and P.P. Toulios, "Integrated circuits for the millimeter through optical frequency range", *Proceedings of MRJ symposium on sub millimeter waves*, ed. J. Fox, Polytechnic Press, New York, 497 (1970).
13. K.S. Chiang, "Dual effective-index method for the analysis of rectangular dielectric waveguides", *Appl. Opt.*, **25**, 2169-2174 (1986).
14. K.S. Chiang, "Analysis of rectangular dielectric waveguides: effective index method with built-in perturbation correction", ", *Electron. Lett.*, **28**, 388-389 (1992).
15. K.S. Chiang, K.M. Lo and K.S. Kwok, "Effective-index method with built-in perturbation correction for integrated optical waveguides", *J. Lightwave Tech.*, **14**, 223-228 (1996).
16. A. Kumar, D.F. Clark and B. Culshaw, "Explanation oferrors present in the effective-index method for analyzing rectangular core waveguides", *Opt. Lett.*, **13**, 1129-1131 (1988).
17. J.E. Goell, "A circular harmonic computer analysis of rectangular dielectric waveguides", *Bell Sys. Tech. J.*, **48**, 2133 (1969).
18. A. Kumar, M.R. Shenoy and K. Thyagarajan, "Modes in anisotropic rectangular waveguides: an accurate and simple perturbation approach", *IEEE Trans. Microwave Theory and Tech.*, **32**, 1415-1418 (1984).
19. A. Kumar, A.N. Kaul and A.K. Ghatak, "Prediction of coupling length in a rectangular-core directional coupler: an accurate analysis", *Opt. Lett.*, **10**, 86-88 (1985).
20. P. Yeh and H.F. Taylor, "Contradirectional frequency-selective couplers for guided-wave optics" *Appl. Optics*, **19**, 2848-2855 (1980).
21. A. Yariv, *Introduction to optical Electronics*, 2nd ed. (Holt, Rinehart and Winston, New York), 391(1976).
22. M. Kuznetsov, "Expressions for the coupling coefficient of a rectangular-waveguide directional coupler", *Opt. Lett.*, **9**, 499-501 (1988).
23. K. Jinguji, N. Takato and M. Kawachi, *Proc. Tech. group Meet. IECE Jpn.*, (1986) (in Japanese).
24. I. Yokohama, K. Okamoto and J. Noda, "Analysis of Fiber-optic Polarizing beam splitters consisting of fused-taprer couplers", *J. Lightwave Tech.*, **4**, 1352-1359 (1986).
25. A. Kumar, R.K. Varshney, "Scalar modes and coupling characteristics of eight-port waveguide couplers", *J. Lightwave Tech.*, **7**, 293-296 (1996).

13

Scalar Variational Analysis for Optical Waveguides

Enakshi Khular Sharma and Sangeeta Srivastava

1. Introduction

Confinement and guidance of electromagnetic waves has been the subject of intensive studies ever since these waves were used as carriers of information. The emergence of semiconductor lasers and the subsequent availability of low loss optical fibers have ushered in an era of optical communication. A miniaturization of the optical fiber system has led to a class of active and passive components that use light guided by films deposited on wafer like substrates. A study of all such components come under the purview of integrated optics, a term coined in 1969 by S. E. Miller.

Dielectric optical waveguides in planar geometry, which provide confinement to lightwaves in one or two dimensions, are the basic building blocks of integrated optical devices. The basic idea is to confine optical waves to a small region and process them in a variety of ways to achieve operations such as coupling, switching and modulation. In order to design these devices, it is imperative that the waveguidance through the basic waveguides is understood and modeled in simple and accurate ways. The simplest optical waveguide is the dielectric planar waveguide, which consists of a dielectric guiding film of index n_2 sandwiched between materials of slightly lower refractive indices n_1 (substrate) and n_3 (cover). A detailed analysis of this slab waveguide is available in most text books[1]. Fabrication techniques[2] such as ion exchange (on glass substrates) and ion exchange or proton exchange (on $LiNbO_3$ substrates), however, result in waveguides with an inhomogeneous refractive index distribution, i.e., graded index waveguides. Based on whether the waveguide provides one-dimensional or two-dimensional confinement, these waveguides are termed as diffused planar or diffused channel graded index waveguides.

Although most practically useful guides are the channel waveguides, the planar waveguides also play an important role. The realization of planar components such as prisms, lenses and gratings[3] require a useful description of the modal fields of planar waveguides. Further, as the planar waveguides are relatively easy to analyze, their modeling gives an insight into the basic guidance phenomenon and all approximate methods for analyzing channel waveguides are based on analyzing suitable planar waveguides. We will discuss methods for analysis of both planar and channel waveguides.

Optical waveguides are dielectric structures and wave propagation through them is defined by Maxwell's equation in a non-conducting, charge free medium with appropriate boundary conditions. However, for 'weakly guiding' structures in which the relative index difference between the dielectric layers is small, the transverse Cartesian components of the electric field satisfy the Helmholtz equation[1], also known as the scalar wave equation. For planar waveguides the Helmholtz equation reduces to an ordinary differential equation and can always be solved as a boundary value problem by direct numerical integration or by the multilayer staircase approximation[4]. Exact analysis for obtaining the modal solutions of the planar waveguides is possible for only a limited number of refractive index profiles[5]. However, for channel waveguides such direct solutions are not possible since the Helmholtz equation is a partial differential equation and one has to use relatively more involved methods based on finite element or finite-difference. These methods are not only numerically intensive and time consuming, they do not lend themselves to simple modeling of devices based on diffused waveguides. Here we discuss semi-analytical techniques based on the variational principle for obtaining the modal solutions. Though not as accurate as numerical methods, these have fairly good accuracy and give analytical forms for the mode fields.

We discuss the scalar variational principle as applied to optical waveguides in Section 2. This is followed by application to diffused planar waveguides in Section 3 and diffused channel waveguides in Section 4. Finally, in Section 5 we discuss the application to directional couplers.

2. Scalar Variational Principle and Optical Waveguides

The refractive index profile of channel waveguides (Fig.1), which provide two-dimensional confinement, can be represented as

$$\begin{aligned} n^2(x,y) &= n_s^2 + 2n_s\Delta n f(y/h)g(x/w) \qquad && y>0 \\ &= n_c^2 && y<0 \end{aligned} \tag{1}$$

where n_s is the substrate refractive index, Δn is the maximum index change from the substrate index, y is the depth from the waveguide surface, $f(y/h)$ models the index variation along the depth and $g(x/w)$ models the index variation along the width. The Helmholtz equation for such a guiding structure is given by

$$\nabla^2\Psi(x,y,z) + k_0^2 n^2(x,y)\Psi(x,y,z) = 0 \tag{2}$$

where $\Psi(x,y,z)$ is one of the transverse Cartesian components of the electric field. Translational invariance of the waveguide along the z-direction, i.e., $n^2(x,y,z) = n^2(x,y)$, enables one to express the field in a separable form as a forward propagating inhomogeneous plane wave

$$\Psi(x,y,z) = \psi(x,y)\exp(-j\beta z) \tag{3}$$

where an implicit time dependence of the form $\exp(j\omega t)$ has been assumed. $\psi(x,y)$ satisfies the scalar wave equation

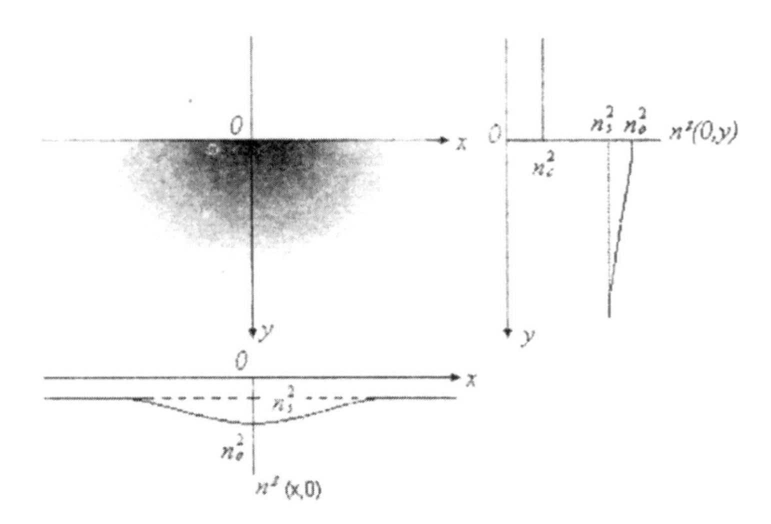

Figure 1 Schematic of the index variation of a diffused channel waveguide. Typical index variation is also plotted.

$$\nabla_t^2\psi + k_0^2\left(n^2(x,y) - n_e^2\right)\psi(x,y) = 0 \tag{4}$$

where, $n_e = \beta / k_0$ is the effective index of the waveguide. It may be noted that the solutions $\psi(x,y)$ of the scalar wave equation and its derivative are continuous everywhere and therefore bounded. Equation (4) is in fact an eigenvalue equation with n_e being the eigenvalue and $\psi(x,y)$ the eigenfunction of the operator H defined as

$$H = \frac{1}{k_0^2}\nabla_t^2 + n^2(x,y) \tag{5}$$

This equation admits only certain discrete solutions called the guided modes and a continuum of solutions called the radiation modes. Depending on the dimensions of the guiding region and the wavelength, a waveguide may support a number of guided modes. However, the most important waveguides are the ones, which support only one guided mode – the so called single-mode waveguides.

The variational principle is based on the integral form of the scalar wave equation. The essence of the principle is the formulation of a stationary expression for a key quantity, the effective index n_e in our case, in terms of the cross sectional integrals involving the fields $\psi(x,y)$. The integral form of the scalar wave equation is

$$n_e^2 = \frac{1}{k_0^2}\iint\psi^*(x,y)\nabla_t^2\psi(x,y)dxdy + \iint n^2(x,y)\left|\psi(x,y)\right|^2 dxdy \tag{6}$$

which can be transformed using the two dimensional form of the divergence theorem to

$$n_e^2 = -\frac{1}{k_0^2}\iint\left|\nabla_t\psi(x,y)\right|^2 dxdy + \iint n^2(x,y)\left|\psi(x,y)\right|^2 dxdy \tag{7}$$

where the modal field $\psi(x,y)$ is assumed to be normalized, i.e.,

$$\iint\left|\psi(x,y)\right|^2 dxdy = 1 \tag{8}$$

All integrals are from $-\infty$ to $+\infty$. The RHS of Eq.(6) and Eq.(7) is referred as the stationary expression for effective index, n_e, since it is stationary[6] with respect to small variations in $\psi(x,y)$, i.e.,

$$\delta n_e^2(\psi + \delta\psi) = 0 \tag{9}$$

to the first order in $\delta\psi$.

Since the modal field is an unknown function and in fact is the function that is sought for as a solution of the propagation problem, one uses an approximation for it as ψ_t, generally referred to as the trial field. This trial field when used in Eq.(7) gives an estimate of the effective index n_{et}. Thus,

$$n_{et}^2 = -\frac{1}{k_0^2}\iint\left|\nabla_t\psi_t(x,y)\right|^2 dxdy + \iint n^2(x,y)\left|\psi_t(x,y)\right|^2 dxdy \tag{10}$$

It is obvious that different forms of the trial field ψ_t would give different expressions and values for n_{et}. An important property of this expression is that all these values of n_{et} would invariably be smaller than the exact value of the effective index (n_e) and the exact value is obtained only when ψ_t is the same as the exact modal field. In order to prove this we consider the Helmholtz equation, Eq.(4), for the diffused waveguide. For each mode (radiation or guided) we get an effective index and the corresponding modal fields. For the nth mode, we can write Eq.(4) in the operator form as

$$H\psi_n(x,y) = n_{en}^2\psi_n(x,y) \tag{11}$$

Multiplying Eq.(11) with ψ_n^* and integrating, we get the effective index of the nth mode as

$$n_{en}^2 = \int\psi_n^* H\psi_n ds \tag{12}$$

where the integration is over the entire cross section and ψ_n are assumed to be normalized. Equation(12) is essentially the same as Eq.(6) with $\psi(x,y)$ as the mode field of the nth mode.

The modal field constitutes a complete set of orthonormal functions, i.e., $\int\psi_n^*\psi_m ds = \delta_{nm}$; thus any function $\phi_t(x,y)$ can be expanded in terms of the modal field as

$$\phi_t(x,y) = \sum_n a_n\psi_n(x,y) \tag{13}$$

In terms of the operator H the integral form of the scalar wave equation, Eq.(6), can be expressed as

$$n_{et}^2 = \frac{\int \phi_t^* H \phi_t ds}{\int \phi_t^* \phi_t ds} \tag{14}$$

where the denominator is the normalisation term. Substituting Eq.(13) in Eq.(14), we get

$$n_{et}^2 = \frac{\int \sum_n a_n^* \psi_n^* H \sum_n a_m \psi_m ds}{\int \sum_n a_n^* \psi_n^* \sum_n a_m \psi_m ds} \tag{15}$$

Using Eq.(11) and the orthonormality condition, Eq.(15) reduces to

$$n_{et}^2 = \frac{\sum_n a_n^* a_n n_{en}^2}{\sum_n a_n^* a_n} \tag{16}$$

Subtracting the fundamental mode index square n_{e0}^2 from both sides of Eq.(16) gives

$$n_{e0}^2 - n_{et}^2 = \frac{\sum_n |a_n|^2 (n_{e0}^2 - n_{en}^2)}{\sum_n |a_n|^2} \tag{17}$$

Since the fundamental mode effective index is always the highest, i.e., $n_{e0} \geq n_{en}$, and $|a_n|^2 \geq 0$ for all n, the RHS of Eq.(17) is always positive, which implies that

$$n_{e0}^2 \geq n_{et}^2 \tag{18}$$

Therefore, any chosen trial field $\phi_t(x, y)$, will always give a value of effective index less than the exact value for the effective index of the fundamental mode. Thus higher the value of n_{et} obtained using the variational expression of Eq.(10), the closer it will be to the exact value of n_e. The corresponding field, $\psi_t(x, y)$, will also be a better approximation to the exact field $\psi(x, y)$.

This principle is employed for analyzing optical waveguides and is implemented as follows: A trial field $\psi_t(x, y; a_1, a_2, ...a_n)$ is set up which involves a number of adjustable variational parameters $a_1, a_2, ...a_n$. The dependence on x and y is chosen in such a way that ψ_t resembles the actual field as far as possible. This field is substituted in Eq.(10), which is then maximized with respect to the variational parameters. The maximum value of n_{et} thus obtained gives the estimate of the effective index and the corresponding $\psi_t(x, y; a_1, a_2, ...a_n)$ with the optimized values of the variational parameters as an approximation for the modal field. Generally, a judicious increase in the number of parameters results in a better trial field. However, a better trial field with minimum number of parameters is always sought for as it simplifies the computations and is easier to use for further modeling of devices involving these waveguides.

3. The Planar Graded Index Waveguide

Variational analyses of planar diffused waveguides based on single function trial fields best illustrates the use of the method and improvement in its accuracy with increase in the number of variational parameters. Planar graded index waveguides (Fig.2) are usually fabricated on $LiNbO_3$ or glass substrate using the process of ion exchange or diffusion. In general, the refractive index profile of the planar graded index waveguide can be expressed as

$$\begin{aligned} n^2(y) &= n_s^2 + 2n_s \Delta n f(y/h) \qquad & y > 0 \\ &= n_c^2 & y < 0 \end{aligned} \tag{19}$$

where, n_s is the substrate refractive index, Δn is the maximum index change from the substrate index, y is the depth from the waveguide surface, h is the depth defined

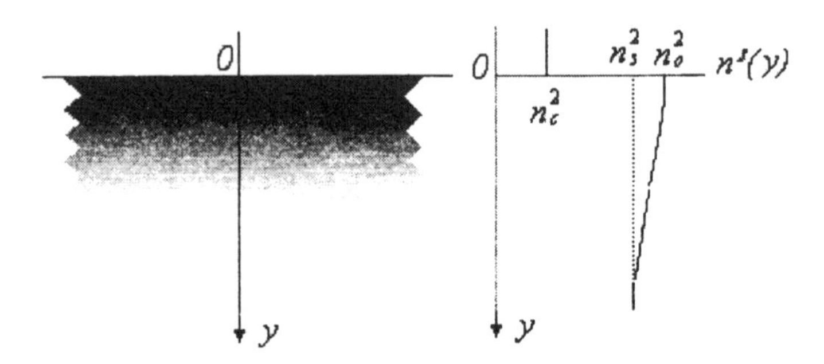

Figure 2 Schematic of the index distribution of a planar waveguide

according to the profile function $f(y/h)$ which best models the index variation obtained by the actual fabrication process. Depending on the diffusion conditions, the typical profile functions obtained in most ion exchange processes can be modeled by

$$\begin{aligned} f(y/h) &= \exp(-y^2/h^2) \qquad & \text{Gaussian Function} \\ &= \text{erfc}(y/h) & \text{Complimentary Error Function} \\ &= \exp(-2y/h) & \text{Exponential Function} \\ &= \exp[-(y-y_0)^2/h^2] & \text{Buried Gaussian} \end{aligned} \tag{20}$$

Often, as in proton exchange guides or in waveguides fabricated by epitaxial growth on GaAs substrates the profile is adequately defined by a step function as $f(y/h) = 1$ for $0 < y/h < 1$ and $f(y/h) = 0$ for $y/h > 1$.

For few specific profiles such as step index, infinite parabolic index and exponential profiles it is possible to obtain analytical solutions. However, in the case of most practical asymmetric graded index profiles of the type described in Eq.(20), analytical solutions are not possible and one has to use numerical or semi-analytical approximate methods. The variational method is a semi-analytical tool which can be used effectively for obtaining the propagation characteristics of guided modes in planar graded index waveguides. For planar waveguides, the variational expressions for effective index given in Eqs.(6) and (7) reduce to the following one dimensional integral equations:

$$n_{et}^2 = \frac{1}{k_0^2}\int \psi_t^*(x)\frac{d^2\psi_t}{dx^2}dx + \int n^2(x)|\psi_t(x)|^2 dx \tag{21}$$

or

$$n_{et}^2 = -\frac{1}{k_0^2}\int \left|\frac{d\psi_t}{dx}\right|^2 dx + \int n^2(x)|\psi_t(x)|^2 dx \tag{22}$$

where all integrals are from $-\infty$ to $+\infty$

In general, trial fields can be constructed in two ways. One is to model the trial field by an appropriate single function involving the variational parameters. Such fields are especially useful for the fundamental mode in single-moded waveguides. Alternatively, the trial field is expressed as an expansion on a suitable finite set of known basis functions as

$$\psi(y) = \sum_{j=0}^{N-1} c_j \phi_j(y) \tag{23}$$

where the coefficients c_j are the undetermined variational parameters. Since, any arbitrary function $\psi(y)$ can be expanded on a complete set of basis functions, when $N \to \infty$, this expansion is exact. However, the series has to be terminated at a convenient value of N for numerical computations, with N being determined by the convergence of results. This is also known as the Rayleigh Ritz[3,7] or Galerkin method. The most important feature in the application of Ritz Galerkin method consists in the choice of basis functions. Usually one chooses a system, which is complete and orthonormal in the region considered and whose members satisfy the required boundary conditions. The Ritz Galerkin approach is not limited to the fundamental mode and can also be used to obtain higher order modes in a multimoded structure.

3.1 Single Function Trial Fields

Single function fields reported in literature can be broadly classified in two categories: i) fields based on the Hermite Gauss[3,8-10] functions for which the integrals in Eq.(21) can be evaluated analytically and ii) fields based on the secant hyperbolic[11] function for which numerical integration is needed. The former are related to the solutions of the scalar wave equation for infinite parabolic profiles while the latter correspond to the sech (y/h) profile function.

3.1.1 Hermite Gauss Function based fields

We first consider the diffused asymmetric planar waveguides. The cover index n_c is usually much lower than the substrate index, i.e., the asymmetry parameter, $p = (n_s^2 - n_c^2)/2n_s\Delta n$, is large and the evanescent field of the guided waves decay very rapidly to zero in the cover region. Hence, as a simplifying assumption, one can assume the fields in the cover to be zero i.e., $\psi(y) = 0$ for $y < 0$ and the following Hermite Gauss (HG) function[3,10] with only one parameter can be used as a trial field:

$$\psi(y) = A\alpha y/h \exp(-\alpha^2 y^2/h^2) \qquad y>0$$
$$= 0 \qquad y<0 \qquad (24)$$

where $A=(8\alpha/h\sqrt{\pi/2})^{1/2}$ is the normalization constant, and α is the variational parameter. Substituting the trial field into the stationary expression for effective index (Eq.10), the normalized effective index $b=(n_e^2-n_s^2)/2n_s\Delta n$ can be expressed in terms of the variational parameter α and normalized parameter $V=k_0h\sqrt{2n_s\Delta n}$ as

$$b=(r-s)/t \qquad (25)$$

where s and t are the profile independent expressions given by

$$s=\frac{3\alpha\sqrt{\pi/2}}{8hV^2}$$
$$t=\frac{h}{8\alpha}\sqrt{\pi/2}$$

and r is the profile dependent term given by

$$r=\frac{h\sqrt{\pi}}{4(2\alpha^2+1)^{3/2}} \qquad \text{for Gaussian profile}$$

$$r=\frac{h}{2\sqrt{\pi}}\left(\frac{\tan^{-1}(\sqrt{2}\alpha)}{\sqrt{2}\alpha}-\frac{1}{2\alpha^2+1}\right) \qquad \text{for Erfc profile}$$

Maximizing Eq.(25) with respect to the variational parameter, α, gives the optimal value of α and hence, the effective index and modal field of the waveguide. A comparison of the optimized HG field with the exact field for a typical Gaussian index profile is shown in Fig.3.

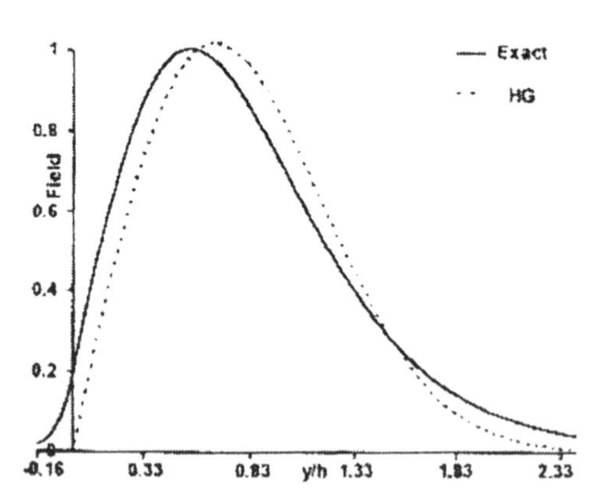

Figure 3 Comparision of HG and exact fields for a Gaussian profile (Δn=0.03, n_s=1.512, n_c=1, λ=0.6328 μm, V=3.6)

In order to take into account the evanescent field in the cover region a simple choice for the trial field is an Evanescent Hermite Gauss (EHG) field[10]. This is now a two variable field with an exponential decay in the cover region given by

$$\psi(y)=A(1+\gamma y/h)\exp(-\alpha^2y^2/h^2) \qquad y>0$$
$$=A\exp(\gamma y/h) \qquad y<0 \qquad (26)$$

where A is the normalization constant and γ and α are the two variational parameters. If we substitute the trial field into the stationary expression for effective index, we again get the closed form expressions for the effective index $b = (n_e^2 - n_s^2)/2n_s\Delta n$ in terms of α, γ, V, and the asymmetry parameter p as

$$b = (r - s)/t \tag{27}$$

where, the profile independent terms are given by

$$s = \frac{4p\alpha}{\gamma} + \frac{1}{V^2}\left(\sqrt{\frac{\pi}{2}}\left((3\gamma^2 + 4\alpha^2) + 4\gamma\alpha\right)\right)$$

$$t = \frac{4\alpha}{\gamma} + 4\sqrt{\frac{\pi}{2}} + \frac{4\gamma}{\alpha} + \frac{\gamma^2}{\alpha^2}\sqrt{\frac{\pi}{2}}$$

and the profile dependent term r is given by

$$r = \frac{2}{(1+2\alpha^2)^{3/2}}\left[2\sqrt{\pi}\alpha(1+2\alpha^2) + \alpha\gamma^2\sqrt{\pi} + 4\alpha\gamma(1+2\alpha^2)^{1/2}\right]$$

for Gaussian profile

$$r = \frac{2}{\sqrt{2\pi}\alpha^2}\left[(4\alpha^2+\gamma^2)\tan^{-1}(\sqrt{2}\alpha) - \frac{\sqrt{2}\alpha\gamma^2}{1+2\alpha^2} + 4\alpha\gamma\sqrt{\frac{\pi}{2}}\left[1-(1+2\alpha)^{-1/2}\right]\right]$$

for Erfc profile

Hence, the optimal values of α and γ are obtained by maximizing b with respect to both the parameters (or by putting $\partial b/\partial \alpha = 0$ and $\partial b/\partial \gamma = 0$) and depend on V and p only. The comparison of the optimized EHG field with the exact field for a typical Gaussian index profile in Fig.4 clearly shows the improvement in the field description for the cover region.

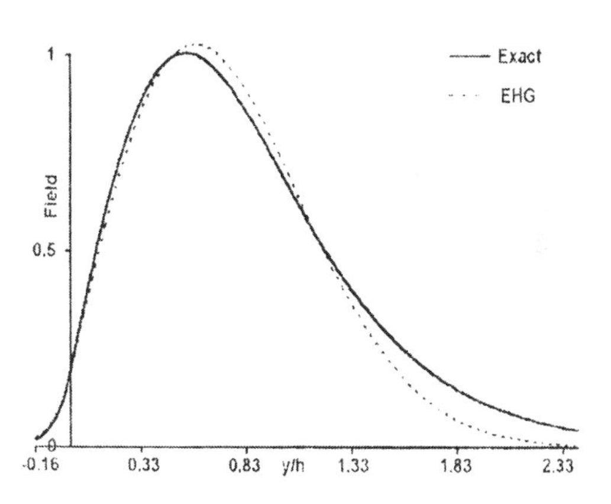

Figure 4 Comparision of EHG and exact fields for a Gaussian profile (Δn=0.03, n_s=1.512, n_c=1, λ=0.6328µm, V=3.6)

In the single mode region, by empirical curve fitting to the curves obtained for these optimal values, simple closed form expressions for α and γ in terms of the waveguide parameters V and p have been obtained[10] as

$$\alpha = b_0 + b_1 p^{-1/2} + b_2 \log V + b_3 p^{-1/2} \log V \tag{28}$$

$$\gamma = a_0 + a_1 p^{1/2} + a_2 V + a_3 V p^{1/2} \tag{29}$$

where the eight constants a_i and b_i (i=1,2,3,4) are a set of numbers for a given profile function $f(y/h)$ and are tabulated in Table-I. Hence, once the parameters V and p and the refractive index profile of the waveguide are known, the optimal α and γ can be easily calculated from Eqs.(28) and (29) to define the field parameters, and the corresponding effective index can be obtained from the analytical expressions.

Another profile that is of interest and models the lateral profile of diffused channel waveguides is the symmetric Gaussian described by

$$n^2(y) = n_s^2 + 2n_s\Delta n \exp(-y^2/h^2) \qquad -\infty < y < \infty \tag{30}$$

For such a profile, the equivalent suitable trial field is a simple Gaussian function of the form

$$\psi(y) = \sqrt{\alpha/(h\sqrt{\pi/2})}\exp\left(-\alpha^2\frac{y^2}{h^2}\right) \tag{31}$$

with only one parameter α. The corresponding b is given by

$$b = -\frac{\alpha^2}{V^2} + \alpha\sqrt{\frac{2}{1+2\alpha^2}} \tag{32}$$

The optimal value of α is obtained by putting $\partial b/\partial\alpha = 0$, which leads to the following transcendental equation relating α and V:

$$\left(1+2\alpha^2\right)^{3/2}\sqrt{2}\alpha = V^2 \tag{33}$$

A closed form expression[10] similar to Eq.(28) for α is obtained by curve fitting as

$$\alpha = a_0 + a_1\sqrt{V} \tag{34}$$

The constants a_0 and a_1 are also tabulated in Table I.

Table-I Empirical constants for various profiles.

Constants	Gaussian	Error Function	Exponential	Step Function	Symmetric Gaussian
a_0	-0.19	-0.19	0.21	-0.40	-0.40
a_1	-0.26	-0.30	-0.54	-0.33	0.84
a_2	0.03	-0.01	-0.12	0.17	-
a_3	1.01	0.97	0.99	1.14	-
b_0	-0.03	-0.66	-0.95	0.49	-
b_1	0.35	0.40	0.52	0.18	-
b_2	0.86	1.23	1.30	0.54	-
b_3	-0.14	-0.06	-0.09	-0.11	-

The trial fields can be further improved by an addition of variational parameters. Another trial field reported in literature is the basic EHG field with a parameter introduced to adjust the peak position of the field[11] distribution (EHGP), defined as

$$\psi(y) = A(1+\gamma y/h)\exp-[\alpha^2(\frac{(y+\tilde{p})^2}{h^2})] \qquad y>0$$

$$= A\exp(\gamma y)\exp(-\frac{\alpha^2(\tilde{p}^2+2\tilde{p})}{h^2}) \qquad y<0 \qquad (35)$$

with $\tilde{p}$, γ and α as the three variational parameters. The expression for effective index can once again be obtained in analytical form. A comparison of the EHGP field with the exact field in Fig.5 shows the improvement in the accuracy of the field.

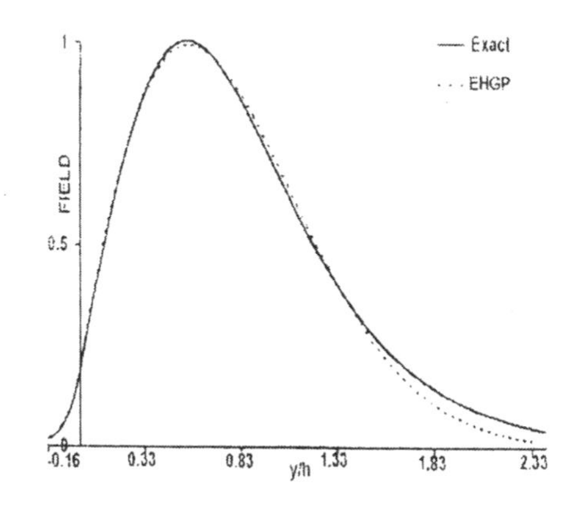

Figure 5 Comparision of EHGP and exact fields for a Gaussian profile (Δn=0.03, n_s=1.512, n_c=1, λ=0.6328 μm , V=3.6)

The above figures show that the approximate fields do not describe the tail region in the substrate accurately. An improved field (IEHG), which takes into account the exponential decay in the tail region, has also been suggested[10]. This trial field is also described by a three parameter function with an exponential decay region starting at different points for different profiles and can be outlined as:

$$\psi(y) = A(1+\gamma a)\exp\left(\alpha^2 a^2 - \frac{\gamma a}{1+\gamma a}\right)\exp\left(-\left(2\alpha^2 a - \frac{\gamma a}{1+\gamma a}\right)\frac{y}{h}\right) \qquad y/h>a$$

$$= A(1+\gamma\frac{y}{h})\exp\left(-\alpha^2\frac{y^2}{h^2}\right) \qquad y/h<a$$

$$= A\exp\left(\gamma\frac{y}{h}\right) \qquad y<0 \qquad (36)$$

The parameter a defines the depth at which the exponential tail starts. Fortunately, it is possible to fix the value of the parameter a for different profiles as

$$\begin{aligned} a &= 1.0 && \text{Gaussian} \\ &= 0.78 && \text{Erfc} \\ &= 0.85 && \text{Exponential} \\ &= 1.2 && \text{Step function} \end{aligned} \qquad (37)$$

which reduces the variational parameters to two. Fig.6 shows a comparison of the fields obtained by EHG, IEHG and exact analysis for typical graded index waveguides. As can be seen the IEHG fields compare extremely well with the exact fields.

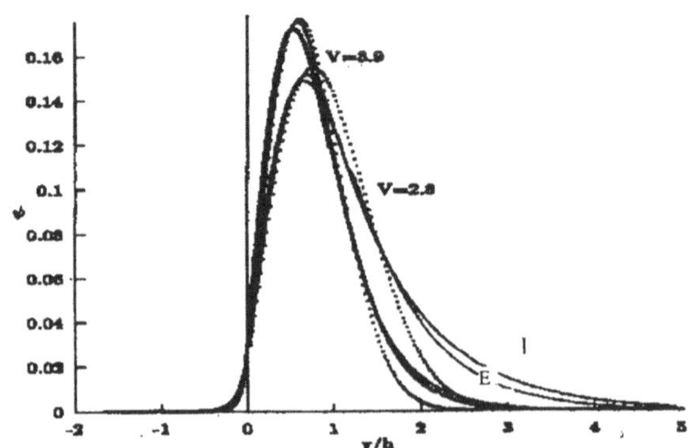

Figure 6 Comparision of EHG, IEHG and exact fields for a Gaussian profile. The dotted curves correspond to EHG fields while the continuous curves corresponding to the exact and IEHG fields almost overlap.

Similarly, for the symmetric Gaussian profile, the field can be improved by introducing a matched exponential tail and will now be given by

$$\psi(y) = A\exp(\alpha^2 a^2)\exp\left(-2\alpha^2 a\frac{y}{h}\right) \qquad \left|\frac{y}{h}\right| > a$$
$$= A\exp\left(-\alpha^2\frac{y^2}{h^2}\right) \qquad -a < y/h < a \qquad (38)$$

3.1.2 Trial Fields based on the Secant Hyperbolic Functions:

A rather unconventional form of the trial field based on the secant hyperbolic functions has been suggested in literature. The one variable Secant Hyperbolic (SH) trial field[11] which assumes the field to be zero in the cover is defined as

$$\psi(y) = A\sinh(y/h)\operatorname{sec}h^{\tau}(y/h) \qquad y \ge 0$$
$$= 0 \qquad y \le 0 \qquad (39)$$

where the exponent τ is the variational parameter. A field comparison is shown in Fig.7.

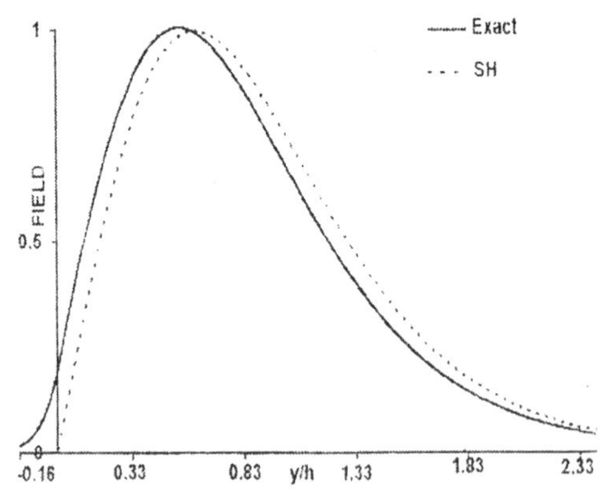

Figure 7. Comparision of SH and exact fields for a Gaussian profile (Δn=0.03, n_s=1.512, n_c=1, λ=0.633 μm , V=3.6)

One can add another parameter to this field to take into account the fields in the cover region also as

$$\psi(y) = A(1+\gamma\sinh(y/h))\operatorname{sec}h^{\tau}(y/h) \qquad y \ge 0$$
$$= A\exp(\gamma y/h) \qquad y \le 0 \qquad (40)$$

where γ and τ are the two variational parameters. This field, also known as the Evanescent Secant Hyperbolic (ESH) field[11] gives extremely accurate results as it accounts for the field variations in all the three regions- cover, guiding and the substrate. A field comparison in Fig.8 shows the excellent accuracy of the field form. However, these field forms do not lead to analytical expressions and numerical integration is required.

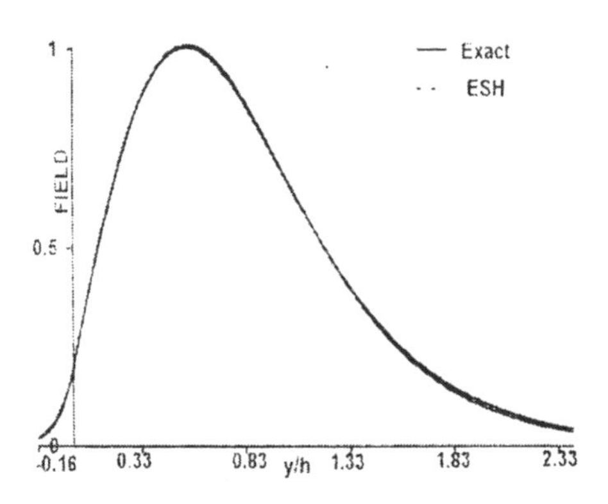

Figure 8 Comparision of ESH and exact fields for a Gaussian profile (Δn=0.03, n_s=1.512, n_c=1, λ=0.6328 μ m, V=3.6)

In addition to the above two categories, there has also been interest in a cosine-exponential (CE)[13,14] form for the trial field. The cosine-exponential fields are solutions of the scalar wave equation for a step refractive index profile and, hence, are useful in defining an "equivalent" step index waveguide.

3.2 The Ritz Galerkin Method

In the Ritz Galerkin[3,7] method the modal field of the waveguide $\psi(y)$ is expanded on a suitable finite set of basis functions $\phi_j(y)$ which satisfy the required boundary conditions, i.e.,

$$\psi(y) = \sum_{j=0}^{N-1} c_j \phi_j(\alpha, y) \tag{41}$$

with the coefficients c_j as the variational parameters. A suitable choice of the basis functions is formed by the normalized Hermite Gauss functions, i.e.,

$$\phi_j(\alpha, y) = \sqrt{\frac{\alpha}{\sqrt{\pi}\, 2^j\, j!}} \exp(-\alpha^2 y^2/2) H_j(\alpha y) \qquad j = 0,\ 1,\ 2.... \tag{42}$$

The chosen basis functions ϕ_j are orthonormal solutions of the Helmholtz equation for an infinitely extended parabolic refractive index, i.e., for $n^2(x) = n_0^2 - \Delta_0 y^2$ with the still arbitrary parameter α related to Δ_0 as

$$\alpha^2 = k_0 \sqrt{\Delta_0} \tag{43}$$

A reference profile that closely resembles the true profile can lead to a rapid convergence of results. Hence, the infinite parabolic profile is selected such that it has the same peak index as the given profile, i.e., $n_0^2 = n_s^2 + 2n_s\Delta n$, and also has the same index as the given graded index profile at the point where the profile function has reduced to half its peak value, i.e., $\Delta_0 = n_s \Delta n / y_{0.5}^2$ where $f(y_{0.5}/h) = 0.5$ and

$$y_{0.5/h} = 0.832 \qquad \text{for Gaussian profile}$$
$$= 0.48 \qquad \text{for Erfc profile.} \tag{44}$$

Hence, the optimal choice of α gets quantified to

$$\alpha^2 = k_0 \sqrt{n_s \Delta n} \,/\, y_{0.5} \tag{45}$$

Substituting for the trial field by the form given in Eq.(41) with $\phi_j(\alpha, y)$ defined by Eq.(42) into the stationary expression for effective index, one obtains:

$$n_e^2 \sum_{m,n} c_m c_n \int \phi_n \phi_m dy = (1/k_0^2) \sum_{m,n} c_m c_n \int \phi_m \frac{d^2\phi_n}{dy^2} dy + \sum_{m,n} c_m c_n \int n^2 \phi_n \phi_m dy \tag{46}$$

Substituting for $d^2\phi_n / dy^2$ from the scalar wave equation for the infinite parabolic index profile and using the orthonormality condition $\int \phi_m \phi_n \, dy = \delta_{mn}$ we can rewrite Eq.(46) as

$$n_e^2 \sum_{m,n} c_m c_n \delta_{mn} = -\sum_{m,n} c_m c_n \delta_{mn} (2n+1) \alpha^2 / k_0^2 + \sum_{m,n} c_m c_n \int \Delta_0 y^2 \phi_m \phi_n dy$$
$$+ \sum_{m,n} c_m c_n \int \Delta_0 n^2(y) \phi_m \phi_n dy \tag{47}$$

Using the stationary condition $\partial n_e^2 / \partial c_k = 0$, one obtains the set of equations given by

$$n_e^2 c_k = -\left[-(2k+1)\alpha^2 / k_0^2 + \int_{-\infty}^{\infty} n^2(y) \phi_k^2 dy + \int_{-\infty}^{\infty} \Delta_0 y^2 \phi_k^2 dy \right] c_k$$
$$+ \sum_{n \neq k} \left[\int_{-\infty}^{\infty} n^2(y) \phi_k \phi_n dy + \int_{-\infty}^{\infty} \Delta_0 y^2 \phi_k \phi_n dy \right] c_n, \quad k = 0,1,2,\cdots(N-1) \tag{48}$$

For the asymmetric index profile of the form in Eq.(19), the above equations can be expressed as the following eigenvalue system:

$$\mathbf{A}_{N \times N} \mathbf{C}_{N \times 1} = n_e^2 \mathbf{C}_{N \times 1} \tag{49}$$

where the matrix $\mathbf{A}$ is a symmetric matrix with the diagonal elements given by A_{kk}

$$A_{kk} = \left(n_c^2 + n_s^2\right)/2 + 2 n_s \Delta n \int_0^{\infty} f(y/h) \phi_k^2 dy - (2k+1)\alpha^2 / \left(2k_0^2\right) \tag{50}$$

and the upper diagonal elements $(k < n)$ given by

$$A_{kn} = \left(n_s^2 - n_c^2\right) T_{kn} + 2 n_s \Delta n \int_0^{\infty} f(y/h) \phi_k \phi_n dy + \sqrt{(k+1)n}\, \delta_{n-1,k+1} \alpha^2 / \left(2k_0^2\right) \tag{51}$$

where

$$T_{kn} = (-1)^{(k+n-1)/2} \sqrt{\frac{k!\,n!}{2\pi}} \sum_{j=0}^{k} t_j$$

$$\frac{t_{j+1}}{t_j} = \frac{-(k-j)(n-j)}{(j+1)(k+n-2j-2)}$$

$$t_0 = \frac{(k+n-2)!}{k!n!}$$

The various eigenvalues of $\mathbf{A}$ correspond to the effective indices n_{em} and the corresponding eigenvectors are the field coefficients c_{jm}; m indicates the mode number. For sufficiently large values of N (~20) the relevant eigenvalues and transverse field distributions so obtained converge to the "exact" value not only for the fundamental mode but also for a few higher order modes. Even if the α values are not optimally chosen, the eigenvalues do converge to exact values provided N is sufficiently large. It may be mentioned that the Ritz-Galerkin method can also be implemented to obtain the complex effective index in optical waveguides with loss/gain[15,16]. However, the coefficients c_i and elements of the matrix $\mathbf{A}$ are now complex and, hence, the numerical effort increases.

For profiles that are symmetric about the axis, the above analysis simplifies considerably, and in fact, results in two distinct eigenvalue equations for the even and odd modes of the structure. The symmetric profile can be described by

$$n^2(y) = n_s^2 + 2n_s\Delta n f(y/h) \qquad -\infty < 0 < \infty$$

where, $f(y/h)$ is a symmetric function of y.

Once again, substituting the trial field given by Eq.(41) into the variational expression for effective index and exploiting the symmetry of the structure, one obtains the matrix elements as

$$A_{kk} = n_s^2 + 4n_s\Delta n \int_0^{\infty} f(y/h)\phi_k^2 dy - (2k+1)\alpha^2/(2k_0^2) \tag{52}$$

and,

$$A_{kn} = 4n_s\Delta n \int_0^{\infty} f(y/h)\phi_k\phi_n dy + \sqrt{n(n-1)}\delta_{k,n-2}\alpha^2/(2k_0^2) \tag{53}$$

for k and n of same parity, and

$$A_{kn} = 0$$

for k and n of opposite parity, Therefore, one obtains two sets of independent eigenvalue equations given by

$$A_{kn}c_k = n_e^2 c_k \tag{54}$$

where $k,n = 0,2,4,\cdots\cdots$ for the symmetric mode and $k,n = 1,3,5\cdots\cdots$ for the antisymmetric mode. In other words, for the symmetric modes of any multi-moded waveguide, only the even numbered coefficients $c_0, c_2, c_4, \cdots\cdots$ of $\phi_0, \phi_2, \phi_4, \cdots\cdots$, respectively, are non-zero and sum up to give the modal fields of the symmetric modes of the structure. This is, as one would expect since $\phi_0, \phi_2, \phi_4, \cdots\cdots$ are all even functions of y, or are symmetric about the y axis. Hence, the total number of terms required to approximate the actual symmetric mode shall be much less. Similarly, for the antisymmetric mode, only the coefficient $c_1, c_3, c_5, \cdots\cdots$ of $\phi_1, \phi_3, \phi_5, \cdots\cdots$ are all nonzero and add up to define the mode of the structure. From

Eq.(41), it is clear that $\phi_1, \phi_3, \phi_5, \cdots\cdots$ are all odd functions of y and hence, a sum over these results is a description of the antisymmetric mode of the structure.

Table-II gives a comparison of the effective index obtained for a typical asymmetric graded index waveguide with a Gaussian profile. It can be observed that the effective index obtained by the single function EHGP and ESH trial fields are the same as those obtained by the converged Ritz Galerkin procedure (N~30) and the exact value.

The single function fields as well as those obtained by Ritz Galerkin procedure have been used to study mode coupling efficiency at component boundaries in planar refractive integrated optical components[3].

Table-II A comparison of the effective indices obtained from the various approximations for a typical asymmetric waveguide. (Δn=0.03, n_s=1.512, n_c=1, h=1.2 μ m, λ =0.6328 μ m, V=3.6)

Procedure	n_e	Maximized Variational Parameters
Exact	1.5230	-
Ritz Galerkin	1.5230	-
HG	1.5213	$\alpha = 1.067$
EHG	1.5228	$\alpha = 1.116,\ \gamma = 12.68$
EHGP	1.5230	$\alpha = 0.892,\ \gamma = 15.059,\ \tilde{p} = 0.526$
SH	1.5215	$\tau = 3.299$
ESH	1.5230	$\tau = 3.493,\ \gamma = 13.131$

4. Channel Waveguides

Although, all characteristics and device performance are well understood in the planar configuration, 'real life' optical circuits require light confinement in both x and y directions. The refractive index distribution of the channel waveguides fabricated by diffusion and ion exchange can be expressed as

$$n^2(x,y) = n_s^2 + 2n_s\Delta n f(y/h)g(x/w) \qquad y>0$$
$$= n_c^2 \qquad y<0 \qquad (55)$$

where $g(x/w)$ models the index variation along the width and is usually represented as

$$g(x/w) = \exp(-x^2/w^2) \qquad (56)$$

where w is the half width of the waveguide and $f(y/h)$ represents the index variation along the depth and is usually modeled by the asymmetric index profile defined by Eq.(19).

For the two dimensional refractive index profile $n^2(x,y)$, the scalar variational expression depends on both x and y coordinates and Eq.(10) is the stationary expression for the fundamental mode of the channel waveguide. A common feature with most methods for channel waveguides is that the modal field are assumed to be seperable in x and y. Hence, the trial field is also approximated as a product of the fields[9,17] along x and y, i.e.,

$$\psi_t(x,y) = \chi(x)\phi(y) \tag{57}$$

Substituting suitably chosen field forms for $\chi(x)$ and $\phi(y)$ into the stationary expression, we obtain closed form expressions for the normalized effective index, $b = \left(n_{eff}^2 - n_s^2\right)/2n_s\Delta n$ in terms of the normalized wave number V_y in y direction, a normalized wave number V_x in x direction and the asymmetry parameter p which are defined as in the planar configuration:

$$\begin{aligned} V_y &= k_0 h\sqrt{2n_s\Delta n} \\ V_x &= k_0 w\sqrt{2n_s\Delta n} \\ p &= \left(n_s^2 - n_c^2\right)/2n_s\Delta n \end{aligned} \tag{58}$$

The refractive index profile along the y-direction is an asymmetric profile while along the x-direction it is symmetric. Hence, the analysis of planar waveguides in Section 3 suggests that a suitable trial field for the channel waveguide is the product of a two variable evanescent Hermite Gauss function (EHG) in the asymmetric y-direction and a Gaussian function in the symmetric x-direction, i.e.,

$$\begin{aligned} \psi(y) &= A(1+\gamma_y y/h)\exp(-\alpha_y^2 y^2/h^2)\exp(-\alpha_x^2 x^2/w^2) \qquad y>0 \\ &= A\exp(\gamma_y y/h)\exp(-\alpha_x^2 x^2/w^2) \qquad y<0 \end{aligned} \tag{59}$$

where A is the normalization constant and γ_y, α_y and α_x are the three variational parameters. Substituting this field into the stationary expression, as in the case of planar waveguides, closed form expression for normalized effective index b in terms of the variational parameters is obtained as

$$b = \left(r_x r_y - s\right)/d_x d_y \tag{60}$$

where

$$d_x = \frac{1}{\alpha_x}\sqrt{\frac{\pi}{2}}, \quad d_y = \frac{1}{2\gamma_y} + \frac{1}{2\alpha_y}\sqrt{\frac{\pi}{2}} + \frac{\gamma_y}{2\alpha_y^2} + \frac{\gamma_y^2}{8\alpha_y^3}\sqrt{\frac{\pi}{2}}$$

$$s = \frac{pd_x}{2\gamma_y} + \frac{1}{V_y^2}s_y d_x + \frac{1}{V_x^2}s_x d_y$$

$$s_x = \alpha_x\sqrt{\frac{\pi}{2}}, \quad s_y = \frac{1}{8\alpha_y}\left(\sqrt{\frac{\pi}{2}}\left(3\gamma_y^2 + 4\alpha_y^2\right) + 4\gamma_y\alpha_y\right)$$

$$r_x = \sqrt{\frac{\pi}{1+2\alpha_x^2}}$$

and the depth profile dependent r_y for a Gaussian profile is

$$r_y = \frac{1}{4(1+2\alpha_y^2)^{3/2}}\left[2\sqrt{\pi}(1+2\alpha_y^2)+\gamma_y^2\sqrt{\pi}+4\gamma_y(1+2\alpha_y^2)^{1/2}\right]$$

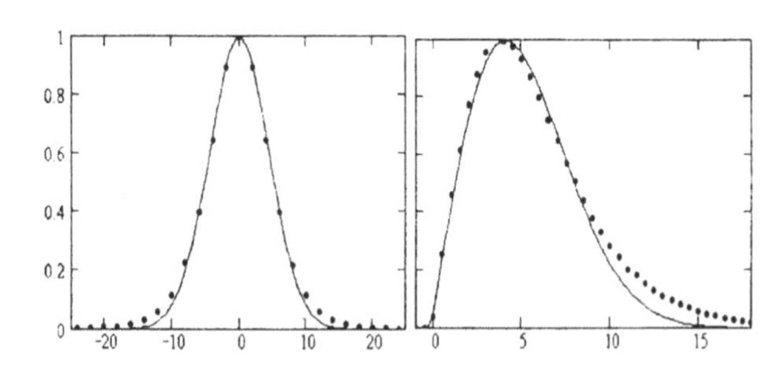

Figure 9 The closed form variational field at wavelength of 1532nm (a) x-variation at y=4.0 μ m. (b) y-variation at x=0.0 μ m, The points correspond to the exact numerical results(BPM) and continuous line to three variable fields (h=6.5 μ m, w=7.5 μ m, n_s=2.297 and Δn=0.0048.)

The optimal field parameters are obtained by maximizing the above expression with respect to the three parameters α_y, γ_y, and α_x. Fig.9 is a comparison[17] of the field forms obtained along the x and y directions from the three variable trial field and exact numerical procedure based on Beam Propagation Method (see Chapter 15). As can be seen the simple field form models the exact numerical field very well. Such simple field descriptions can be conveniently used for analyzing fiber to waveguide coupling[17], amplification in Er-doped waveguides[18] and directional couplers.

As in the case of planar waveguides, the accuracy of the fields can be improved by increasing the number of variational parameters in the trial field. For example, to improve the field in the tail region along the y-direction, the IEHG[17] field can be used instead of the EHG field while the field in the x-direction can be improved by adding a matched exponential tail.

5. Variational Analysis and Directional Couplers

One of the key components of integrated optic circuits is the directional coupler[1]. In its simplest configuration it consists of two closely spaced waveguides interacting through their evanescent fields. The typical configuration of a identical waveguides directional coupler is shown in Fig.10. The refractive index profile of the directional coupler can be expressed as the sum of the individual waveguide profiles as

$$\begin{aligned} n^2(x,y) &= n_s^2 + 2n_s\Delta n f(y/h)\left[g((x+s)/w)+g((x-s)/w)\right] && y>0 \\ &= n_c^2 && y<0 \qquad (61) \end{aligned}$$

with the individual waveguide refractive indices, denoted by $n_1(x,y)$ and $n_2(x,y)$, given by

$$n_{1,2}^2(x,y) = n_s^2 + 2n_s\Delta n f(y/h) g((x\pm s)/w) \qquad y>0$$

$$= n_c^2 \qquad\qquad y < 0 \tag{62}$$

The modal fields $\psi_1(x,y)$, $\psi_2(x,y)$ of the two individual waveguides individually satisfy the scalar wave equation

$$\frac{d^2\psi_j}{dx^2} + \frac{d^2\psi_j}{dy^2} + k_0^2\left(n_j^2(x,y) - n_{ej}^2\right)\psi_j = 0 \qquad j=1,2 \tag{63}$$

where $n_{e1} = n_{e2} = n_e$ denotes the effective index of each individual waveguide.

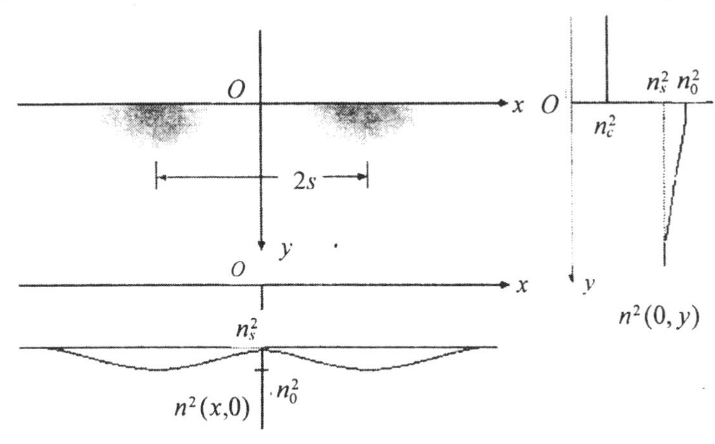

Figure 10 Schematic of the refractive index distribution of a direction coupler made up of diffused channel waveguides.

We now apply the variational principle to obtain the fields of the coupled waveguide configuration. If the trial fields $\varphi_t(x,y)$ are not normalized the integral form of the scalar equation is given by

$$n_{et}^2 = \frac{\frac{1}{k_0^2}\iint \phi_t^*(x,y)\nabla_t^2\phi_t(x,y)dxdy + \iint n^2(x,y)\left|\varphi_t(x,y)\right|^2 dxdy}{\iint\left|\varphi_t(x,y)\right|^2 dxdy} \tag{64}$$

The variational trial field of the coupler is chosen as a linear combination of the individual waveguide fields. Thus

$$\phi_t(x,y) = a_1\psi_1(x,y) + a_2\psi_2(x,y) \tag{65}$$

with a_1 and a_2 as the two variational parameters. Substituting Eq.(65) into Eq.(64) we obtain

$$\tilde{n}_e^2 \iint (a_1\psi_1 + a_2\psi_2)^2 dxdy = \frac{1}{k_0^2}\iint (a_1\psi_1^* + a_2\psi_2^*)(a_1\nabla_t^2\psi_1 + a_2\nabla_t^2\psi_2)dxdy$$
$$+ \iint n^2\left(a_1\psi_1 + a_2\psi_2\right)^2 dxdy \tag{66}$$

For the normalized fields ψ_1 and ψ_2, neglecting the field overlap integral (i.e., assuming $\iint \psi_1\psi_2 dxdy = 0$), the above expression reduces to

$$\tilde{n}_e^2\left(a_1^2 + a_2^2\right) = a_1^2\left(n_e^2 + 2n_e\kappa_{11}\right) + a_2^2\left(n_e^2 + 2n_e\kappa_{22}\right) + 2n_e a_1 a_2\left(\kappa_{12} + \kappa_{21}\right) \tag{67}$$

where κ_{ij} are defined as

$$\kappa_{ij} = \iint \frac{\left(n^2 - n_j^2\right)\psi_i \psi_j}{2n_{ej}} dxdy \tag{68}$$

The perturbation correction term κ_{ii} and coupling coefficient κ_{ij} have been defined to be consistent with the definitions used in coupled mode theory[1], which is frequently used to analyze coupled waveguide configurations[1]. Since the waveguides are identical, $\kappa_{11} = \kappa_{22} = \widetilde{\kappa}$ and $\kappa_{12} = \kappa_{21} = \kappa$, Eq.(67) can be simplified to

$$\widetilde{n}_e^2\left(a_1^2 + a_2^2\right) = \left(a_1^2 + a_2^2\right)\left(n_e^2 + 2n_e\widetilde{\kappa}\right) + 4n_e a_1 a_2 \kappa \tag{69}$$

Using the stationary condition $\partial\widetilde{n}_e^2 / \partial a_1 = \partial\widetilde{n}_e^2 / \partial a_2 = 0$, one obtains the following homogeneous equations:

$$(\widetilde{n}_e^2 - n_e^2 - 2n_e\widetilde{\kappa})a_1 - 2n_e\kappa a_2 = 0$$
$$-2n_e\kappa a_1 + (\widetilde{n}_e^2 - n_e^2 - 2n_e\widetilde{\kappa})a_2 = 0 . \tag{70}$$

For non-trivial solutions of Eq.(70) the determinant of the coefficient matrix should be zero; this gives the following analytical expressions for the effective indices of the two modes of the coupled waveguide configuration:

$$\widetilde{n}_e = (n_e^2 + 2n_e\widetilde{\kappa} \pm 2n_e\kappa)^{1/2} \tag{71}$$

It may be noted that $\widetilde{\kappa}$ and κ are both small compared to n_e and, hence, by a binomial expansion of the RHS of Eq.(71), retaining only the first order terms, $\widetilde{n}_e$ can be expressed as

$$\widetilde{n}_e = (n_e + \widetilde{\kappa}) \pm \kappa \tag{72}$$

This is the form given in most text books[1] where the directional coupler is usually analyzed by coupled mode theory in the 'slowly varying approximation'.

From Eqs.(70) it may also be noted that $a_1 = \pm a_2$. Therefore, the fields of the coupled waveguide configuration are

$$\varphi(x,y) = a_1(\psi_1(x,y) \pm \psi_2(x,y)) \tag{71}$$

where the constant $a_1 = 1/\sqrt{2}$ is essentially the normalization constant. The fields $\psi_1(x,y)$ and $\psi_2(x,y)$ of the two waveguides can be obtained in simple analytical forms by use of the variational analysis detailed in the previous sections.

As an illustrative example, we consider a planar graded index directional coupler with identical Gaussian profile waveguides. The refractive index profile of the coupler is defined as

$$n^2(x) = n_s^2 + 2n_s\Delta n[\exp(-(x+s)^2 / w^2) + \exp(-(x-s)^2 / w^2)] \tag{72}$$

with the index profile of the individual waveguides given by

$$n_{1,2}^2(x,y) = n_s^2 + 2n_s\Delta n f(y/h)\exp(-(x \pm s)^2 / w^2) \tag{73}$$

The individual waveguide fields, from the variational analysis in Section 3, can be written as

$$\psi_{1,2}(x) = \sqrt{\alpha / (w\sqrt{\pi / 2})}\, \exp\left(-\alpha^2 \frac{(x \pm s)^2}{w^2} \right) \tag{74}$$

With these simple field forms $\tilde{\kappa}$ and κ can be evaluated analytically as

$$\tilde{\kappa} = \frac{n_s \Delta n \alpha}{n_e h} \sqrt{\frac{\pi}{\delta}} \exp\left(\frac{\mu^2}{4\delta} - s\mu \right) \tag{75}$$

$$\kappa = \frac{n_s \Delta n \alpha}{n_e h} \sqrt{\frac{\pi}{\delta}} \exp\left(\frac{\mu_1^2}{4\delta} - s\mu_1 \right) \tag{76}$$

where

$$\delta = \frac{2\alpha^2}{h^2} + \frac{1}{w^2}, \quad \mu = \frac{4s}{w^2},$$

$$\mu_1 = 4s\left(\frac{\alpha^2}{h^2} + \frac{1}{w^2} \right)$$

A comparison of the fields obtained from the variational analysis and the exact fields is shown in Fig.11 for the two modes of the directional coupler. It is evident that the two-step variational procedure outlined above offers sufficiently good accuracy.

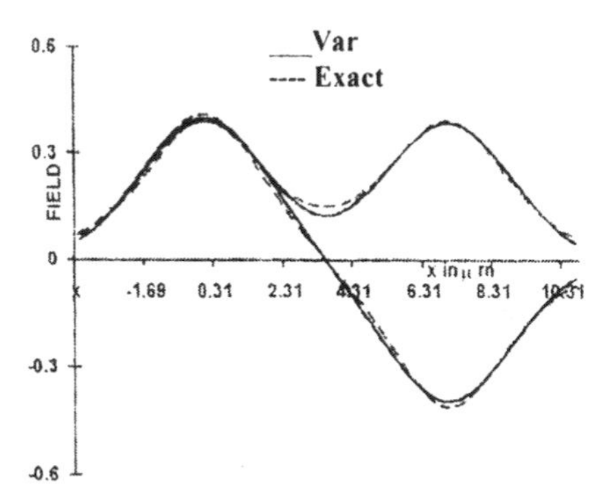

Figure 11 Variation of field along x for a directional coupler with $w_1 = w_2 = 1.9\mu m$, $\lambda = 1.03\mu m$, $\Delta n_1 = \Delta n_2 = 0.0076$, $n_s = 1.50709$, $s = 3.5\mu m$

6. Summary

In summary, the scalar variational principle has been discussed as a useful semi-analytical tool for the analysis of diffused planar and channel waveguides. Various single function variational field forms and their accuracies have been compared. For simple estimation of modal fields in typical diffused planar waveguides closed form expressions for the field parameters have been given. The analysis has also been applied to directional couplers. Also included is the Ritz-Galerkin form of the variational analysis.

Acknowledgements: We express our gratitude to the Fiber Optics Group at IIT Delhi and all co-workers of our Group at UDSC for their contribution to our better understanding of waveguidance.

References

1. A.K. Ghatak and K. Thyagarajan, *Optical Electronics*, Cambridge University press, Cambridge, England (1989).
2. H. Nishiara, M. Haruna and T. Suhura, *Optical Integrated Circuits*, McGraw Hill Book Company, New York (1989).

3. E. K. Sharma, S. Srivastava and J. P. Meunier, "Mode coupling analysis for the design of refractive ion exchange integrated optical components," *IEEE J. Select. Top Quantum. Electron.* **2**, 165 (1996).
4. E.K. Sharma and M.P. Singh, "Multilayer waveguide devices with absorbing layers: an exact analysis," *J. Opt. Commun.* **14**, 134 (1993).
5. D. Marcuse, "TE modes of graded index slab waveguides," *IEEE J. Quantum. Electron.* **QE-9**, 1000 (1973).
6. M.J. Adams, *An Introduction to Optical Waveguides,* John Wiley, Chichester (1981).
7. J.P. Meunier, J. Pigeon and J.N. Massot, "A numerical technique for the determination of propagation characteristics of inhomogeneous planar optical waveguides," *Opt. Quant. Electron.* **15**, 77 (1983).
8. S.K. Korotoky, W.J. Minford, L.L. Buhl, M.D. Divino and R.C. Alferness, "Mode size and method of estimating the propagation constant of single mode Ti:$LiNbO_3$ strip waveguides," *IEEE J. Quantum Electron.* **QE-18**, 1796 (1982).
9. A. Sharma and P. Bindal, "Analysis of diffused planar and channel waveguides," *IEEE J. Quantum Electron.* **QE-29**, 150 (1993).
10. A.K. Taneja, S. Srivastava and E.K. Sharma, "Closed form expressions for propagation characteristics of diffused planar optical waveguides," *Microwave Optical Technol. Lett.* **5**, 305 (1997).
11. A. Sharma and P. Bindal, "Variational analysis of diffused planar and channel waveguides and directional couplers", *J. Opt. Soc. Am. A* **11**, 2244 (1994)
12. W.H. Tsai, S.C. Chao and M.S. Wu, "Variational analysis of single mode inhomogeneous waveguides," *J. Lightwave Technol.* **10**, 747 (1992).
13. P.K. Mishra and A. Sharma, "Analysis of singlemode inhomogeneous planar waveguides", *J. Lightwave Technol.* **4**,204 (1986).
14. M. Saini and E.K. Sharma, "Equivalent refractive index of multiple quantum well waveguides by variational analysis', *Opt. Lett.* **20**, 2081 (1995).
15. E.K. Sharma, M.P. Singh and A. Sharma, "Variational Analysis of optical fibers with loss or gain", *Opt. Lett.* **18**, 2096 (1993).
16. Sunanda and E.K. Sharma, "Field variational analysis for modal gain in Er-doped fiber amplifiers", *J. Opt. Soc. Am. B* **16**, 1344 (1999).
17. G. Jain, A.K. Taneja, E.K. Sharma, "Closed form modal field expressions in diffused channel waveguides," *Proc. SPIE* **5349**, 163 (2004)
18. G. Jain and E.K. Sharma, "Gain calculation in Erbium doped $LiNbO_3$ channel waveguides by defining a complex index profile," *Opt. Eng.* **43**, 1454 (2004).

14

Optimal Variational Method for Rectangular and Channel Waveguides

Anurag Sharma

1. Introduction

Modal analysis of integrated optical waveguides, both rectangular waveguides and diffused channel waveguides, has been a subject of study ever since the beginning of integrated optics in 1969 and has been included in Chapters 12 and 13. Marcatili's method[1] was followed by the effective index method[2], the perturbation method[3] (see Chapter 12), the variational method[4-7] (see Chapter 13), the weighted index (WI) method,[8] the transverse resonance method,[9,10] the moment method[11,12] and several modifications and improvements suggested to some of these methods. Most of these techniques have been comprehensively reviewed by Chiang[13] and Benson and Kendall.[14] All these approximate methods for rectangular waveguides are based on defining effective or equivalent planar waveguides and obtaining their modes. The propagation characteristics of optical waveguides are obtained by solving the Helmholtz equation, which is a partial differential equation and one has to use relatively more involved methods based on finite-elements or finite difference. These numerical methods are not only numerically intensive, they do not easily lend themselves to simple modelling of the devices based on the diffused waveguides such as directional coupler, power dividers, filters, modulators and switches. Therefore, a number of simple approximate models have been developed which, though are not as accurate as the numerical methods, have fairly good accuracy, give analytical or semi-analytical forms for the modal fields and, of course, are much quicker to use. All these models are based on the variational principle and therefore, we discuss briefly the basic features of the variational principle, which has been discussed in Chapter 13. Analytical field models in the variational framework for channel waveguides have also been discussed in Chapter 13. In this chapter, we discuss a method, which we term as the optimal variational method (VOPT), since the method does not require any *a priori* assumption for the form of the field and the field is generated, generally numerically, by the method itself. Thus, the method results in an optimal field for the given waveguiding structure under the assumption of separability.

2. The Optimal Variational (VOPT) Method[15-18]

Modes of a waveguide characterized by $n^2(x, y)$ satisfy the wave equation

$$\frac{\partial^2\psi}{\partial x^2}+\frac{\partial^2\psi}{\partial z^2}+[k_0^2 n^2(x,z)-\beta^2]\,\psi(x,y)=0 \tag{1}$$

where β is the propagation constant and $k_0=\omega/c$ is the free space wave number. The integral form of this equation can be written as[19-21]

$$\beta^2 = \iint k_0^2 n^2(x,y)\left|\psi(x,y)\right|^2 dxdy - \iint\left|\nabla\psi\right|^2 dxdy \tag{2}$$

where both the integrals are over $-\infty$ to ∞ and it is assumed that the modal field is normalised

$$\iint\left|\psi(x,y)\right|^2 dxdy = 1 \tag{3}$$

The right hand side of Eq.(2) is usually referred to as the stationary expression for the propagation constant, β since it is stationary with respect to variations in $\psi(x,y)$. Since the modal field is an unknown function and, in fact, is the function that is sought for as the solution of the propagation problem, one uses an approximation for it as $\psi_t(x,y)$, generally referred to as the trial field. This trial field when used in Eq. (2) gives an estimate of the propagation constant, say β_t. Thus, we have

$$\beta_t^2 = \iint k_0^2 n^2(x,y)\left|\psi_t(x,y)\right|^2 dxdy - \iint\left|\nabla\psi_t\right|^2 dxdy \tag{4}$$

As discussed in Chapter 13, an important property of this expression is that all these values of β_t would invariably be smaller than the exact value of β and the exact value is obtained *only* when $\psi_t(x,y)$ is same as the exact modal field. Thus, higher the value of β_t obtained using Eq. (4), closer it is to the exact value of β and better is the corresponding $\psi_t(x,y)$ as an approximation for the modal field. Therefore, a value of β_t obtained through the variational expression necessarily represents a better approximation for β than any other approximation which gives a smaller value. The same could be concluded about the corresponding approximations for the modal field. This property is extremely useful in developing simple models for the mode of a given waveguide. The procedure for setting up a trial function as well as a number of such trial functions for channel waveguides have been discussed in Chapter 13.

Most of the analytical trial functions assume the separability of the field in the x and y directions:

$$\psi_t(x,y)=X(x)\,Y(y) \tag{5}$$

Most methods differ in their assumption for the functional forms of $X(x)$ and $Y(y)$. Some of these are included in Chapter 13. Thus, the accuracy of the method is governed by two assumptions: the separability and the functional forms for the factors in Eq.(5). We discuss an iterative variational method in which the separability is still assumed but no functional forms are assumed for $X(x)$ and $Y(y)$. These are generated numerically without any approximation. Thus, these two one-dimensional fields evolve with the iterations within the framework of the variational method and the accuracy obtained is the *best* under the assumption of separability. Therefore, we term the method as the optimal variational (VOPT) method.

With $\psi_t(x,y)$ substituted from Eq. (5) into the variational expression, Eq. (4) becomes

$$\beta_t^2 = \iint k_0^2 n^2(x,y)\,|X(x)|^2\,|Y(y)|^2\,dxdy - \int\left|\frac{dX}{dx}\right|^2 dx - \int\left|\frac{dY}{dy}\right|^2 dy \tag{6}$$

where it is assumed that both $X(x)$ and $Y(y)$ are normalised

$$\int|X(x)|^2\,dx = 1 = \int|Y(y)|^2\,dy \tag{7}$$

The method is iterative and we assume, to start with, a planar index distribution $n_{\bar{y}}^2(y)$ (it could as well be $n_{\bar{x}}^2(x)$). We introduce this index distribution into the variational expression of Eq. (6), which can be rewritten as

$$\begin{aligned}\beta_t^2 = &\int k_0^2 n_{\bar{y}}^2(y)\,|Y(y)|^2 - \int\left|\frac{dY}{dy}\right|^2 dy \\ &+ \int k_0^2\left[\int\{n^2(x,y) - n_{\bar{y}}^2(y)\}\,|Y(y)|^2\,dy\right]|X(x)|^2\,dx - \int\left|\frac{dX}{dx}\right|^2 dx.\end{aligned} \tag{8}$$

Equations (6) and (8) are identical since the terms containing $n_{\bar{y}}^2(y)$ cancel out exactly. However, in writing the equation in this manner, we have separated the RHS in two terms (written on two separate lines in Eq.18). We will show in the following that each of these terms is positive and can be individually maximized giving, thus, the maximum value of β_t. The first term,

$$\int k_0^2 n_{\bar{y}}^2(y)|Y(y)|^2\,dy - \int\left|\frac{dY}{dy}\right|^2 dy$$

is the variational expression for the planar index profile, $n_{\bar{y}}^2(y)$, and hence, is exactly equal to the square of the propagation constant, $\beta_{\bar{y}}^2$, of its mode. It is thus positive and has a maximum value $\beta_{\bar{y}}^2$ (see Chapter 13, Sec.3). This value and the corresponding modal field $Y(y)$ can be obtained exactly using a standard numerical method. The modal field $Y(y)$ thus obtained can be normalised to satisfy the condition of Eq.(7). We have thus obtained the maximum value of the first term of Eq.(8) and have in the process generated the function $Y(y)$.

The second term of Eq.(8) is also in the form of the variational expression for a planar index distribution, $n_{\bar{x}}^2(x)$, which is defined as

$$n_{\bar{x}}^2(x) = \int\{n^2(x,y) - n_{\bar{y}}^2(y)\}|Y(y)|^2\,dy \tag{9}$$

which can be easily evaluated using $n_{\bar{y}}^2(y)$ and $Y(y)$ of the first term. Thus, the second term is also positive and its maximum value is $\beta_{\bar{x}}^2$, where β_x is the propagation constant of the waveguide defined by $n_{\bar{x}}^2(x)$ of Eq.(9). The value of β_x and the corresponding modal $X(x)$ can be easily obtained numerically. The field $X(x)$ can then be normalised as required by Eq.(7).

Next, we use $n_{\bar{x}}^2(x)$ generated above to rewrite the variational expression of Eq.(6) as

$$\beta_l^2 = \int k_0^2 n_{\bar{x}}^2(x)|X(x)|^2 dx - \int \left|\frac{dY}{dx}\right|^2 dx + \int k_0^2 |Y(y)|^2 \left[\int \{n^2(x,y) - n_{\bar{x}}^2(x)\} |X(x)|^2 dx\right] - \int \left|\frac{dY}{dy}\right|^2 dy \tag{10}$$

The first term on the RHS of the above equation is exactly the same as the second term of Eq.(8) and has already been maximized. The second term, in Eq.(10), is again a variational expression for the index profile $n_{\bar{y}}^2(y)$, which is now defined as

$$n_{\bar{y}}^2(y) = \int \{n^2(x,y) - n_{\bar{x}}^2(x)\} |X(x)|^2 dx. \tag{11}$$

This completes one cycle of iteration; starting from an arbitrary $n_{\bar{y}}^2(y)$, we have generated a new $n_{\bar{y}}^2(y)$ through the variational expression for the given index profile $n^2(x,y)$. This $n_{\bar{y}}^2(y)$ is the starting point for the next cycle of iteration. The quantity $\beta_{\bar{x}}^2 + \beta_{\bar{y}}^2$ gives an estimate for the propagation constant β^2 of the mode of the given channel waveguide $n^2(x,y)$. At the end of each cycle of iteration, one checks for convergence in this quantity and the iterations are stopped when the convergence to a required accuracy is achieved. In most cases, one requires 2-3 iterations to obtain typical convergence.

The VOPT method can be implemented as an iterative procedure and various steps required are given in Table-I. For translation of this procedure in to a computational scheme, one requires the following three elements:

1. *Computation of the propagation constant of a planar waveguide.* We have used the Riccati transformation and have solved the resulting first order differential equation using the predictor-corrector method.
2. *Computation of the modal field.* We have used the predictor-corrector method for the wave equation directly.
3. *Integration over the field to normalize it and to obtain the index distribution.* We have used Bode's 4-point formula for evaluation of integrals, since the truncation error of this formula is of the same order as that of the predictor-corrector method.

Table-I: Steps for Implementation of the VOPT Method

STEP 1: Choose an $n_{\bar{y}}^2(y)$. A good choice is $n_{\bar{y}}^2(y) = n^2(x=0,y)$

STEP 2: Obtain $\beta_{\bar{y}}^2$ and $Y(y)$ numerically. Normalize $Y(y)$.

STEP 3: Obtain $n_{\bar{x}}^2(x)$ using Eq.(9).

STEP 4: Obtain $\beta_{\bar{x}}^2$ and $X(x)$ numerically for $n_{\bar{x}}^2(x)$. Normalize $X(x)$.

STEP 5: Obtain $n_{\bar{x}}^2(x)$ using Eq.(11).

STEP 6: Obtain $\beta_{\bar{y}}^2$ and $Y(y)$ numerically for $n_{\bar{y}}^2(y)$. Normalize $Y(y)$.

STEP 7: Compute $\beta_l^2 = \beta_{\bar{x}}^2 + \beta_{\bar{y}}^2$. Check for convergence in β_l^2.

IF Converged, GO TO STEP 8 ELSE GO TO STEP 3.

STEP 8: β_l^2 and $\psi_l(x,y) = X(x)Y(y)$ are the required propagation constant and modal field.

3. Equivalent 1-*D* Profiles for Channel Waveguides

In the VOPT method, two 1-*D* index profiles $n_x^2(x)$ and $n_y^2(y)$ are generated for the given 2-*D* profile $n^2(x,y)$. These 1-*D* profiles can be used to obtain *equivalent* 1-*D* profiles for the given 2-*D* profile in the sense that the propagation constant of its fundamental mode and the variation of its modal field along $x-$ (or, $y-$) direction match those of the given waveguide. These equivalent profiles can be used to simulate or model interactions between different waveguides in the $x-z$ (or, the $y-z$) plane. The equivalent waveguides can also be used to obtain, for example, the approximate vector modes of the given waveguide by obtaining the TM mode of the equivalent waveguide (see Sec.8). Propagation characteristics of directional couplers can also be modelled using equivalent waveguides (see Sec.7).

The refractive index distributions of the equivalent waveguide in the $x-$ and $y-$ directions obtained through the VOPT method are defined, respectively, by

$$n_{x_{eq}}^2(x) = \int \{n^2(x,y) - n_y^2(y)\} |Y(y)|^2 \, dy + \beta_t^2 / k_0^2 \, , \tag{12}$$

and

$$n_{y_{eq}}^2(y) = \int \{n^2(x,y) - n_x^2(x)\} |X(x)|^2 \, dx + \beta_t^2 / k_0^2 \, . \tag{13}$$

The index profiles $n^2(x,y)$ and $n_x^2(x)$, for example, are equivalent in the sense that the modes for both the profiles have the same propagation constant, β_t, and have identical $x-$ variation of the modal fields. In other words, the propagating modal field $X(x)e^{-i\beta_t z}$ of the equivalent profile $n_{x_{eq}}^2(x)$ is a projection, on the $x-z$ plane, of the propagating modal field $\psi_t(x,y)e^{-i\beta_t z}$ of the given waveguide $n^2(x,y)$. We shall return to applications of the equivalent waveguides in Sec.7 and Sec.8.

4. Rectangular Waveguides

The waveguides formed in semiconductor material systems, such as the GaAs/GaAlAs, are piecewise homogeneous layered structures. These are normally referred to as rectangular waveguides as opposed to diffused channel waveguides in which the index is graded in the main guiding layer. In this section, we consider the former class of waveguides and would return to the latter class in Sec.7.

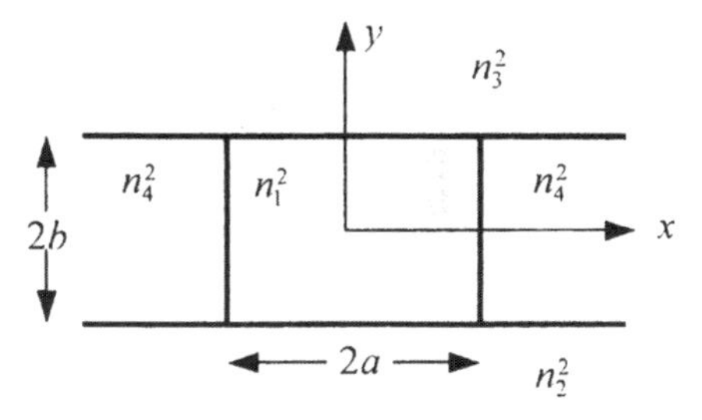

Figure 1 Schematic of a rectangular waveguide

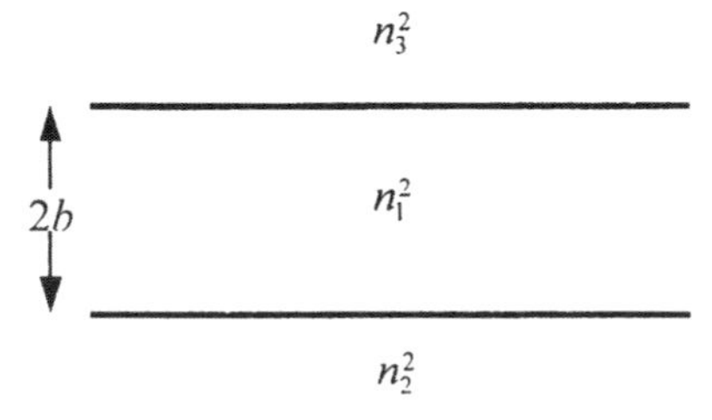

Figure 2 Schematic of the initial planar waveguide for the VOPT method

As an example, we consider a rectangular-core waveguide shown in Fig.1. For simplicity, we are considering only a 3-layer structure, although the methods discussed, and the conclusions drawn, here are valid for structures with an arbitrary number of layers. The structure shown in Fig.1 becomes a rectangular waveguide with $n_4^2 = n_3^2 = n_2^2$, a channel waveguide with $n_4^2 = n_2^2$, or a strip waveguide with $n_4^2 = n_3^2$. To implement the VOPT method, we begin with an arbitrary $n_y^2(y)$, which could be defined as the given waveguide at the center (at $x = 0$, see Fig. 2):

$$n_y^2(y) = \begin{cases} n_1^2 & |y| < b & \text{core} \\ n_2^2 & y < -b & \text{substrate} \\ n_3^2 & y > b & \text{cover.} \end{cases} \tag{14}$$

We obtain the modal field $Y(y)$ and the propagation constant β_l^2 of this waveguide. The field would be in term of sinusoidal functions in the core and exponential in the cover and the substrate as discussed in Chapter 12. Next, we obtain an equivalent $x-$waveguide using Eq.(12). The index profile of this $x-$waveguide, in the first iteration, would be given by

$$n_x^2(x) = \begin{cases} n_e^2 & |x| < a \\ n_e^2 + P_{co}^y(n_4^2 - n_1^2) \equiv n_4^2 + \gamma(n_e^2 - n_1^2) & |x| > a \end{cases} \tag{15}$$

where

$$\gamma = 1 - (1 - P_{co}^y)\left(\frac{n_1^2 - n_4^2}{n_1^2 - n_e^2}\right), \tag{16}$$

$n_e^2 = \beta_l^2 / k_0^2$ and P_{co}^y is the fraction of power in the core of the $y-$waveguide. Marcatili's method[1] and the effective-index method[2] have been discussed in Chapter 12 for such waveguides; it can be shown that for these methods, $\gamma = 1$ and $\gamma = 0$, respectively. In all the three methods, the next step is to obtain the modal field, $X(x)$ and a new value for the propagation constant β_l^2. This completes one cycle of iteration of the VOPT method. The other two methods stop here, while the VOPT method continues further. For the next cycle, we obtain an improved $n_y^2(y)$ using Eq.(13) as

$$n_y^2(y)\Big|_{\text{improved}} = \int\{n^2(x,y) - n_x^2(x)\}|X(x)|^2\, dx + \beta_l^2 / k_0^2 , \tag{17}$$

which is also a 3-layer piecewise homogeneous waveguide much like the one defined in Eq.(14), however, the values of refractive indices of the layer would now be different.

5. Relationship with Other Methods

It is easy to see that for each step of the iteration, the form of the modal fields $X(x)$ and $Y(y)$, are in terms of cosine/sine and exponential functions. In fact, the first variational method for rectangular waveguides was based on these functions and was called the cosine-exponential (CE) method.[4-6] Subsequently, an iterative version of this method was developed and it was named as the CEVAR method.[7] It was also shown[7] that the weighted index (WI) method[8] was exactly equivalent to the CEVAR method. The CEVAR method, when applied to general channel waveguides, both rectangular as well as diffused waveguides (and even fibers), is the VOPT method.[7,16]

Recently, a new variant of the effective-index method with built in perturbation correction (EIMPC) has been developed.[22-24] It has been shown that this method gives much improved accuracy and is based on finding an effective planar waveguide such that the perturbation correction due to the given rectangular waveguide on this effective waveguide is zero. It has however been shown[25] that the EIMPC is exactly the same method as the VOPT method or the CEVAR method if the iterations in the VOPT method are stopped after one cycle and are not taken to the state of convergence. For convenience, we will denote by VOPT* the method in which VOPT method is stopped after one cycle. It is obvious that the accuracy after the first cycle would be lower as compared to that in the converged state. Considering once again the rectangular waveguide of Fig.1, in the EIMPC method, one first solves the planar waveguide defined in Eq.(14). This gives the field and the effective-index n_e^2. Next, one defines an effective-index waveguide as

$$\tilde{n}_x^2(x) = \begin{cases} n_e^2 & |x| < a \\ n_4^2 + \tilde{\gamma}(n_e^2 - n_1^2) & |x| > a \end{cases} \tag{18}$$

which is similar to one given in Eq.(15), but now $\tilde{\gamma}$ is a parameter yet to be determined. As mentioned above, in the conventional effective-index method, $\tilde{\gamma} = 0$. In the EIMPC, the value of $\tilde{\gamma}$ is obtained from the condition that the perturbation correction to the obtained propagation constant vanishes when the given waveguide (Fig.1) is treated as perturbation to the equivalent composite waveguide $[n_y^2(y) + \tilde{n}_x^2(y) - n_e^2(y)]$ for which the fields are now known. This parameter $\tilde{\gamma}$ can be obtained analytically[23] and we have $\tilde{\gamma} = \gamma$ where γ is given by Eq.(16). Thus, the EIMPC method and the VOPT* method are exactly the same.

It may be worth mentioning that the perturbation method, the variational method and the perturbation method with built-in correction are intimately connected to each other, in general, independent of the problem to which these are applied. This relationship has been explored in the Appendix.

In order to have a numerical appreciation of the performance of various methods, we consider a specific example of a rectangular $n_4^2 = n_3^2 = n_2^2$ and $b/a = 0.5$. The results for the normalised propagation constant,

$$B = \frac{(\beta / k_0)^2 - n_2^2}{n_1^2 - n_2^2} \tag{19}$$

of the fundamental scalar mode are plotted as a function of

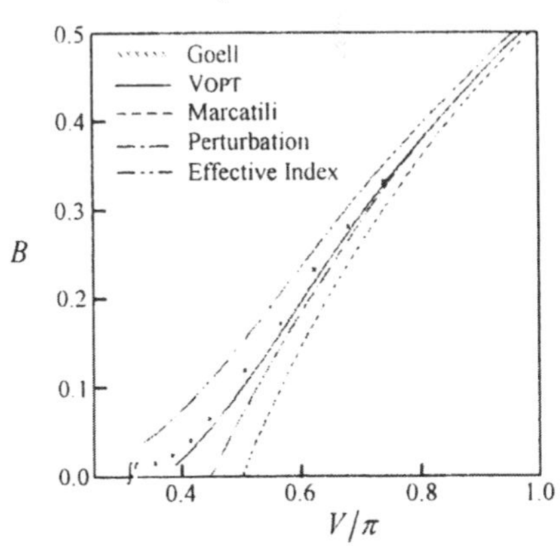

Figure 3 Normalised propagation constant for a rectangular waveguide using the methods of Goell[26] and Marcatili,[1] the effective index method, the perturbation method (Chapter 12) and the VOPT method.

$$V = k_0 a \sqrt{n_1^2 - n_2^2} \tag{20}$$

in Fig.3 for different methods including the numerical method based on expansion in circular harmonics,[26] which can be considered as being closest to the exact value. In any case, as shown in Chapter 13, the variational method gives a lower bound on the propagation constant, which means the highest value given by any variational method is closest to the exact value. The conventional effective method, not being a variational method, overestimates the value of B and is not very accurate. The figure shows that the VOPT method gives the best results.

Next, we consider a channel waveguide ($n_4^2 = n_2^2$) such that $\Delta_2 = 0.005$ and $\Delta_2 = 0.4$ where $\Delta_i = (n_1^2 - n_i^2)/2n_1^2$. The results for B using the VOPT and the EIMPC methods have included Table-II. In the case of VOPT method, we have continued the iterations till the convergence in the four decimal places is obtained. The number of iterations ($N_{iterations}$) necessary for this is also shown in the table. These results show that the first iterative cycle itself gives very good results if the mode is sufficiently far from its cutoff. As the mode draws near to its cutoff, more iterations are required. However, in wave-guides with 4 or 5 layers, such as the strip loaded waveguide shown in the inset, the number of iterations required is more even when the mode is not close to its cutoff and the first iteration (and hence, the EIMPC) is not sufficiently accurate as shown by the result in Table-II.

Table-II Convergence of Iterations in VOPT

a/b	V	Mode	EIMPC/ VOPT*	VOPT	$N_{iterations}$
2	1.2	E_{11}	0.0973	0.0995	2
2	1.7	E_{11}	0.4141	0.4142	2
2	1.7	E_{21}	0.0701	0.0731	2
2	2.0	E_{11}	0.5369	0.5369	1
2	2.0	E_{21}	0.2587	0.2594	2
1	1.4	E_{11}	0.0445	0.0547	3
1	1.6	E_{11}	0.1740	0.1777	2
1	3.0	E_{11}	0.6559	0.6559	1
1	3.0	E_{21}	0.2357	0.2367	2
See the figure below		E_{11}	0.5397	0.5492	4

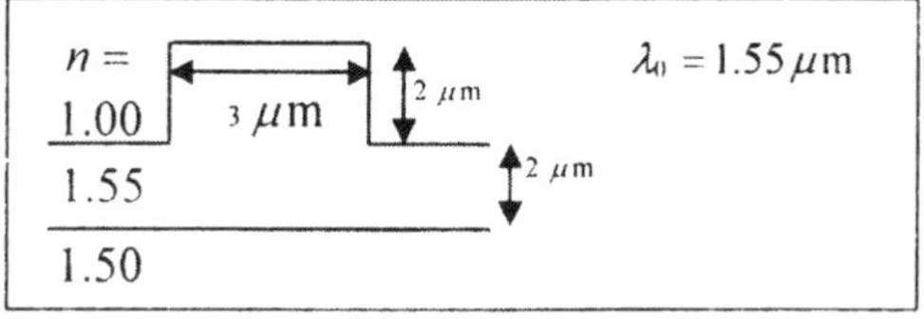

$n =$ 1.00
3 μm
2 μm
$\lambda_0 = 1.55\,\mu$m
1.55
2 μm
1.50

6. Diffused Channel Waveguides

Diffused channel waveguides are formed by ion diffusion and ion exchange in substrates like $LiNbO_3$ and glass. The guiding layer of such waveguides has a continuously varying refractive index, the exact nature of variation being dependent on the conditions of diffusion. Diffused waveguides, being graded-index, are more difficult to analyze than the piecewise homogeneous rectangular waveguides. Variational methods and other numerical methods have played an important role in the theory and design of these waveguides. The VOPT method once again gives the best results for such waveguides under the assumption of separability of fields, as we shall see through examples.

The refractive index profile of a diffused channel waveguide (see Fig.1 of Chapter 13 for a schematic) is usually written as

$$n^2(x,y) = \begin{cases} n_s^2 + 2n_s\,\Delta n\, f(x)\,g(y) & y > 0 \\ n_c^2 & y < 0 \end{cases} \tag{21}$$

where n_s and n_c are the indices of the substrate and the cover, respectively, Δn is the grading parameter and $f(x)$ and $g(y)$ are the grading profiles along the lateral and the depth directions, respectively. As can be seen from the figure, the index profile is symmetric along the lateral direction (x-axis), while it is highly asymmetric along the depth (y-axis). In fact, usually the cover is air ($n_c = 1$) and the index contrast at the interface ($y = 0$) is so large that the field drops very sharply from a maximum under the surface to a very small value at the interface leaving a very small evanescent field in the cover. A variety of functional forms for $f(x)$ and $g(y)$ has been used in the literature. The common ones are

$$f(x) = \begin{cases} \exp\left(-x^2/W^2\right) & \text{Gaussian} \\ \dfrac{\operatorname{erf}\left(\frac{x+W}{D}\right) - \operatorname{erf}\left(\frac{x-W}{D}\right)}{2\operatorname{erf}(W/D)} & \text{error function} \end{cases} \tag{22}$$

and

$$g(y) = \begin{cases} \exp\left(-y/D\right) & \text{exponential} \\ \exp\left(-y^2/D^2\right) & \text{Gaussian} \\ \operatorname{erfc}(y/D) & \text{Complementary error function} \end{cases} \tag{23}$$

where D is the diffusion depth and W is the half-width of the initial metal strip used for diffusion. Several results are discussed elsewhere,[19] which show the performance of the VOPT method in comparison of other methods. We include here one example of the Gaussian and Gaussian profile, which has been studied earlier using the variational finite difference (VFD)[27] and magnetic field based finite difference (HFD)[28] methods. The results are given in Fig.4, which shows that the VOPT method gives the highest value of B for any given value of $V = k_0 D\sqrt{2n_s\Delta n}$. Since these are obtained using a variational method, the corresponding exact value of B must be somewhat larger than this value and the exact value curve would lie above the one corresponding to the VOPT method. Therefore, among the results shown in the figure, those obtained using the VOPT method are the most accurate and the finite-difference methods give very poor accuracy. The low accuracy of the VFD- and HFD-results can be ascribed to rather small size, 14×14 and 20×20, respectively, of the grid for sampling the field in the transverse cross-section. In the case of VOPT method, in which one has to consider only a one-dimensional sampling of the field at a time, the field is sampled, in effect, on a grid of size 200×200. In addition, in the finite difference methods, one assumes that the field vanishes at the boundaries of a window whose size is kept large enough in order to keep the effect of this approximation at a negligible level. However, larger the size of the window, more are the grid points required to sample

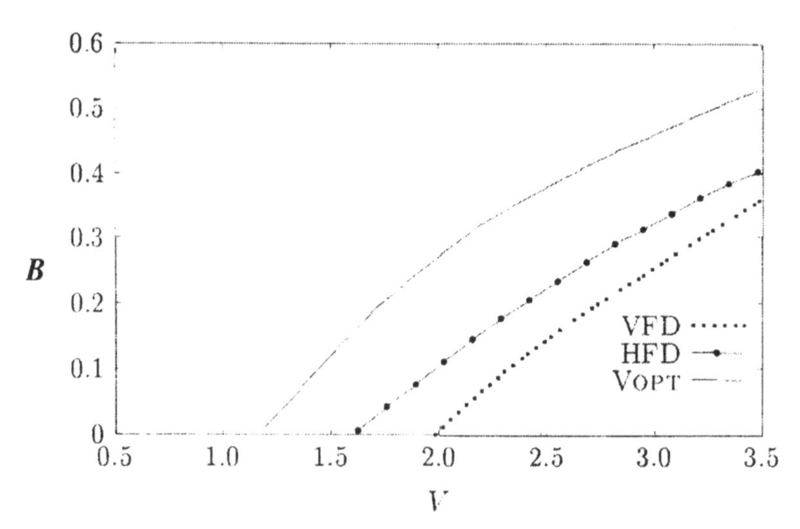

Figure 4 Normalised propagation constant, *B*, for a diffused channel waveguide Gaussian-Gaussian profile

the field so that these are close enough to approximate its variation adequately, particularly in the guiding region. On the other hand, in VOPT method, the field is assumed to decay exponentially out side the computational grid. These two aspects of finite-difference (and also finite-element) methods limit the accuracy rather severely unless very large computer memory and time are at disposal. The grid sizes of 20×20, used in HFD, and 14×14, used in VFD, are highly inadequate as shown by above results.

7. Directional Couplers

Directional couplers are the basis for a variety of integrated optical devices such as switches, modulators, and power dividers and have been introduced in earlier chapters. A directional coupler consists of two identical waveguides placed parallel to each other along the z – axis separated by a constant distance (see Fig.10 of Chapter 13) (in some special applications, the waveguides may be non-identical and/or the separation between them may not be constant). The modes of the two waveguides, due to the overlap of their evanescent fields, get coupled to each other and exchange power between them as they propagate along the z – axis.

Diffused waveguide directional couplers have been widely studied, both experimentally[29-31] and theoretically.[32-37] Most earlier studies either neglected the variation of refractive index in the direction normal to the plane of the directional coupler[30] or used the effective index method to reduce the 2-*D* profile to a 1-*D* Profile.[29,34-36] The first analysis taking full 2-*D* index profile into account was carried out by Jain *et al.*[32] Later, Hawkins and Goll[37] presented a variational analysis modifying the Hermite-Gauss (HG) trial field (see Chapter 13). Another approach was developed by Fiet and Fleck,[33] which was computationally intensive.

The refractive index profile for a directional coupler made up of two diffused channel waveguides can be written as (*cf.* Eq.61 of Chapter 13)

$$n^2(x,y) = \begin{cases} n_s^2 + 2n_s\,\Delta n\; g(y)[f(x-s)+f(x+s)] & y>0 \\ n_c^2 & y<0 \end{cases} \quad (24)$$

where $2s$ is the separation between the centres of the two constituent channel waveguides. The composite waveguiding structure thus formed has two modes one of which is symmetric and the other is antisymmetric. The propagation constant of these modes, β_s and β_a, respectively, depend on the separation parameter, s, and are such that $\beta_s \geq \beta \geq \beta_a$ (β being the propagation constant of single isolated waveguide). Due to this difference in the propagation constant a continuous exchange of power occurs between the two waveguides, and the distance after which the maximum power is transferred from one waveguide to the other is called the coupling length, l_c and is given by

$$l_c = \frac{\pi}{\beta_s - \beta_a} \quad (25)$$

This coupling length is the main parameter of a directional coupler and different methods have been used to obtain its value for a given refractive index profile. We discuss here briefly the use of the VOPT method directly as well as the use of the 1-*D* waveguide in obtaining the coupling length of a directional coupler. We also include comparison with available experimental results as well as those obtained other theoretical methods.

In the direct approach, the VOPT method is used to obtain the propagation constants of the symmetric and the antisymmetric modes of the composite waveguide that the directional coupler is. The coupling length of the coupler is then computed using these propagation constants. In the other approach, we use the equivalent waveguide n_{xeq}^2 given by Eq.(12) to define an equivalent 1-*D* directional coupler with constituent equivalent waveguides placed at $x = s$ and $x = -s$. Thus, the equivalent 1-*D* directional coupler has the following refractive index distribution

$$n_x^2 = n_{xeq}^2(x-s) + n_{xeq}^2(x+s) - n_{seq}^2 \quad (26)$$

where n_{seq} is the index for substrate region of the equivalent waveguide. The equivalent 1-*D* directional coupler with $n_x^2(x)$ is used to obtain the propagation constants, β_s and β_a, of its first two modes. The coupling length is obtained from Eq.(25). As an example, we consider the directional couplers reported by Korotky and Alferness.[31] The profile assumed is error function-Gaussian with parameters $n_c = 1.0$, $n_s = 2.1398$, $W = 4.0\mu$m, $D = 3.8\mu$m and $\lambda_0 = 1.32\mu$m. The metal strip thickness is 0.068μm, which controls the value of Δn. The results are shown in Fig.5, which also includes the theoretical results of Hawkins and Goll[37] along with the results obtained using VOPT method. The figure shows that the results obtained using VOPT method are in very good agreement with the experimental data. Another example, that we consider here, concerns the directional couplers reported by Noda *et al.*[30] The profile assumed is Gaussian-Gaussian with parameters $n_c = 1.0$, $n_s = 2.152$, $W = 4.0\mu$m, $D = 5.0\mu$m and $\lambda_0 = 1.153\mu$m. Results are given in

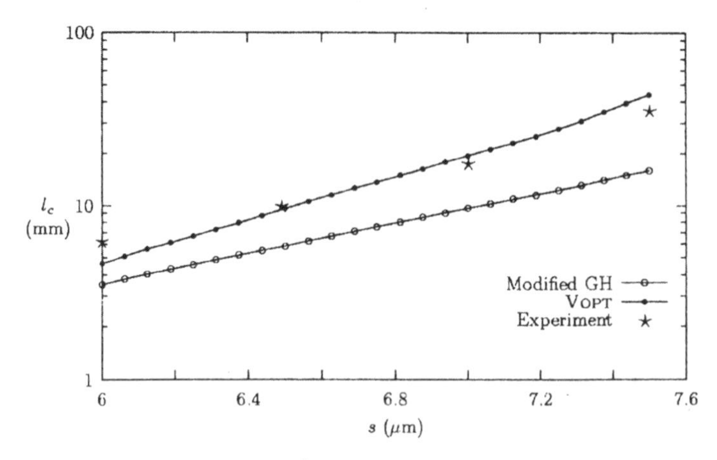

Figure 5 Coupling length, l_c, for the directional coupler reported by Korotky and Alferness[31]

Table-III. The table shows that the 1-*D* equivalent waveguide method[38] agrees very well with the VOPT method and that the values obtained using both these methods are in good agreement with the experimental values for the coupling length. On the other hand, the values obtained using the effective index method have substantial error ranging from 6% to 30%.

Table-III Coupling length of diffused channel waveguide directional couplers obtained using different methods

Waveguide parameters	Coupling length, l_c (mm)			
	VOPT method	1-*D* equivalent waveguide method	Effective index method	Experimental Result[33]
$\Delta n = 0.006, s = 6.08\mu m$	5.7965	5.8019	7.9054	5.28
$\Delta n = 0.006, s = 6.56\mu m$	10.5211	10.5360	11.6874	10.60
$\Delta n = 0.004, s = 6.08\mu m$	3.6956	3.6892	4.5916	3.40
$\Delta n = 0.004, s = 6.56\mu m$	5.5469	5.5463	5.8601	5.28

8. Vector Modes

As has been discussed in Chapter 12, a single mode channel waveguide which supports a single scalar mode would allow two orthogonally polarized vector modes, namely, E_{11}^x and E_{11}^y modes which have slightly different propagation properties. These modes, if coupled by unavoidable irregularities, may cause undesirable effects when a polarization sensitive device is used at the end of the waveguide. Therefore, it is important to study vector modal properties of such waveguide. We know that the two fundamental vector modes E_{nm}^x and E_{nm}^y are hybrid in nature and all the components of E and H are nonzero. However, E_{nm}^x mode has strong E_x and H_y, and other components are such that $E_x, H_y >> E_z, H_z >> E_y, H_x$ and hence, it behaves like a TM_n mode in equivalent (1-*D*) x – waveguide and a TE_m mode in the equivalent (1-*D*) y – waveguide. Similarly, E^y_{nm} mode behaves like a TM_m mode in the equivalent y – waveguide and a TE_n mode in the equivalent x – waveguide. It is thus possible to use the equivalent waveguide obtained in the VOPT method to obtain the approximate E^x_{nm} and E^y_{nm} vector modes of the given 2-*D* waveguide. Since for 1-*D* waveguides, the TE-modes and the scalar modes (with appropriate polarization given to the scalar field) are identical, one has to obtain the TM mode only of the appropriate equivalent waveguide to finally obtain the desired vector mode. For example, to obtain the E^x_{11} mode, one has to solve the mode problem only for the TM_0 mode of the equivalent x – waveguide. In the actual procedure, one uses the VOPT method to obtain the scalar modal field $\psi_{scalar}(x, y) = X(x)Y(y)$ and the propagation constant β_t. Since the mode behaves like a TE mode in the equivalent (1-*D*) y – waveguide, $Y(y)$ gives the y – dependence of the dominant field E_x of the vector mode; however, to obtain the correct x – dependence and the propagation constant, one has to solve for the TM mode of the equivalent (1-*D*) x – waveguide, defined by Eq. (12). This would give a solution for the transverse electric field component (x – component), say $\widetilde{X}(x)$ and a corrected value of the propagation constant $\widetilde{\beta}_t$. Thus, the approximate field for the E^x_{nm} mode would be $E_x(x, y) = \widetilde{X}(x)Y(y)$ and the propagation constant would be $\widetilde{\beta}_t$.

As an example, we consider a rectangular waveguide (see Fig.1) with $n_1 = 1.5$, $n_2 = n_3 = n_4 = 1.0$ and $b/a = 0.5$. Results are shown in Fig.6, where we have compared our results with exact results of Goell[29], and those obtained by Marcatili's method.[1] We can see that in case of step-index rectangular waveguide, our results are much closer to exact results of Goell than those of Marcatili's method.

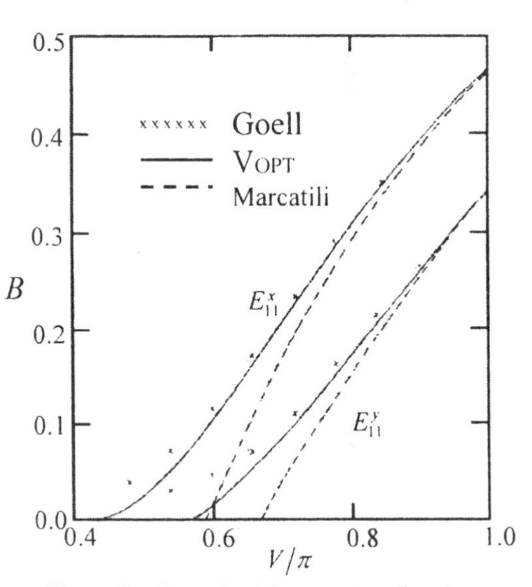

Figure 6 Normalised Propagation of vector modes of a rectangular waveguide

9. Summary

We have discussed in this chapter, the optimal variational (VOPT) method for obtaining modes of two-dimensional waveguides. Applications of the method to rectangular waveguides and diffused channel waveguides have also been discussed. The method also leads to the definition of one-dimensional equivalent waveguides, which have been used to analyze directional couplers and to obtain vector modes. Numerical results have also been included, which show that the VOPT method better results than other methods.

Appendix: The Variational and the Perturbation Methods

In this Appendix, we show that the perturbation method can be derived from the expression which is generally used for the variational method and that there exists a close relationship between the two methods. The mathematical framework used is that of quantum mechanics since the problem of scalar modes in waveguide is very similar to the problem of quantum mechanical states for a given potential well[39]. Further, we use for brevity, the notations of bra and ket used in quantum mechanics.[40-42]

We consider a uniform waveguide with index profile given by $n^2(x, y)$ which is isotropic and lossless. The scalar modes of such a waveguide are given as the solution of the scalar wave equation, which can be written as

$$H|\phi\rangle = E|\phi\rangle \tag{A1}$$

where $H \equiv \nabla_t^2 + k_0^2 n^2(x, y)$ is the operator (generally referred to as the Hamiltonian), the eigenavalue, E , of which is the propagation constant(β^2) and the eigenfunction, $|\phi\rangle$, represents the modal field $\psi(x, y)$ of the waveguide mode. We will consider here only the fundamental mode of the given waveguide. Further, we assume that the modal field is normalized:

$$\iint|\psi(x, y|^2 \, dx \, dy = 1 = \langle\phi|\phi\rangle. \tag{A2}$$

A. The Variational Method

In this method, the Rayleigh quotient, $\langle\phi|H|\phi\rangle$, is used to obtain an estimate for the eigenvalue when the exact solution of the eigenvalue problem is not possible. A suitably chosen function $|\phi_t(p)\rangle$ is used as a trial function. Here p represent an adjustable parameter (or a group of parameters) in the function. The trial function is substituted in the Rayleigh quotient to obtain E_t :

$$E_t(p) = \langle\phi_t(p)|H|\phi_t(p)\rangle. \tag{A3}$$

The best estimate of the eigenvalue is obtained by maximizing the value of the quotient with respect to p, i.e., we obtain the solution of

$$\frac{\partial E_t}{\partial p} = 0 \tag{A4}$$

which gives an optimum value of p , say p_{opt} . The value $E_t(p_{\text{opt}})$ gives an estimate for the eigenvalue while $|\phi_t(p_{\text{opt}})\rangle$ is an approximation for the field. The accuracy of the estimated eigenvalue and the approximate field depends on the closeness in form of the trial field to the exact field.

B. The Perturbation Method

In this method, we look for a Hamiltonian, H_0, which is close to H and is solvable exactly. Thus, its eigenvalue E_0 and the corresponding field $|\phi_0\rangle$ are known exactly. Thus,

$$H_0|\phi_0\rangle = E_0|\phi_0\rangle, \quad \text{or} \quad E_0 = \langle\phi_0|H_0|\phi_0\rangle \tag{A5}$$

where $|\phi_0\rangle$ is assumed to be normalized. In the first order perturbation method, the approximate eigenvalue for the Hamiltonian H is given by

$$E_1 = E_0 + \langle\phi_0|H - H_0|\phi_0\rangle \tag{A6}$$

On using the value of E_0 from Eq.(A5), this equation becomes

$$E_1 = \langle\phi_0|H_0|\phi_0\rangle + \langle\phi_0|H - H_0|\phi_0\rangle = \langle\phi_0|H|\phi_0\rangle \tag{A7}$$

which shows that the first order perturbed eigenvalue is the same as the Rayleigh quotient with the unperturbed field as trial field.

C. The Perturbation Method with Built-in Correction

In this method, an unperturbed eigenvalue problem is sought such that the first order correction is made to vanish. In this method, a suitable unperturbed Hamiltonian is set-up with a built-in parameter by varying which the first order perturbation correction is made zero. Let $H_t(p)$ be such a Hamiltonian where p is the adjustable parameter. Further, let the corresponding unperturbed field be $|\phi_t(p)\rangle$. The perturbed eigenvalue for the Hamiltonian H then is given by (*cf.* Eq. A7)

$$E_t(p) = \langle\phi_t(p)|H|\phi_t(p)\rangle \tag{A8}$$

Now the aim is to obtain a value of p, say p_0, such that the first order correction to the eigenvalue is zero. We can develop a Taylor expansion for $E_t(p)$ around $p = p_0$ to obtain

$$E_t(p = p_0 + \Delta p) = E(p_0) + \Delta p \left.\frac{\partial E_t}{\partial p}\right|_{p=p_0} \tag{A9}$$

where the second term on the right hand side is the first order perturbation correction to the eigenvalue and is a function of p_0. To make the first order perturbation correction vanish, we must have

$$\left.\frac{\partial E_t}{\partial p}\right|_{p=p_0} = 0, \tag{A10}$$

which gives the required value of p_0 and the corresponding value of $E_t(p_0)$ as the eigenvalue estimate.

Equation (A10) is the same as Eq.(A4) showing that the built-in correction method is exactly the same as the variational method if the forms used for the field in the two methods are the same.

References

1. E.A.J. Marcatili, "Dielectric rectangular waveguide and directional coupler for integrated optics," *Bell Syst. Tech. J.* **48**, 2071-2102 (1969).
2. G.B. Hocker and W.K. Burns, "Mode dispersion in diffused channel waveguides by the effective index method," *Appl. Opt.* **16**, 113-118 (1977).
3. A. Kumar, K. Thyagarajan and A.K. Ghatak, "Analysis of rectangular-core dielectric waveguides: an accurate perturbation approach," *Opt. Lett.* **8**, 63-65 (1983).
4. A. Sharma, P.K. Mishra and A.K. Ghatak, "Analysis of single mode waveguides and directional couplers with rectangular cross section," *Proceedings of 2nd European Conference on Integrated Optics* (Florence, Italy), October 17-18, 1983, pp.9-12 (1983).
5. P.K. Mishra, A. Sharma, S. Labroo and A.K. Ghatak, "Scalar variational analysis of single-mode waveguides with rectangular cross-section," *IEEE Trans. Microwave Theory Tech.* **MTT-33**, 282-286 (1985).
6. A. Sharma, P.K. Mishra and A.K. Ghatak, "Single-mode optical waveguides and directional couplers with rectangular cross section: a simple and accurate method of analysis," *J. Lightwave Technol.* 6, 1119-1125 (1988).
7. A. Sharma, "On approximate theories of single mode rectangular waveguides," *Opt. Quantum Electron.* **21**, 517-520 (1989).
8. P.C. Kendall, M.J. Adams, S. Ritchie and M.J. Robertson, "Theory for calculating approximate values of the propagation constant of an optical rib waveguide by weighting the refractive index," *IEE Proc.* A **134**, 699 (1987).
9. F.P. Payne, "A new theory of rectangular optical waveguides," *Opt. Quantum Electron.* **14**, 525 (1982).
10. F.P. Payne, "A generalized transverse resonance model for planar optical waveguides," *Tenth European Conference on Optical Communications*, Stuttgart (Germany), Sept.3-6, 1984, pp.96-97 (1984).
11. D.A. Roberts and M.S. Stern, "Accuracy of method of moments and weighted index method," *Electron. Lett.* **23**, 784 (1987).
12. She Shouxian, "Analysis of rectangular-core waveguide structures and directional couplers by an iterated moments method," *Opt. Quantum Electron.* **20**, 125 (1988).
13. K.S. Chiang, "Review of numerical and approximate methods for the modal analysis of general optical dielectric waveguides," *Opt. Quantum Electron.* **26**, S113-S134 (1994).
14. T.M. Benson and P.C. Kendall, "Variational techniques including effective and weighted index methods," *Progress in Electromagnetic Research* **10**, 1-40 (1995) [EMW Publishing, Cambridge, Mass (USA)].
15. A. Sharma and P. Bindal, "An optimal approximate method for the analysis of diffused channel waveguides", *SPIE International Conference on Emerging Optoelectronic Technologies (CEOT)*, Bangalore (India), December 18-20, 1991. (SPIE Proceedings, vol. 1622).

16. A. Sharma and P. Bindal, "An accurate variational analysis of single mode diffused channel waveguides," *Opt. Quantum Electron.* **24**, 1359-1371 (1992).
17. A. Sharma and P. Bindal, "Modelling of single mode diffused channel optical waveguides", *Journal of the Indian Institute of Science* (India) **76**, 183-199 (1996).
18. A. Sharma and P. Bindal, "Solutions of the 2-*D* Helmholtz equation for optical waveguides: semi-analytical and numerical variational approaches", *LAMP Series Report*, **LAMP/92/2**, 1-29 (1992) (published by the International Centre for Theoretical Physics, Trieste,Italy).
19. H. Kogelnik, "Theory of Optical Waveguides" in *Integrated Optics*, T. Tamir (Ed.), Springer Verlag, Berlin, 1975, p.44.
20. A.K. Ghatak and K. Thyagarajan, *Optical Electronics*, Cambridge University Press, Cambridge, 1989.
21. A.W. Snyder and J.D. Love, *Optical Waveguide Theory*, Chapman Hall, London, 1983.
22. K.S. Chiang, "Analysis of rectangular dielectric waveguides: effective-index method with built-in perturbation correction," *Electron. Lett.* **28**, 388-389 (1992).
23. K.S. Chiang, K.M. Lo and K.S. Kwok, "Effective-index method with built-in perturbation correction for integrated optical waveguides," *J. Lightwave Technol.* **14**, 223-228 (1996).
24. K.S. Chiang, K.S. Kwok and K.M. Lo, "Effective-index method with built-in perturbation correction for the vector modes rectangular-core optical waveguides," *J. Lightwave Technol.* **17**, 716-722 (1999).
25. A. Sharma, "Analysis of integrated optical waveguides: variational method and effective-index method with built-in perturbation correction," *J. Opt. Soc. Am.-A* **18**, 1383-1387 (2001).
26. J.E. Goell, "A circular-harmonic computer analysis of rectangular dielectric waveguides," *Bell Syst. Tech. J.* **48**, 2133-2160 (1969).
27. R.K. Lagu and R.V. Ramaswamy, "A variational finite difference method for analyzing channel waveguides with arbitrary index profiles," *IEEE J. Quant. Electron.* **QE-22**, 968-976 (1986).
28. N. Schulz, K. Bierwirth, F. Arndt, and U. Koster, "Finite difference method without spurious solutions for the hybrid-mode analysis of diffused channel waveguides," *IEEE Trans. Microwave Theory Tech.* **MTT-38**, 722-729 (1990).
29. R.C. Alferness, R.V. Schmidt, and E.H. Turner, "Characteristics of Ti-diffused lithium niobate optical directional couplers," *Appl. Opt.* **18**, 4012-4015 (1979).
30. J. Noda, M. Fukuma and O. Mikami, "Design calculations for directional couplers fabricated by Ti-diffused $LiNbO_3$ waveguides," *Appl.Opt.* **20**, 2284-2291 (1981).
31. S.K. Korotky and R.C. Alferness, "The Ti:$LiNbO_3$ integerated-optic technology: fundamentals, design considerations and capabilities," in *Integrated Optical*

Circuits and Components, L.D. Hutcheson (Ed.), Marcel Dekker, New York, 1987.

32. U. Jain, A. Sharma, K. Thyagarajan and A.K. Ghatak, "Coupling characteristics of a diffused channel-waveguide directional coupler," *J. Opt. Soc. Am.* **72**, 1545-1549 (1982).
33. M.D. Fiet and J.A. Fleck, Jr., "Comparison of calculated and measured performance of diffused channel-waveguide couplers," *J. Opt. Soc. Am.* **73**, 1296-1310 (1983).
34. A. Sharma, E. Sharma, I.C. Goyal and A.K. Ghatak, "Variational analysis of directional couplers with graded-index profile," *Opt. Commun.* **34**, 39-42 (1980).
35. J. Ctyroky, M. Hofman, J. Janta, and J. Schrofel, "3-D analysis of $LiNbO_3$:Ti channel waveguides and directional couplers," *IEEE J. Quant. Electron.* **QE-20**, 400-408 (1984).
36. C.M. Kim and R.V. Ramaswamy, "WKB analysis of asymmetric directional couplers and its application to optical switches," *J. Lightwave Technol.* **6**, 1109-1118 (1988).
37. R.T. Hawkins and J.H. Goll, "Method for calculating coupling length of Ti:$LiNbO_3$ waveguide directional couplers," *J. Ligtwave Technol.* **6**, 887-891 (1988).
38. P. Bindal and A. Sharma, "Modelling of Ti:$LiNbO_3$ directional couplers," *IEEE Photon. Technol. Lett.* **4**, 728-731 (1992).
39. R.J. Black and A. Ankiewicz, "Fiber-optic analogies with mechanics," *Am. J. Phys.* **53**, 554-563 (1985).
40. P.A.M. Dirac, *Principles of Quantum Mechanics*, Oxford University Press, 1958.
41. L.I. Schiff, *Quantum Mechanics*, 3rd edition, McGraw-Hill, New York, 1968.
42. A.K. Ghatak and S. Loknathan, *Quantum Mechanics: Theory and Applications*, MacMillan Press, New Delhi, 1984

15

Numerical Methods for Scalar Wave Propagation

Anurag Sharma

1. Introduction

The development of optoelectronic devices involves a time consuming cycle of design, fabrication, characterization and possible redesign. The role of computerized modeling and simulation tools is important in reducing the time and costs involved in this cycle. These tools provide an alternative to trial fabrication for arriving at the design of a new device as well as in redesign of an existing device for new conditions. Coupled with advances in device technology itself, much progress has been made in the field of modeling and simulation techniques used to model devices, a description of some of these modeling techniques is given in the literature (see, e.g.,Refs. 1-8). Through continuous efforts, new techniques are being devised and existing techniques are being improved. Based on such techniques, commercial simulation tools and software have also become available and are popular in the market amongst researchers, engineers and manufacturers. These tools have to be accurate, fast, robust, easy-to-use with minimum computation and memory requirement as far as possible.

Modeling techniques may be classified as semi-analytical or numerical in nature. The former include the effective index method[9] (see Chapter 12), the variational method[10] (see Chapters 13 and 14), the WKB analysis (see Chapter 11) and the coupled mode analysis[11-13]. These semi-analytical methods work well for uniform optical waveguides or with coupled waveguides carrying few modes. Often, however, the device has a complicated, non-uniform structure that is difficult to model accurately with such approaches and it becomes inevitable to use numerical methods. Modeling techniques can be vectorial, semi-vectorial or scalar in capability. In a dielectric waveguide, in which quasi TE or TM modes are possible, the scalar approaches neglect the polarization effects, and the derivative of the relative permittivity with respect to the transverse directions are assumed to be negligible. In the semi-vectorial approach, the two transverse components are not coupled via the boundary though the derivative with respect to the transverse directions is not neglected. Finally, in the vectorial approach, the transverse components are coupled due to the effect of the boundary and the full polarization effects are taken into consideration. For most practical waveguiding structures, the relative variation in the refractive index is small enough to allow the scalar wave approximation. Under this assumption the wave equation can be written in the form of the Helmholtz equation, the solution of which, termed as the scalar field, represents one of the transverse components of the electric or magnetic field.

Various scalar numerical techniques have been developed for modeling of optical waveguides, which treat the total field, including the guided and the radiation modes together. These include the fast-Fourier transform based beam propagation method, FFT-BPM[14-17], finite-difference beam propagation method FD-BPM[17-19], finite element beam propagation method, FE-BPM[6,20] and the collocation method.[21-26] In this Chapter, we will discuss the basic beam propagation methods and their advanced versions for wide-angle propagation.

2. The Propagation Problem

For simplicity, we shall confine our discussions in this paper to two-dimensional waveguides; however, the methods discussed can be extended to three-dimensional structures. A 2-*D* waveguide structure is defined by its refractive index distribution $n^2(x,z)$, which contains all information regarding the interfaces also. The electromagnetic fields that propagate through such a dielectric structure must satisfy Maxwell's equations. However, in a majority of practical waveguiding structures (we will confine our discussion to such cases), the relative variation of the refractive index is sufficiently small to allow the scalar wave approximation[11-13]. It, then, suffices to consider instead a much simpler Helmholtz equation:

$$\frac{\partial^2\Psi}{\partial x^2}+\frac{\partial^2\Psi}{\partial z^2}+k_0^2 n^2(x,z)\Psi(x,z)=0 \tag{1}$$

where $\Psi(x,z)$ represents one of the Cartesian components of the electric field (generally referred to as the scalar field). The time dependence of the field has been assumed to be $\exp(i\omega t)$ and $k_0=\omega/c$ is the free space wave number.

The problem that is addressed in this paper is then to obtain the solutions $\Psi(x,z)$ of Eq.(1) given the field $\Psi(x,z=0)$ at a plane $z=0$. Thus, we are dealing with an initial value problem with respect to the variable z (which is generally taken as the overall direction of propagation). However, the presence of the partial derivative with respect to x makes this problem much more complex and one has to devise special methods even to obtain numerical solutions. If the waveguide is uniform in the direction of propagation, the problem of wave propagation can be solved in terms of the modes of the waveguide[11-13]. These modes, which have well defined transverse field patterns and have specific phase constants, form an orthogonal set of basis functions. An incident field at $z=0$ can be expressed as a linear combination of these modes, which then propagate along the waveguide with their phases changing in a definite way but differently, in general, for different modes. Therefore, at any other value of z, these would combine to give a field, which represents the field after propagation through the length z of the waveguide. Thus, once the modes are known for a given uniform waveguide, one can, in general, propagate a given field through any length of the waveguide. It must, however, be mentioned that these modes - the guided and the radiation - are infinite in number and, in fact, the radiation modes form a continuum of modes and, hence, the summation/integration over these modes has to be truncated leading to inaccuracy in any propagation. The approach of total field propagation, that we discuss next, can therefore also be useful even for uniform waveguides.

A straight and uniform waveguide is, however, an idealization and practical waveguides do have variations along their lengths, such as bends in optical fibers or refractive index and/or geometric variations in fibers/integrated optic (IO) waveguides. In addition and more importantly, there are devices, which inherently have structures that are nonuniform. These include, *e.g.*, tapers, y-junctions,. couplers, etc. Wave propagation through such guided-wave structures cannot, in general, be modelled in terms of the modes, although in some structures it can be treated via coupled mode theory[11-13] (which is usually limited to very few modes). This has led to the development of the approach in which the total fields or beams are propagated by numerically solving the Helmholtz equation directly, and lately, a number of methods have been developed for this purpose.

There are several advantage of treating total fields, rather than individual modes; for instance, the analysis is not restricted to uniform or near uniform waveguides and one can handle, in principle, arbitrary index and/or geometry variations along the direction of propagation. Further, the radiation modes are included in the total field propagating along the waveguiding structure and it is not necessary to obtain them explicitly, which is often not practical. These methods can be used to obtain the evolution of fields in a structure that is not uniform along the direction of propagation (see Fig. 1, for example). Some of the methods are the FFT-BPM[14-17], the FD-BPM[17-19] and the collocation method[21-26]

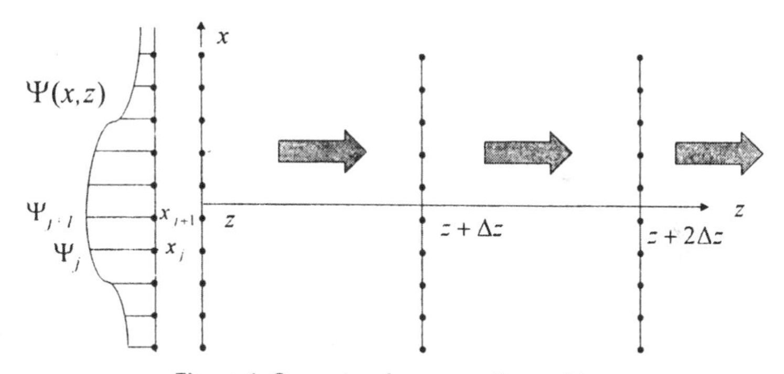

Figure 1 Geometry of a propagation problem

All these methods solve the propagation problem starting from a plane, say $z = 0$, on which the transverse field is known, in small successive steps of Δz upto the desired plane, say $z = z_f$. Further, in order to propagate the field over a single step, the transverse field in discretized by sampling over N sample points along the transverse axis (or, $N \times M$ sample points on a plane in case of a 3-*D* propagation problem). The geometry of the propagation problem is depicted schematically in Fig.1. The three methods mentioned above differ mainly in the technique to effect the propagation over a single step. In this article, we present the basic principles of these three methods and discuss some examples of each giving their advantages and limitations.

3. Propagation Equations

The wave equation, Eq.(1), describes the propagation of a field $\Psi(x,z)$ completely, in all the directions, both forward and backward. Presence of second derivatives with respect to all the space variables is a manifestation of this fact. However, as mentioned in the previous section, most practical propagation problems involve propagation in one dominant direction, which we have designated as the z direction. When the wave propagation is considered only along the $+z$ direction and the waves in the $-z$ direction are either non-existent or neglected, we term it as unidirectional propagation and when the propagation along both the directions is considered, it is bi-directional propagation. Although all structures, except longitudinally uniform waveguides, generate reflections, these are usually negligible since the variations in index or dimensions of a structure are usually very small on the scale of a wavelength. On the other hand, in several structures such as those involving step changes in index or dimension along the direction of propagation or those with periodic variation (*e.g.*, fiber gratings; see Chapter 9), the reflections cannot be neglected and one must consider bi-directional propagation. We shall, however, restrict our discussions to unidirectional propagation.

In order to investigate the nature of the wave equation, we write Eq.(1) in the operator form as

$$(\mathsf{D}^2 + G)\Psi(x,z) = 0 \tag{2}$$

where we have used the following operators

$$\mathsf{D} \equiv \frac{\partial}{\partial z}; \quad G \equiv \frac{\partial^2}{\partial x^2} + k_0^2 n^2(x,z)\ . \tag{3}$$

Equation 2 can be formally factorized as

$$\mathsf{D}_-\mathsf{D}_+\Psi(x,z) = \left(\mathsf{D} - i\sqrt{G}\right)\left(\mathsf{D} + i\sqrt{G}\right)\Psi(x,z) \tag{4}$$

where $\mathsf{D}_\pm$ represent the propagation operator for directions $\pm z$, respectively. It should be noted here that the factorization in Eq.(4) is strictly valid only for propagation though uniform (z-independent structure), since otherwise $\mathsf{D}_\pm$ and G do not commute. However, since the numerical propagation is carried out in steps of very small length Δz, we can assume that the structure is uniform over one step and consider Eq.(4) valid over the step.

Unidirectional propagation, or the beam propagation, is then given by $\mathsf{D}_+\chi = 0$, which can be written as

$$\frac{\partial \Psi}{\partial z} = -i\sqrt{G}\,\Psi\ . \tag{5}$$

The operator G effectively represents the divergence of the beam through the index variation and the derivative with respect to x. The presence of the square root in the propagation equation makes its solution difficult and a number of ways have been devised to overcome or reduce this difficulty.

It can be easily recognized that the field represented by $\Psi(x,z)$ varies rapidly on account of its phase factor. However, when one assumes that the propagation is dominantly in the $+z$ direction, one can factor out an average phase term and write

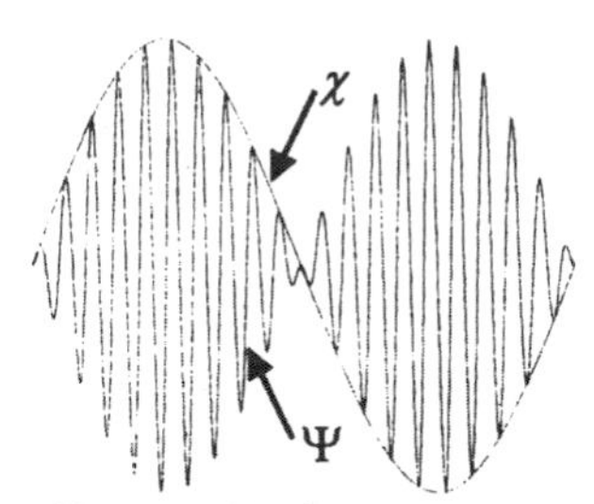

Figure 2 The field, $\Psi(x,z)$, and its envelope, $\chi(x,z)$.

$$\Psi(x,z) = \chi(x,z)\ \exp(-ikz)\,, \tag{6}$$

where $\chi(x,z)$ is the envelope of the wave which varies with much less rapidity (see Fig. 2) and $k = k_0 n_r$ with n_r being the index of a reference medium. Substituting from Eq.(6) into Eq.(5), we obtain the equation satisfied by the envelope $\chi(x,z)$ as

$$\frac{\partial \chi}{\partial z} = -i\left(\sqrt{G} - k\right)\chi(x,z)\,. \tag{7}$$

Further, we define the index variation $n^2(x,z)$ of the structure as the sum of the reference index and a variation from it:

$$n^2(x,z) = n_r^2 + \Delta n^2(x,z)\,, \tag{8}$$

so that we can now rewrite the operator G as

$$G = \frac{\partial^2}{\partial x^2} + k_0^2\ \Delta n^2(x,z) + k_0^2 n_r^2 \equiv G_r + k^2 \tag{9}$$

and Eq.(7) then becomes

$$\frac{\partial \chi}{\partial z} = -i\left(\sqrt{G_r + k^2} - k\right)\chi\,. \tag{10}$$

If we assume that the magnitude of the operator G_r is much smaller than k, we can make a binomial expansion of the square root in Eq.(10) and retain the linear term in G_r to obtain the paraxial wave equation,

$$\frac{\partial \chi}{\partial z} = -i\frac{G_r}{2k}\chi\,, \tag{11}$$

which can be written as

$$2ik\frac{\partial \chi}{\partial z} = \frac{\partial^2 \chi}{\partial x^2} + k_0^2[n^2(x,z) - n_r^2]\ \chi(x,z)\,. \tag{12}$$

This approximation is known as the Fresnel or the paraxial approximation and is valid for a medium, in which the index is not varying very rapidly along the z direction and one can neglect the second derivative of $\chi(x,z)$. The condition of validity can be expressed as

$$\left|\frac{\partial^2 \chi}{\partial z^2}\right| << \left|2k_0 n_r \frac{\partial \chi}{\partial z}\right| \tag{13}$$

The paraxial approximation has been extensively used in the study of waveguides and is, in fact, an essential approximation for the FFT-BPM and the FD-BPM, and in the collocation method, it is optional, as we would see later. These basic methods of beam propagation have been discussed in the following sections.

Soon after the development of paraxial beam propagation methods, it became apparent that these could be applied only in limited situations of practical importance, since, in most cases, the beams propagate at angles well beyond the paraxial region. A number of wide-angle methods have been developed to model such situations. The essence of most of methods lies in the ways to treat the square root operator in Eq.(7) or (10). The most successful method[27,28] is to approximate the square root by the Padé approximant. Thus, we write

$$\frac{\partial \chi}{\partial z} = -i\frac{N}{D}\chi, \tag{14}$$

where N and D are polynomials in of order n and d for the approximant of order (n,d). In this scheme, the approximant of order (1,0) corresponds to the paraxial approximation, and the approximant of order (1,1) gives $N = G_r/2k$ and $D = 1 + (G_r/4k^2)$ leading to the following propagation equation

$$(G_r + 4k^2)\frac{\partial \chi}{\partial z} = -2ikG_r\chi. \tag{15}$$

In this equation, the operator G_r is to be used twice and hence, the computation effort would be at least twice that involved in the paraxial propagation. Higher order approximations similarly require computation effort equivalent to several paraxial propagations. Table-I gives the expressions for the first few higher order approximations. It may be noted that in these expressions G_r^2 means that the propagator G_r is to be used twice in succession. Further, one also has to determine the required order of the Padé approximant for a given problem.

Table-I First few orders of Padé approximants and the explicit form of the polynomials

Padé Order (n,d)	Expression for Propagator
1,0	$-\frac{iG_r}{2k}$
1,1	$-i\frac{G_r/2k}{1+G_r/4k^2}$
2,1	$-i\frac{G_r/2k+G_r^2/8k^3}{1+G_r/2k^2}$
2,2	$-j\frac{G_r/2k+G_r^2/4k^3}{1+3G_r/4k^2+G_r^2/16k^4}$

4. The FFT-BPM [14-17]

This was the first method developed for beam propagation through optical waveguiding structures and was simply called the beam propagation method (BPM).[14-17] The method is based on the paraxial wave equation, Eq.(12), and further assumes that the relative index variation in the waveguiding structure is small

$$\Delta n^2 \equiv [n^2(x,z) - n_r^2] << n_r^2. \tag{16}$$

The paraxial wave equation can then be written as

$$\frac{\partial \chi}{\partial z} = \frac{1}{2ik}\left[\frac{\partial^2}{\partial x^2} + k_0^2\,\Delta n^2\right]\chi(x,z) \equiv H(x,z)\chi(x,z), \tag{17}$$

where $H(x,z)$ is an operator representing the effect of propagation and of the phase change due to index variation (or the "lens action"). Thus, we can write

$$H(x,z) = H_1(\text{propagation}) + H_2(\text{lens action})$$
$$= \left[\frac{1}{2ik}\frac{\partial^2}{\partial x^2}\right] + \left[\frac{1}{2ik}k_0^2\,\Delta n^2\right] \tag{18}$$

so that Eq.(17) becomes

$$\frac{\partial \chi}{\partial z} = H(x,z)\chi(x,z) = (H_1 + H_2)\chi(x,z)\,. \tag{19}$$

A formal solution of the above equation over a small extrapolation interval Δz can be written as

$$\chi(x, z+\Delta z) = e^{(H_1+H_2)\Delta z}\chi(x,z)\,, \tag{20}$$

where the inherent assumption is that Δz is so small that $H_1(x,z)$ and $H_2(x,z)$ are constant with respect to z over this interval. However, since $H_1(x,z)$ and $H_2(x,z)$ do not commute ($H_1H_2 \neq H_2H_1$), one has to invoke the so-called symmetrized splitting of the operator to increase the accuracy. Thus, we have

$$\chi(x,z+\Delta z) = \exp\left[\left(\frac{H_1}{2} + H_2 + \frac{H_1}{2}\right)\Delta z\right]\chi(x,z)$$
$$= e^{H_1\Delta z/2}e^{H_2\Delta z}e^{H_1\Delta z/2}\chi(x,z) + O\left((\Delta z)^3\right), \tag{21}$$

which can be written as

$$\chi(x,z+\Delta z) = P\,Q(z)\,P\,\chi(x,z) + O\left((\Delta z)^3\right) \tag{22}$$

with

$P = e^{H_1\Delta z/2}$: Propagation by $\Delta z/2$ through medium n_r^2

$Q(z) = e^{H_2\Delta z}$: Phase change due to $\Delta n^2(x,z)$ over Δz (*lens action*).

Thus, a single propagation step of length Δz, represented by Eq.(12), involves propagation in a uniform medium n_r from z to $z + {}^{\Delta z}\!/_2$ (operator P), followed by a lens action of the index variation over Δz (operator $Q(z)$), and finally another propagation in a uniform medium n_r from $z + {}^{\Delta z}\!/_2$ to $z + \Delta z$ (operator P). Such successive steps propagate the beam through a desired distance, say from $z = 0$ to z_f. The scheme of propagation is depicted schematically in Fig. 3. The accuracy of propagation depends on the size of the step, Δz. The propagation in uniform space is carried out through the plane wave expansion method in which the known field at

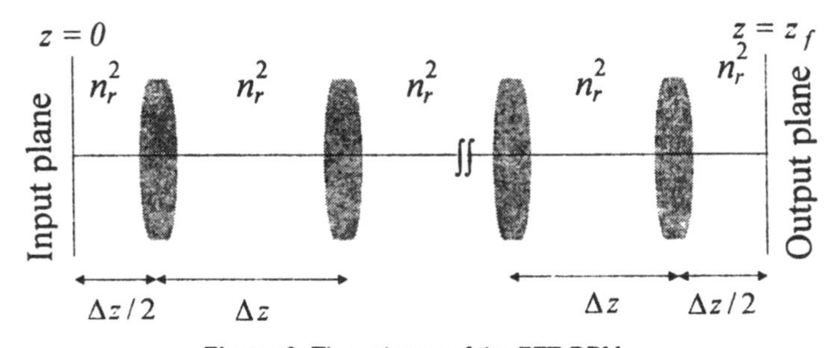

Figure 3 The scheme of the FFT-BPM.

z is expressed as a linear combination of plane waves that are then simply propagated over the propagation step, and finally, these waves are superposed to get the propagated field. The expansion and superposition is done through fast Fourier transform (FFT) and inverse FFT and hence, the method is termed as the FFT-BPM. For implementation of the FFT algorithms, the field is sampled at specified number of equidistant points along the x-axis. The accuracy, obviously, depends on the number of these samples.

5. The FD-BPM [17-19]

In the finite-difference (FD) beam propagation method, the paraxial wave equation, Eq.(12), is solved using the standard Crank-Nicholson method for parabolic partial differential equations. In this method, the partial derivatives in the equation are replaced by finite-difference approximations and the wave equation reduces to a set of simultaneous linear equation for the field at the sampled points, called nodes. The entire computation domain is covered with a mesh of unit cell $\Delta x \times \Delta z$ and the field is computed at the nodes, which are at the corners of these cells (see Fig.4). The nodes along the x and z-axes are counted through indices i, j, respectively. Thus, we use the approximations for the derivatives

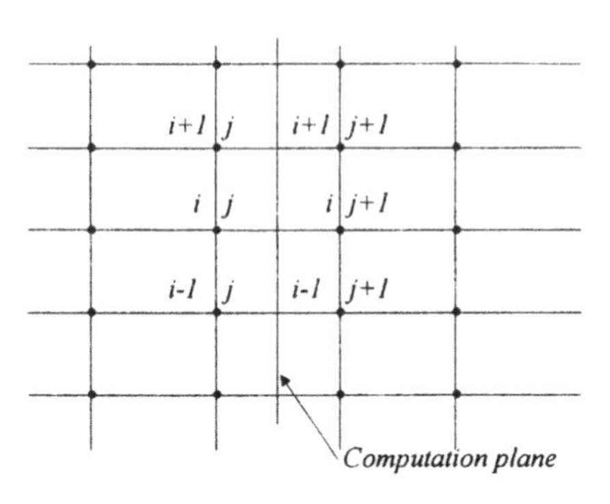

Figure 4 Geometry of computation for the FD-BPM.

$$\left.\frac{\partial \chi}{\partial z}\right|_{z+\frac{1}{2}\Delta z} = \frac{\chi(z+\Delta z)-\chi(z)}{\Delta z},$$
$$\left.\frac{\partial^2 \chi}{\partial x^2}\right|_{x} = \frac{\chi(x+\Delta x)-2\chi(x)+\chi(x-\Delta x)}{(\Delta x)^2} \tag{23}$$

and evaluate Eq.(12) at the computation plane situated between the nodal planes j and $j+1$ (Fig. 4). Now, defining χ_i^j to represent the field envelope $\chi(x,z)$ at the node (i,j), we obtain the finite-difference approximations for Eq.(12) as

$$\left.\frac{\partial \chi}{\partial z}\right|_{i}^{j+\frac{1}{2}} = \frac{\chi_i^{j+1}-\chi_i^j}{\Delta z}$$
$$= \frac{1}{2ik}\left[\frac{1}{2}\left(\left.\frac{\partial^2 \chi}{\partial x^2}\right|_i^j + \left.\frac{\partial^2 \chi}{\partial x^2}\right|_i^{j+1}\right) + k_0^2 \Delta n^2(x_i, z_j + \Delta z/2)\left(\chi_i^{j+1}+\chi_i^j\right)/2\right], \tag{24}$$

where the second derivative with respect to x is evaluated as

$$\left.\frac{\partial^2 \chi}{\partial x^2}\right|_i^j = \frac{\chi_{i+1}^j - 2\chi_i^j + \chi_{i-1}^j}{(\Delta x)^2}. \tag{25}$$

It may be noted that the values of χ and $\partial^2\chi/\partial x^2$ at the plane $j+\frac{1}{2}$ are evaluated by averaging the respective quantities at j and $j+1$. With Eq.(25), the finite difference equation, Eq.(24), can be written as a set of linear equations for the unknown values χ_i^{j+1} in terms of values of χ_i^j (for all values of i):

$$A_{i+1}^{j+1}\chi_{i+1}^{j+1} + B_i^{j+1}\chi_i^{j+1} + C_{i-1}^{j+1}\chi_{i-1}^{j+1} = D_i \tag{26}$$

where A, B, C and D can be expressed in terms of known quantities including χ_i^j. Thus, we have

$$\begin{aligned} &A_{i+1}^{j+1} = C_{i-1}^{j+1} = -T, \qquad B_i^{j+1} = 1 + 2T - k_0^2 \Delta n^2 T \\ &D_i = T\chi_{p+1}^l + (1 - 2T + k_0^2\Delta n^2)\chi_p^l + T\chi_{p-1}^l, \end{aligned} \tag{27}$$

where $T = \Delta z / 4ik\Delta x^2$. Equation 26 represents a tridiagonal matrix equation, which can be easily solved to obtain the unknown χ_i^{j+1}. Thus, each step of propagation involves solving a linear matrix equation of size $N \times N$, where N is the number of samples (nodes) in the transverse direction. Evidently, in case of a 3-*D* propagation problem in which the field envelope is sampled on a grid of $N \times M$ nodes, the size of the matrix would be $NM \times NM$, which could be rather formidable. For this reason, usually one tries to reduce a 3-*D* problem to a 2-*D* problem using some procedure of equivalence such as the effective index method[9] or a variational method[10] see Chapter 14).

6. The Collocation Method[21-26]

In the collocation method, it is not necessary to invoke the Fresnel (paraxial) approximation and one can solve the second order wave equation directly. Thus, we now consider Eq.(1) and express the unknown field as a linear combination over a set of orthogonal basis function, $\phi_n(x)$:

$$\Psi(x,z) = \sum_{n=1}^{N} c_n(z)\,\phi_n(x), \tag{28}$$

where $c_n(z)$ are the expansion coefficients, n is the order of the basis functions and N is the number of basis functions used in the expansion. The choice of $\phi_n(x)$ depends on the boundary conditions and the symmetry of the guiding structure. For the one-dimensional expansion, we have used the Hermite-Gauss functions as well as sinusoidal functions. Here we shall use the former functions and in Sec. 10, we shall use the latter functions. The coefficients of expansion $c_n(z)$ are unknown and represent the variation of the field with z. In the collocation method, these coefficients are effectively obtained by requiring that the differential equation, Eq.(1), be satisfied *exactly* by the expansion, Eq.(28), at N collocation points x_j, $j = 1,2,\ldots, N$, which are chosen such that these are the zeroes of $\phi_{N+1}(x)$. Thus, using this condition and with some algebraic manipulations[21-23], one converts the wave equation, Eq.(1), into a matrix ordinary differential equation:

$$\frac{d^2\mathbf{\Psi}}{dz^2} + [\mathbf{S}_0 + \mathbf{R}(z)]\mathbf{\Psi}(z) = 0, \tag{29}$$

where $\mathbf{\Psi}(z) = col.\{\Psi(x_1,z) \quad \Psi(x_2,z) \quad \cdots\cdots \quad \Psi(x_N,z)\}$ is a column vector, $\mathbf{R}(z) = k_0^2 \times diag.\{n^2(x_1,z) \quad n^2(x_2,z) \quad \cdots\cdots \quad n^2(x_N,z)\}$ is a diagonal matrix and $\mathbf{S}_0$ is a constant known matrix defined by the basis functions. For the Hermite-Gauss basis functions, the matrix $\mathbf{S}_0$ is given as

$$\mathbf{S}_0 = \mathbf{D}_1 - \mathbf{A}\mathbf{D}_2\mathbf{A}^{-1},$$

where $\mathbf{A} = \{A_{ij} : A_{ij} = \phi_j(x_i) \equiv \eta_{j-1} H_{j-1}(\alpha x_i) e^{-\alpha^2 x_i^2/2}\}$ with η being the normalization factor, α a scaling factor, $\mathbf{D}_1 = \alpha^4 \times diag.[x_1^2 \quad x_2^2 \ldots x_N^2]$, $\mathbf{D}_2 = \alpha^2 \times diag.[1 \quad 3 \quad 5 \quad \cdots \quad (2N-1)]$ and H_j is the Hermite polynomial of order j.

We refer to Eq.(29) as the *collocation equation*. In deriving this equation from the wave equation, Eq. 1, no approximation has been made except that N is finite and Eq.(29) is exactly equivalent to Eq.(1) as $N \to \infty$. Thus, the accuracy of the collocation method improves indefinitely as N increases. The collocation equation is a matrix ordinary differential equation and can be solved as an initial value problem using any standard method such as the Runge-Kutta method or the predictor-corrector method.

In the collocation method, one can either solve the collocation equation directly or invoke the paraxial approximation, if valid, to obtain the equation for the envelope:

$$\frac{d\boldsymbol{\chi}}{dz} = \frac{1}{2ik}\left[\mathbf{S}_0 + \mathbf{R}(z) - k^2\mathbf{I}\right]\boldsymbol{\chi}(z), \tag{30}$$

where $\boldsymbol{\chi}(z) = col.[\chi(x_1,z) \quad \chi(x_2,z) \quad \cdots \quad \chi(x_N,z)]$ and $\mathbf{I}$ is a unit matrix. This equation can be solved directly using, *e.g.*, the Runge-Kutta method, or using the operator method like in the case of FFT-BPM. The latter procedure has been shown to be unconditionally stable numerically.[12]

A unique feature of the collocation method is that one obtains an equation as a result which can be solved or modified in a variety of ways. It can be solved as an initial value problem using any standard method such as the Runge-Kutta method [10] or the predictor-corrector method. In the paraxial form, it can also be solved using matrix operator methods based on the approach of symmetrized splitting of the sum of two non-commutating operators.[24] One could also use a suitable transformation of the independent and/or dependent variable to an advantage. Indeed, we have shown[25] that a transformation could be used to redistribute the collocation points (which are the field sampling points in the transverse cross-section) in such a way that the density of points increases in and around the guiding region, and the transverse extent, covered by the sampled field, also increases.

7. Performance of Paraxial Methods

In this section, we include some example to compare the performance of the paraxial methods discussed above. In our examples, we have compared the FD-BPM and the collocation method with the FFT-BPM as the latter was the standard method when the other two methods were developed.

In the first example[18], we consider an asymmetric step index planar waveguide with core width 4 μm, core index 3.38, substrate index 3.377 and the cover being air. The wavelength of the propagating wave is 1.15 μm. The fundamental mode of this waveguide was propagated through a straight waveguide and numerical loss was computed for the FFT-BPM and the FD-BPM. The result as a function of the step-size, Δz, is shown in Fig. 5a. A comparison of the computation efficiency in terms of the computation time per propagation step is shown in Fig. 5b. This example shows that the FD-BPM is better than the FFT-BPM with respect to both the accuracy of computation as well as the efficiency of computation.

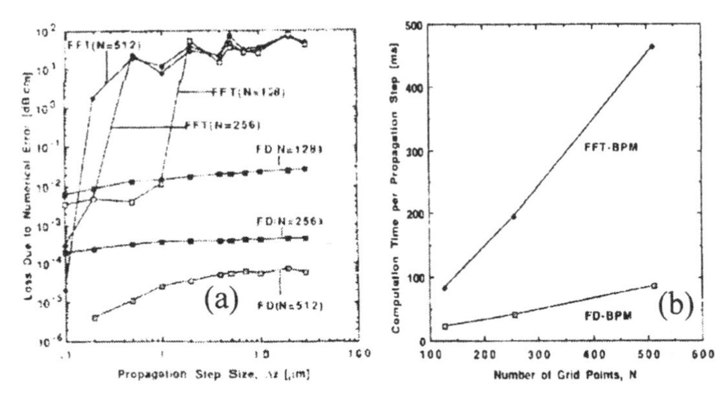

Figure 5 Comparison of the performance of the FFT-BPM and the FD-BPM for an asymmetric step-index waveguide: (a) comparison of accuracy in terms of numerical loss in dB/cm as a function the step-size; (b) comparison of computation time per propagation step as a function of step-size.
[After Chung and Dagli, *IEEE J. Quantum Electron.* **QE-26**, 1335 (1990).; © IEEE]

Next, we consider an example[21-23] of a graded-index waveguide with refractive index profile given by

$$n^2(x) = n_2^2 + (n_1^2 - n_2^2)\,\mathrm{sech}^2(x/a)\,, \tag{31}$$

where $n_1 = 1.45$, $n_2 = 1.4476$ and $a = 3\ \mu$m. The fundamental mode is propagated at $\lambda_0 = 1.31\ \mu$m. As a measure of accuracy we have computed the correlation factor (CF) of the propagating field at $z = z_f = 100\mu$m with the incident field:

$$CF = \frac{\int \Psi^*(x,z=0)\Psi^*(x,z=z_f)\,dx}{\sqrt{\int |\Psi^*(x,z=0)|^2\,dx}\sqrt{\int |\Psi^*(x,z=z_f)|^2\,dx}} \tag{32}$$

The absolute value of the correlation factor should be unity, since only the phase changes as the mode propagates through the waveguide. Thus, $E_R = 1 - |CF|$ gives the error in the method used for computing the propagated field. This quantity is plotted in Fig. 6. For comparison, we have also included the results obtained using the FFT-BPM with a grid of 128 points in the computation window of $-50\ \mu$m to $50\ \mu$m. The figure shows the error as a function of the number of collocation points, N, for the collocation method for which Eq.(30) was solved using the Runge-Kutta method with the extrapolation interval Δz. The figure shows the improvement in the performance as N increases. The method has better accuracy than the FFT-BPM for much smaller values of N. Further, the figure also includes a comparison of

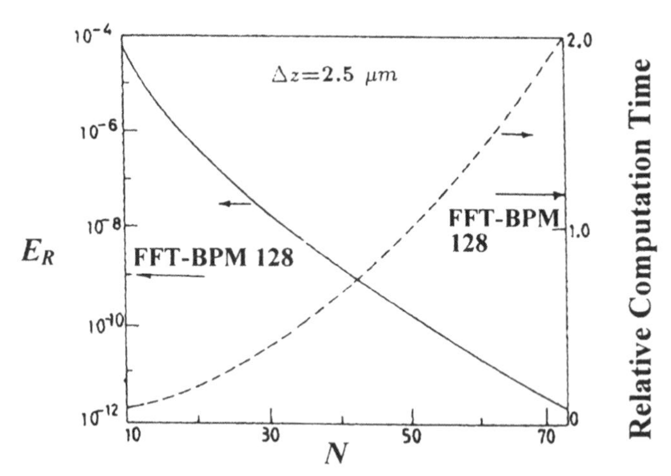

Figure 6 Comparison of the collocation method and the FFT-BPM in terms of the error in propagation and the computation time.

computation time, which also shows that for a given accuracy the collocation method is more efficient computationally.

The two representative examples show that both the FD-BPM and the collocation method represent improvement over the FFT-BPM. The collocation method, however, has a further advantage that it can solve the second order (non-paraxial) wave equation also; while in the FD-BPM (or the FFT-BPM) require special iterative means to deal with propagation beyond the paraxial approximation. We discuss this aspect briefly in the latter sections.

8. Non-paraxial Propagation: Direct Numerical Solution

As mentioned earlier, one should solve the wave equation, Eq.(1), directly to obtain a non-paraxial solution. However, it is difficult to solve and hence, various approximations have been used. We have further seen in Sec. 6 that the collocation method can be directly used for this equation and one obtains the collocation equation, Eq.(29), the accuracy of which increases indefinitely as the number of collocation points, N, increases. The collocation equation is an ordinary matrix equation and can be solved sing a standard numerical method such as the Runge-Kutta method. This method has been implemented[29] successfully and its accuracy has been demonstrated for various examples. It may be mentioned that the computation effort in this case would obviously be more than that involved in solving the first order equation, Eq.(30), but there is no approximation, and no additional convergence check as in the case of the Padé approximants are needed.

We include here two examples of wide-angle propagation using the collocation method[29]. As the first example, we consider the propagation of a Gaussian beam at 45° to the z-axis for a distance of $10\ \mu m$. The width of the Gaussian beam intensity was $2.828\ \mu m$ and the wavelength $1.06\ \mu m$. The computation was done with 120 collocation points and the width of the numerical window was about $80\ \mu m$. The result is shown in Fig. 7 in which the paraxially propagated beam is also included. This result matches well with that obtained by Hadley.[28]

In the next example we show the propagation of the TE_1 mode of a tilted step-index waveguide.[30] In the waveguide, $n_{core}=1.002$, $n_{clad}=1.000$, $w=15.092$ μm and $\lambda=1.0\ \mu$m. We have used 500 points for this computation. The field is launched at an angle of 45° with respect to the direction of propagation and moves form left to right. Figure 8 shows the propagation of the TE_1 mode. The mode propagates undistorted and there is very little power loss. The mode is displaced in the x-direction by the exact amount that traveling at 45° for a distance of 100 μm would entail, showing the accuracy of propagation.

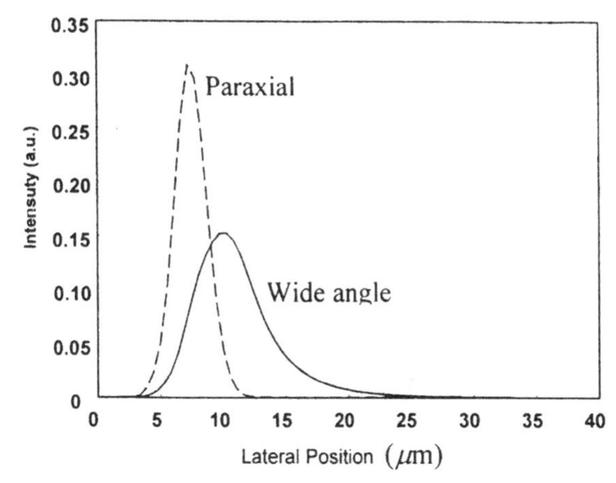

Figure 7 Intensity (a.u.) of a Gaussian beam obtained using paraxial (dashed curve) and non-paraxial (continuous curve) propagation. The beam is phase tilted at 45 degrees with respect to the direction of propagation.

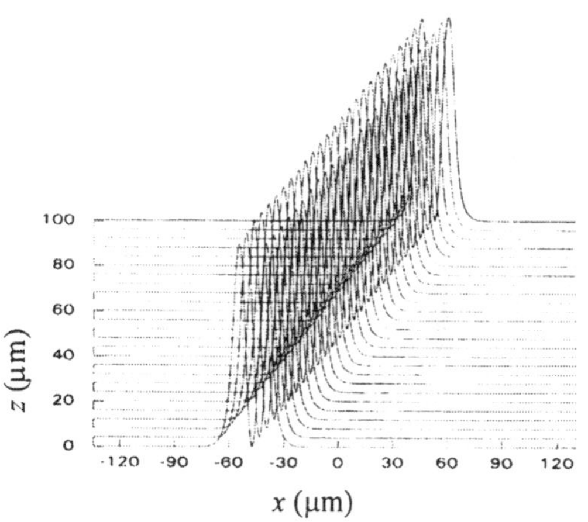

Figure 8 Propagation of the TE_1 mode of the step index waveguide, launched at 45°.

In the above discussions, we have considered wave propagation along the z direction and no reflections have been considered. There are, however, a number of cases where reflections are important and cannot be neglected. Bragg gratings, for example, are components that operate on the principle of reflection (see Chapter 9). These have recently attracted considerable attention because of their applications as reflectors and filters with narrow bandwidth and low side-lobe level in WDM systems, and for flattening of the gain spectrum in erbium-doped fiber amplifiers (see Chapters 1 and 2). To model propagation through such components, one has to use methods which are bi-directional and are capable of propagating waves in both directions $+z$ and $-z$. Since the method described here is based on the direct solution of the wave equation, propagation in both – forward and backward – is automatically taken into account[31,32].

9. Non-paraxial Propagation: Split-Step Formulation

The splitting of the propagation operator has been used extensively in propagation methods; however, it has been restricted to the paraxial equation. Recently, it has been applied successfully to the non-paraxial equation.[33] In this section, we briefly discuss the basics of this splitting procedure and discuss its implementation in the collocation method and the FD-BPM in the following sections. The splitting is very much similar to the one used in the FFT-BPM; each propagation step is split into two operations: the propagation in a uniform reference medium and the effect of the actual index variation in the form of a phase factor. The operation of propagation through the uniform medium is identical in each step and therefore, it is to be computed only once and then used in each step.

We start again with the scalar wave equation, Eq.(1), which can be written as two simultaneous first order (in z) differential equations and can in turn be represented in a matrix form as

$$\frac{\partial \mathbf{\Phi}}{\partial z} = \mathbf{H}(z)\mathbf{\Phi}(z), \tag{33}$$

where

$$\mathbf{\Phi}(z) = \begin{bmatrix} \Psi \\ \frac{\partial \Psi}{\partial z} \end{bmatrix}, \qquad \mathbf{H}(z) = \begin{bmatrix} 0 & 1 \\ -\frac{\partial^2}{\partial x^2} - k_0^2 n^2 & 0 \end{bmatrix}. \tag{34}$$

The operator $\mathbf{H}$ can be written as a sum of two operators, one representing the propagation through a uniform medium of index, say n_r, and the other representing the effect of the index variation of the guiding structure; thus,

$$\begin{aligned} \mathbf{H}(z) &= \mathbf{H}_1 + \mathbf{H}_2(z) \\ &= \begin{bmatrix} 0 & 1 \\ -\frac{\partial^2}{\partial x^2} - k_0^2 n_r^2 & 0 \end{bmatrix} + \begin{bmatrix} 0 & 0 \\ k_0^2 (n_r^2 - n^2) & 0 \end{bmatrix} \end{aligned} \tag{35}$$

A formal solution of Eq.(35), after using the symmetrized splitting of summation of operators as in Eq.(35), can be written as (cf. Eq.22)

$$\mathbf{\Phi}(z + \Delta z) = \mathbf{P}\,\mathbf{Q}(z)\,\mathbf{P}\,\mathbf{\Phi}(z) + O\big((\Delta z)^3\big), \tag{36}$$

$$\mathbf{P} = e^{\frac{1}{2}\mathbf{H}_1 \Delta z}, \quad \mathbf{Q}(z) = e^{\mathbf{H}_2 \Delta z}. \tag{37}$$

The operator $\mathbf{P}$ represents propagation in the uniform medium, n_r, over a distance of $\Delta z/2$, and hence, can be evaluated using any method like the collocation, finite-difference or FFT methods. The operator $\mathbf{Q}(z)$ can also be easily evaluated due to the specific form of the matrix and it can be easily seen by expanding the exponential that

$$\mathbf{Q}(z) = \begin{bmatrix} 1 & 0 \\ k_0^2 (n_r^2 - n^2) & 1 \end{bmatrix}, \tag{38}$$

since,

$$(\mathbf{H}_2)^m = \mathbf{0}, \qquad m \geq 2 \tag{39}$$

due to the special form of the matrix $\mathbf{H}_2$. It may be noted that for lossless propagation the matrix $\mathbf{P}$ would be Hermitian, while the matrix $\mathbf{Q}$ always has a determinant value equal to unity.

The method given above can be implemented with any of the numerical methods employed to solve the wave equation, *e.g.*, the FFT-BPM, FD-BPM, or the collocation method. In the following sections, we discuss the implementation using the collocation method and the FD-BPM.

10. Split-Step Non-paraxial (SSNP) Implementation in the Collocation Method[33]

We shall use the sinusoidal basis functions for this implementation. Although any set of basis function can be used depending on the symmetry and boundary conditions of the structure to be analyzed, these functions lead to simpler formulation and efficient computation, as we shall see later.

The sinusoidal functions are chosen such that they vanish at a far off boundary situated at $x = \pm L$. Thus, we have

$$\begin{aligned} \phi_n(x) &= \cos(\nu_n x) \quad \text{for } n = 1,3,5,\ldots, N-1 \\ &= \sin(\nu_n x) \quad \text{for } n = 2,4,6,\ldots, N \end{aligned} \tag{40}$$

where $\nu_n = n\pi/2L$ and the collocation points, being zeroes of $\cos(\nu_{N+1}x)$, are

$$x_j = (\frac{2j}{N+1} - 1)L, \qquad j = 1,2,\ldots, N \tag{41}$$

Further, it can be shown that

$$\mathbf{S}_0 = \mathbf{A}\,\mathbf{H}\,\mathbf{A}^{-1} \qquad \text{and} \qquad \mathbf{A}^{-1} = \left(\frac{2}{N+1}\right)\mathbf{A}^T \tag{42}$$

where $\mathbf{A} = \{A_{ij} : A_{ij} = \phi_j(x_i)\}$ and $\mathbf{H} = diag.(-\nu_1^2 \; -\nu_2^2 \; -\nu_3^2 \cdots -\nu_N^2)$. Applying the procedure described in Sec. 9 to Eq.(29), we will obtain the formal solution as Eq.(36) with the matrices $\mathbf{\Phi}$, $\mathbf{P}$ and $\mathbf{Q}$ now being block matrices defined as

$$\mathbf{\Phi}(z) = \begin{bmatrix} \mathbf{\Psi} \\ \dfrac{d\mathbf{\Psi}}{dz} \end{bmatrix}, \quad \mathbf{P} = \exp\left\{\frac{\Delta z}{2}\begin{bmatrix} \mathbf{0} & \mathbf{I} \\ -\mathbf{S}_0 - k_0^2 n_r^2 \mathbf{I} & \mathbf{0} \end{bmatrix}\right\}, \quad \mathbf{Q}(z) = \begin{bmatrix} \mathbf{I} & \mathbf{0} \\ -\Delta\mathbf{R}(z) & \mathbf{I} \end{bmatrix}, \tag{43}$$

where $\Delta\mathbf{R}(z) = \mathbf{R}(z) - k_0^2 n_r^2 \mathbf{I}$. Using the expressions for the sinusoidal basis functions, the matrix $\mathbf{P}$ can be obtained analytically as

$$\mathbf{P} = \left(\tfrac{2}{N+1}\right)\begin{pmatrix} \mathbf{A} & 0 \\ 0 & \mathbf{A} \end{pmatrix}\begin{bmatrix} c_1 & 0 & \cdots & 0 & s_1 & 0 & \cdots & 0 \\ 0 & c_2 & \cdots & 0 & 0 & s_2 & \cdots & 0 \\ \vdots & \vdots & \ddots & \vdots & \vdots & \vdots & \ddots & \vdots \\ 0 & 0 & \cdots & c_N & 0 & 0 & \cdots & s_N \\ \tilde{s}_1 & 0 & \cdots & 0 & c_1 & 0 & \cdots & 0 \\ 0 & \tilde{s}_2 & \cdots & 0 & 0 & c_2 & \cdots & 0 \\ \vdots & \vdots & \ddots & \vdots & \vdots & \vdots & \ddots & \vdots \\ 0 & 0 & \cdots & \tilde{s}_N & 0 & 0 & \cdots & c_N \end{bmatrix}\begin{pmatrix} \mathbf{A}^T & 0 \\ 0 & \mathbf{A}^T \end{pmatrix} \tag{44}$$

where $c_i = \cos(\sqrt{\Lambda_i}\,\Delta z/2)$, $\quad s_i = \sin(\sqrt{\Lambda_i}\,\Delta z/2)\big/\sqrt{\Lambda_i}$, $\tilde{s}_i = -\Lambda_i\, s_i$ and $\Lambda_i = k_0^2 n_r^2 - \nu_i^2$. In cases where Λ_i is negative, the quantities c_i, s_i and $\tilde{s}_i$ remain real and sine and cosine functions are evaluated through the corresponding hyperbolic functions. However, the hyperbolic functions are exponentially increasing in nature and thus, such values of Λ_i correspond to evanescent waves which lead to instability in the propagation. By adjusting the window size and hence the values of v_i^2, and by choosing a suitable value of the reference refractive index (such that Λ_i has only positive values), we can easily avoid these evanescent waves. Each propagation step thus requires 12 multiplications of an $N \times N$ square matrix with a column matrix except at the first and the last steps where 8 such multiplications are additionally required. We would like to emphasize that using the sinusoidal basis functions in the collocation method here has an advantage, since no FFT, matrix inversion or matrix diagonalization need be done for propagation through the uniform medium and all matrices involved are obtained analytically. As an example, we consider the propagation of the TE_{10} mode in the benchmark step-index waveguide[34] with n_{co}=3.3, n_{clad}=3.17, w=8.8 μm and λ=1.55 μm. We have obtained the power remaining in the guide after propagation of 100 μm at a tilt angle of 20°. Table-II compares the SSNP method with other methods. It is quite obvious from the table that with fewer points, the SSNP method shows higher accuracy. The method is faster than the DNS, taking only about half the time. It is also much easier to implement. Here, N_x and N_z represent the average number of discretizations points along the transverse direction and the longitudinal direction, respectively.

Table II: Comparison of propagation to 100 micron in the benchmark step index waveguide for TE_{10} modes using different methods.

Method	N_z	N_x	Power in the waveguide at 20°
SSNP-Coll	1000	800	~0.96
DNS	1000	800	~0.90
AMIGO[34]	1429	1311	~0.95
FD2BPM[34]	1000	2048	~0.95
FTBPM[34]	1000	256	~0.55
LETI-FD[34]	200	1024	~0.15

11. Split-Step Non-paraxial (SSNP) Implementation in the Finite-Difference (FD) Method[35]

The implementation using the FD-method differs mainly in the propagation procedure through the uniform medium. Thus, only the evaluation of **P** is done using the finite-difference approximations of the differential as discussed in Sec. 5, while the evaluation of the matrix $\mathbf{Q}(z)$ is straightforward. Thus, the matrix **P** is the solution of Eq.(33) over a distance of $\Delta z/2$ for the uniform medium of index n_r.

$$\mathbf{\Phi}(z + \Delta z/2) = e^{\mathbf{H}_1 \Delta z}\mathbf{\Phi}(z) \equiv \mathbf{P}\mathbf{\Phi}(z) \tag{45}$$

For the uniform medium, we can obtain the solution in terms of sinusoidal functions directly as

$$\Psi(z+\tfrac{1}{2}\Delta z)=\cos(\tfrac{1}{2}\sqrt{\mathbf{S}}\Delta z)\Psi(z)+\frac{1}{\sqrt{\mathbf{S}}}\sin(\tfrac{1}{2}\sqrt{\mathbf{S}}\Delta z)\Psi'(z)$$
$$\Psi'(z+\tfrac{1}{2}\Delta z)=-\sqrt{\mathbf{S}}\sin(\tfrac{1}{2}\sqrt{\mathbf{S}}\Delta z)\Psi(z)+\cos(\tfrac{1}{2}\sqrt{\mathbf{S}}\Delta z)\Psi'(z) \tag{46}$$

where the matrix **S** now is a finite-difference representation of the operator $\partial^2/\partial x^2+n_r^2$. The evaluation of cosine and sine of the matrices can be done either by diagonalization of the matrix or by series expansion of these functions. In the latter case, we obtain

$$\mathbf{P}=\begin{pmatrix} \mathbf{I}-\frac{\Delta z^2}{8}\mathbf{S}+\frac{\Delta z^4}{544}\mathbf{S}^2 & \Delta z\mathbf{I}-\frac{\Delta z^3}{6}\mathbf{S}+\frac{\Delta z^5}{3840}\mathbf{S}^2 \\ -\Delta z\mathbf{S}+\frac{\Delta z^3}{6}\mathbf{S}^2-\frac{\Delta z^5}{3840}\mathbf{S}^3 & \mathbf{I}-\frac{\Delta z^2}{8}\mathbf{S}+\frac{\Delta z^4}{544}\mathbf{S}^2 \end{pmatrix} \tag{47}$$

The error in this case is $O(\Delta z^6)$. In the simplest case, the FD-representation of **S** is based on the three-point central difference defined in Eq.(23); thus

$$\mathbf{S}=k_0^2n_r^2\mathbf{I}+\mathbf{D}_0/\Delta x^2 \tag{48}$$

where $\mathbf{D}_0$ is a tridiagonal matrix with diagonal elements being -2 and off diagonal elements being 1. This three-point approximation is however not sufficient and we have to have a much better approximation for **S**. This can be obtained by a series approximation of $\partial^2/\partial x^2$ which in the matrix form can be written as

$$\mathbf{S}=k_0^2n_r^2\mathbf{I}+\frac{1}{\Delta x^2}\sum_{m=1}^{\infty}b_m\,\mathbf{D}_0^m \tag{49}$$

where

$$b_m=\frac{1}{4^{m-1}}\sum_{l=1}^{m}a_l a_{m-l+1} \qquad \text{with} \qquad a_{l+1}=-\frac{(2l-1)^2}{2l(2l+1)}a_l \tag{50}$$

The lowest power, $m=1$ corresponds to the 3-point formula and as the power m increases, more and more $(2m+1)$ transverse grid points are used in approximating $\partial^2/\partial x^2$. The matrix **S** does not remain tridiagonal, becomes increasingly dense. However, this matrix and the matrix **P** are to be calculated only once before the beginning of the propagation and the latter matrix is used in each step. It may be noted that the size of the matrix remains unaffected by the value of m and hence, the computation effort does not change significantly.

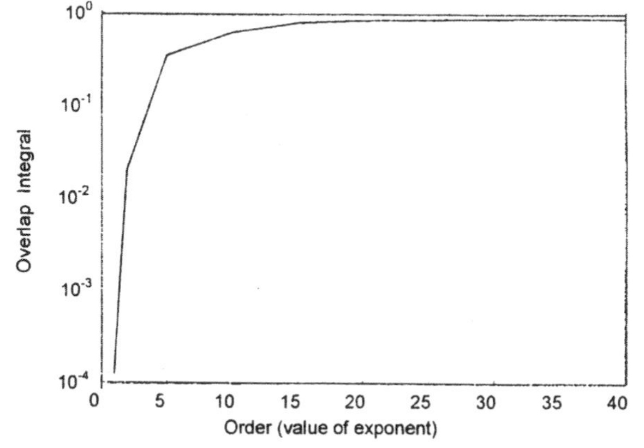

Figure 9 Overlap integral with order of exponent for propagation of TE_1 mode in step-index waveguide[30] at 50° for a distance of 100 μm.

Figure 9 shows the overlap integral of the expected field and the computed field as a function of the exponent m. The figure clearly shows the improvement in accuracy as the order of exponent is increased.

12. Summary

In this article, we have discussed three commonly used methods for numerically propagating waves through waveguiding structures have been discussed: the FFT-BPM, the FD-BPM and the collocation method. Examples have also been included and the advantages and limitations of these methods have been discussed. Both the FD-BPM and the collocation method are better, both in terms of accuracy obtainable and the computational efficiency, than the FFT-BPM. The collocation has the additional advantage that it can treat both wide-angle and the bi-directional propagation in an non-iterative manner. Examples on wide angle propagation have also been included.

Acknowledgments: The author would like to thank Arti Agrawal for help in preparation of this chapter.

References

1. S.M. Saad, S. M. (1985), "Review of numerical methods for the analysis of arbitrarily-shaped microwave and optical dielectric waveguides", *IEEE Trans. On Microwave Theory and Technol.* **MTT-33**, 894-899.
2. Yevick, D. (1994), "A guide to electric field propagation techniques for guided-wave optics", (Special Issue on Optical Waveguide Modeling,) *Optical and Quantum Electronics* **26**, S185-S197.
3. Chiang, K. S. (1994), "Review of numerical and approximate methods for the modal analysis of general optical dielectric waveguides", (Special Issue on Optical Waveguide Modeling), Opt. Quant. Electron., **26**, S113-S134.
4. R. März, "Integrated optics: design and modeling", Artech House, Boston, U.S.A. (1995).
5. W. P. Huang, "Methods for Modeling and Simulation of Guided-Wave Optoelectronic Devices, EMW Publishing, Cambridge, Massachussttes, U.S.A. (1995).
6. M. Koshiba and Y. Tsuji, "Design and modeling of microwave photonic devices", *Optical and Quantum Electronics* **30**, 995-1003 (1998).
7. R. Scarmozzino, A. Gopinath, R. Pregla and S. Helfert, "Numerical techniques for modeling guided-wave photonic devices", *IEEE J. Selected Topics Quant. Electron.* **6**, 150-162 (2000).
8. K. Kawano and T. Kitoh *Introduction to optical waveguide analysis*, John Wiley and Sons Inc., New York, U.S.A. (2001).
9. G.B. Hocker and W.K. Burns, "Mode dispersion in diffused channel waveguides by the effective index method," *Appl. Opt.* **16,** 113 (1977).

10. A. Sharma and P. Bindal, "An accurate variational analysis of single mode diffused channel waveguides," *Opt. Quantum Electron.* 24, 1359-1371 (1992).
11. A.K. Ghatak and K. Thyagarajan, *Optical Electronics* (Cambridge University Press, Cambridge), 1989.
12. A.K. Ghatak and K. Thyagarajan, *An Introduction to Fiber Optics* (Cambridge University Press, Cambridge), 1998.
13. A.W. Snyder and J.D. Love, *Optical Waveguide Theory*, (Chapman & Hall, London), 1983.
14. M.D. Feit and J.A. Fleck Jr., "Light propagation in graded index optical fibers", *Appl. Opt.* **17**, 3990-3998 (1978).
15. M.D. Feit and J.A. Fleck Jr., "Calculation of dispersion in graded-index multimode fibers by a propagating beam method", *Appl. Opt.* **18**, 2843-2851 (1979).
16. M.D. Feit and J.A. Fleck Jr., "Computation of mode eigen-functions in graded-index optical fiber waveguides by a propagating beam method", *Appl. Opt.* **19**, 1154-1166 (1980).
17. D. Yevick and B. Hermansson, "Efficient beam propagation techniques", *IEEE J. Quant. Electron.* **26**, 109-112 (1990).
18. Y. Chung and N. Dagli, "An Assessment of finite difference beam propagation method", *IEEE J. Quant. Electron.* **26**, 1335-1339 (1990).
19. C.L. Xu and W.P. Huang, in *Methods for Modeling and Simulation of Optical Guided-Wave Devices*, (ed: W Huang) (EMW Publishers, Cambridge, Mass.) vol. 11, pp. 1 (1995).
20. T. Koch, J. Davies and D. Wickramasinghe, "Finite element/finite difference propagation algorithm for integrated optical devices", *Electron. Lett.* 25, 514-516 (1989).
21. A. Sharma and S. Bannerjee, " Method for propagation of total fields or beams through optical waveguides", *Opt. Lett.* **14**, 94-96 (1989).
22. A. Sharma and S. Bannerjee, "Propagation characteristics of optical waveguiding structures by direct solution of the Helmholtz equation for total fields", *J. Opt. Soc. Am. A.* **6**; 1884-1894 (1989); Errata: **7**, 2156 (1990).
23. A. Sharma, "Collocation method for wave propagation through optical waveguiding structures" , in *Methods for Modeling and Simulation of Guided-Wave Optoelectronic Devices*, W.P.Huang, ed. (EMW Publishing, Cambridge, Mass., U.S.A.), vol. 11, pp. 143-198 (1995).
24. A. Sharma and A. Taneja, "Unconditionally stable procedure to propagate beams through optical waveguides using the collocation method", *Opt. Lett.* **16**, 1162-1164 (1991).
25. Sharma and A. Taneja, "Variable transformed collocation method for field propagation through optical waveguiding structures", *Opt. Lett.* **17**, 804-806 (1992).

26. Taneja A and A. Sharma, "Propagation of beams through optical waveguiding structures: comparison of the beam propagation method (BPM) and the collocation method," *J. Opt. Soc. Am. A* **10,** 1739-1745 (1993).
27. D. Yevick and M. Glasner, "Forward wide-angle light propagation in semiconductor rib waveguides", *Opt. Lett.* **15** 174-176 (1990).
28. G.R. Hadley, "Multistep method for wide-angle beam propagation", *Opt. Lett.* **17**, 1743-1745 (1992).
29. A. Sharma and A. Agrawal, "Wide angle and bi-directional beam propagation using the collocation method for the non-paraxial wave equation," *Opt. Commun.* **216,** 41-45 (2003).
30. J. Yamauchi, J. Shibayama and H. Nakano, "Modified finite-difference beam propagation method based on the Generalised Douglas scheme for variable coefficients", *IEEE Photon. Technol. Lett.* 7, 661-663 (1995).
31. A.K. Taneja and A. Sharma, "Modelling of guided-wave bragg gratings and grating sensors using the collocation method," *Opt. Quant. Electron.* **32**, 1033-1046 (2000).
32. A. Sharma, "Collocation method for numerical scalar wave propagation through optical waveguiding structure," *Asian J. Phys.*, **12**, 143-152 (2003).
33. A. Sharma and A. Agrawal, "New method for non-paraxial beam propagation," *J. Opt. Soc. Am. A* **21**, 1082-1087 (2004).
34. H.-P. Nolting, and R. März, "Results of benchmark tests for different numerical BPM algorithms", *J. Lightwave Technol.*, **13**, 216-224 (1995).
35. A. Agrawal, *Paraxial and Non-paraxial Wave Propagation through Optical Waveguides*, Ph.D. Thesis, India Institute of Technology Delhi, New Delhi (2004).

Index